Cultural Histories of Noise, Sound and Listening in Europe, 1300–19

Cultural Histories of Noise, Sound and Listening in Europe, 1300–1918 presents a range of historical case studies on the sounding worlds of the European past. The chapters in this volume explore ways of thinking about sound historically, and seek to understand how people have understood and negotiated their relationships with the sounding world in Europe from the Middle Ages through to the early twentieth century. They consider, in particular: sound and music in the later Middle Ages; the politics of sound in the early modern period; the history of the body and perception during the *Ancien Régime*; the sounds of the city in the nineteenth century and sound and colonial rule at the *fin de siècle*.

The case studies also range in geographical orientation to include considerations not only of Britain and France, the countries most considered in European historical sound studies in English-language scholarship to date, but also Bosnia-Herzegovina, British Colonial India, Germany, Italy and Portugal. Out of this diverse group of case studies emerge significant themes that recur time and again, varying according to time and place: sound, power and identity; sound as a marker of power or violence; sound, physiology and sensory perception and technologies of sound, consumption and meaning.

Ian Biddle is a cultural theorist and musicologist, working on a range of topics in music- and sound-related areas. He is co-founder and co-ordinating editor of the journal *Radical Musicology* and is co-convenor, with Beate Müller, of the Newcastle University Genocide Research Group. His work deals with: the cultural history of music and masculinity; music's intervention in communities and subjectivities; sound, soundscapes and urban experience; the politics of noise. He has interests in Holocaust studies, memory studies, sound studies, Italian workerist and autonomist theory and psychoanalysis. He is author of *Music, Masculinity and the Claims of History: The Austro-German Tradition from Hegel to Freud* (Ashgate, 2011), editor of *Music and Identity Politics* (Ashgate, 2012), co-editor, with Kirsten Gibson, of *Masculinity and Western Musical Practice* (Ashgate, 2009) and co-editor of *Sound, Music Affect: Theorizing Sonic Experience* (2013).

Kirsten Gibson's research has focused on John Dowland's printed songbooks, Elizabethan court politics, early modern musical print culture and discourses of music and masculinity in sixteenth- and seventeenth-century England. Her work appears in *Early Music History*, *Journal of the Royal Musical Association*, *Renaissance Studies* and *Early Music* as well as in various essay collections. She is co-editor, with Ian Biddle, of *Masculinity and Western Musical Practice* (Ashgate, 2009).

Cultural Histories of Noise, Sound and Listening in Europe, 1300–1918

Edited by
Ian Biddle and Kirsten Gibson

Routledge
Taylor & Francis Group
LONDON AND NEW YORK

First published 2017 by Routledge

2 Park Square, Milton Park, Abingdon, Oxfordshire OX14 4RN
52 Vanderbilt Avenue, New York, NY 10017

Routledge is an imprint of the Taylor & Francis Group, an informa business

First issued in paperback 2019

British Library Cataloguing-in-Publication Data
A catalogue record for this book is available from the British Library

Library of Congress Cataloging-in-Publication Data
Cultural histories of noise, sound and listening in Europe, 1300–1918 /
edited by Ian Biddle and Kirsten Gibson.
pages cm
Includes bibliographical references and index.
ISBN 978-1-409-44439-8 (hardcover: alk. paper)
1. Music—Social aspects—Europe—History. 2. Sound—Social aspects—Europe—History. 3. Noise—Social aspects—Europe—History. I. Biddle, Ian D. II. Gibson, Kirsten, 1978–
ML3917.E85C85 2015
306.4′842094—dc23 2015019017

ISBN: 978-1-4094-4439-8 (hbk)
ISBN: 978-0-367-23051-7 (pbk)

Typeset in Times New Roman
by codeMantra

Contents

Figures

Contributors

Olivier Balaÿ studied at the School of Architecture of Grenoble from 1984 to 2007 and the School of Architecture of Saint Etienne from 1996 to 1998. He completed his PhD at the Institute of Urban Studies, Grenoble, in 1992 and his Habitation à diriger de recherches in 2002. In 1983 he became a researcher at CRESSON (the Centre de recherche sur l'espace sonore et l'environnement urbain) and he also began his career as a professional architect in Lyon (first as a freelancer and then as a partner from 1993 at SARL Balaÿ Boinay Pierron). He is now a professor at the Ecole Nationale Supérieure d'Architecture de Lyon, and remains affiliated with CRESSON. He is author of *L'espace sonore de la ville au XIXe siècle* (A la Croisée, 2003).

Ian Biddle is a cultural theorist and musicologist, working on a range of topics in music- and sound-related areas. His work ranges from the cultural history of music and masculinity, theorising music's intervention in communities and subjectivities, sound, soundscapes and urban experience, and the politics of noise. He has interests in memory studies, sound studies, Italian workerist and autonomist theory, psychoanalysis and theoretical approaches to 'affective' states. He is co-founder and co-ordinating editor (with Richard Middleton) of the journal *Radical Musicology*. He is author of *Music, Masculinity and the Claims of History* (Ashgate, 2001), co-editor (with Marie Thompson) of *Sound, Music, Affect* (Bloomsbury, 2013), co-editor (with Kirsten Gibson) of *Masculinity and Western Musical Practice* (Ashgate, 2009) and editor of *Music and Identity Politics* (Ashgate, 2013).

Alexander Bonus is Assistant Professor of Music at Bard College, where he teaches courses on music history, theory, composition and performance practices. In his capacities as continuo player and conductor, Dr Bonus directs the Bard Baroque Ensemble from the organ and harpsichord. From 2011 to 2012, Dr Bonus taught at Duke University, where he led the vocal and instrumental collegium. His recent publications include the 'Handbook to the Metronome' in *Oxford Handbooks Online* and 'A Musical Treatise from Lost Times, Part I' (*Current Musicology*, 2014). He has also contributed numerous entries to the latest *Grove Dictionary of Musical Instruments*. As a professional early-brass specialist and keyboard player, Dr Bonus regularly performs across North America. He has worked with ensembles including the Boston Early Music Festival, Apollo's Fire, Chicago Opera Theater, the Newberry Consort, the Folger Consort and Tragicomedia. In 2011, the American Council of Learned Societies awarded him a New Faculty Fellowship, due in part to his scholarship regarding musical time and clockwork technologies. In 2008, he received an Arts and Sciences Fellowship from Case Western Reserve University. Alexander Bonus holds a PhD in Musicology from Case Western Reserve University in addition to MM and BM degrees from the Eastman School of Music.

Helen Coffey is Lecturer in Music at the Open University. Her research interests primarily concern musical patronage in German, Austrian and British cities and courts from the fifteenth to the eighteenth century. She has published on various aspects of the employment of civic and court musicians as a reflection of social and political developments and has a particular interest in the role of music in civic and court rituals, as well as the movement of musicians between institutions across Europe. Helen is co-editor of the five-volume work *George Frideric Handel: Collected Documents* and is a council member of the Handel Institute.

Lisa Colton is Director of the Centre for the Study of Music, Gender and Identity at the University of Huddersfield, where her research focuses on early and contemporary music from various critical perspectives. Her journal articles include studies of women's roles in the cultivation of fourteenth-century polyphony (*Early Music*), seventeenth-century keyboard music (*Music and Letters*), gender and virginity in motets relating to nuns (*Journal of the Royal Musical Association*) and of the music of Judith Weir (*Contemporary Music Review*). Her forthcoming work includes the monograph: *Angel Song: Medieval English Music in History*.

Veit Erlmann is an anthropologist/ethnomusicologist and the Endowed Chair of Music History at the University of Texas at Austin. His main areas of interest are music and popular culture in South Africa and Indonesia; sound studies; and the anthropology of intellectual property law. His most recent publication is *Reason and Resonance. A History of Modern Aurality* (Zone Books, 2010). Currently he is working on a book on intellectual property in the South African music industry that will be published by Duke University Press. He also serves as co-editor of *Sound Studies. An Interdisciplinary Journal*.

Daniele V. Filippi studied musicology in Cremona and in Heidelberg. His scholarly interests include early modern music and spirituality, historical soundscapes and intertextuality. He has published books, articles (*Polifonie*, *Journal of the Alamire Foundation*, *Revista de musicología*) and book chapters on Gaffurio, Palestrina, Victoria, Marenzio, de Monte, and G.F. Anerio. In 2012–14 he developed a project entitled 'The Soundscape of Early Modern Catholicism' thanks to a research fellowship from the Jesuit Institute at Boston College (see http://www.sonicexperience.org/); a book with the same title is forthcoming.

Kirsten Gibson is Lecturer in Music at Newcastle University, UK. Her research has focused on John Dowland's printed songbooks, Elizabethan court politics, early modern musical print culture and discourses of music and masculinity in sixteenth- and seventeenth-century England. Her work appears in *Early Music History*, *Journal of the Royal Musical Association*, *Renaissance Studies* and *Early Music* as well as in various essay collections. She is co-editor, with Ian Biddle, of *Masculinity and Western Musical Practice* (Ashgate, 2009).

Adam Hansen is Senior Lecturer in English at Northumbria University. He is author of numerous articles and chapters on early modern literature and culture with a particular focus on London, as well as the monograph *Shakespeare and Popular Music* (Continuum, 2010). He is co-editor of *Litpop: Writing and Popular Music* (Ashgate, 2014) and *Shakespearean Echoes* (Palgrave, 2015).

Emily Laurance is Professor and Chair of Music History and Literature at the San Francisco Conservatory of Music. Her PhD dissertation was a study of Gustave Charpentier's opera *Louise* (1900) and its relationship to realist aesthetics in the operatic tradition. In addition to French opera and aesthetics, Dr Laurance has research interests in nineteenth- and early twentieth-century American music, the French vocal romance of the late eighteenth and early nineteenth centuries and the early pedal harp. She co-founded the duo DoubleAction (with tenor Thomas Gregg), an ensemble that specialises in harp-accompanied song from 1770–1840, which they perform using an 1829 single-action pedal harp. In 2005–06 she held a postdoctoral fellow at the John W. Kluge Center at the Library of Congress. Dr Laurance previously taught at UNC-Chapel Hill and Duke University. She holds Bachelor's degrees from Oberlin College, an MM from the New England Conservatory and a Doctorate in Musicology from the UNC-Chapel Hill.

Hedy Law is Assistant Professor of Musicology at the University of British Columbia. She has written for *Cambridge Opera Journal*, *Musique et Geste en France*, *CENTER: Architecture and Design in America*, *Oxford Handbook of Music Censorship* and *Oxford Handbook of Music and Disability Studies*.

Risto Pekka Pennanen – ethnomusicologist and historian of ideas specialising in the Balkans – is an adjunct professor of ethnomusicology at University of Tampere, Finland. Pennanen received his PhD from University of Tampere and has been an Alexander von Humboldt Research Fellow at the Department of Musicology at Georg-August-University in Göttingen, Germany; a research associate at the School of Oriental and African Studies (SOAS), University of London; and a research fellow at Helsinki Collegium for Advanced Studies. He has published articles on various Balkan musical styles, especially those of Greece, Bulgaria and Bosnia-Herzegovina, and on discography, the recording industry and the canons of music history in the peninsula. His current research project is *Music, Musicians, Soundscape and Colonial Policies in Habsburg Bosnia-Herzegovina, 1878–1918*.

João Silva completed his PhD in ICMuS (Newcastle University). He studied at the Universidade Nova de Lisboa and teaches in several music schools in Portugal. An active researcher in INET-MD since 2003, he was involved with editorial responsibilities in the *Enciclopédia da Música em Portugal no Século XX* project. His main research interests focus on late nineteenth-century musical cultures and their relations with modernity and technology. He published an article on the historiography of recording in *Radical Musicology* and *Entertaining Lisbon: Music Theater, and Modern Life in the Late 19th Century*, a monograph about music and theatre in Lisbon (Oxford University Press, 2016).

Jim Sykes is an Assistant Professor of Music at the University of Pennsylvania. An ethnomusicologist who focuses on the interrelations between sound, personhood, aesthetics, religion and modernity in the Indian Ocean region, he finished his PhD at the University of Chicago (2011) and was subsequently a postdoctoral research fellow at King's College London (2011–13). His fieldwork in Sri Lanka, India and Singapore has been funded by the Mellon Foundation, Fulbright-Hays DDRA, the Wenner-Gren Foundation and the European Research Council.

Acknowledgements

As editors of this volume, we would like to thank, for their support in compiling this volume, our generous colleagues at the International Centre for Music Studies and, for their diligence and unending kindness, the staff of the Robinson Library, Newcastle University.

We are extremely grateful to Heidi Bishop and her team (Katie Macdonald, Nicole Norman, Laura Macy and Sarah Charters) at Ashgate and Katharine Bartlett for helping to bring this project to fruition and for their careful and good-natured proofing of the manuscript.

In the end, the writing and editing of a book is a complex and demanding process, but one which our loved ones make easier and more bearable: Mark and Darren, we thank you for your love and support.

Introduction

Ian Biddle and Kirsten Gibson

To historicize sound is to raise some challenging ontological and epistemological questions about the nature of sound production and its reception by hearing and listening subjects. In this sense, aurality, as an 'object' of historical enquiry, is radically dispersed: its materials are distributed, especially in the pre-phonographic era, across numerous media, locked into the interstitial spaces between texts, images, scores, buildings and instruments (both musical and technical) for the production of sound. The aural residues of the past can be traced in a vast array of primary source materials including literature, treatises, scientific writings, newspapers, letters, diaries, account books, legislative documents, musical scores, civic records and parish registers. Sources like these – though silent, material objects – bear witness to there having been sound, and encourage us to listen out for the now faint echoes of fleeting, intangible, aural worlds of the past: the cacophony of street noises and the hum of the market; the hushed murmurs of the court; the raucous strains of the ballad seller on market day; the sociable musical gatherings of amateur musicians; the carefully engineered soundworlds of civic and religious festivities; the sounds of public worship; the ringing of church bells; the sounding of fusillades and cannon salvos; and the changing soundscapes of urban spaces through architectural developments, the introduction of new sound reproduction technologies and the encroachment of industrial modernity. These residues do more, however, than simply open up sonic windows on reverberations from the past. They engage us in narratives of how sound has shaped past lives: how it has carried meaning for individuals and institutions; how it has intervened in and reflected individual and corporate identities; and how, in various ways, it has been manipulated, or has inadvertently acted, as a mechanism of power (both hegemonic and subversive); and how, in turn, such sonic articulations of power and its subversion have been manipulated and controlled. Stories of sound are, indeed, stories of subjectivity, sociability, power and politics. The complexity of thinking aurality (a term we use here to emphasize the interchange of agents that contribute to the making, marking and reception of sound) in the pre-phonographic era is, however, always about a certain gap between an imagined plenitude that the historian brings to soundscapes past and the partiality (and dispersed nature) of materials and discourses that enable any such thinking.

For historians of sound, one might say, the question that presses itself on the scholar most regularly is not one of sufficiency per se, but one of how to manage the relationship between wild speculation, on the one hand, and the demands of rigorous historical scholarship, on the other. When, in other words, does the incompleteness of the materials that bear witness to there having been sound enable or interdict speculation? What are the disciplinary terms on which that speculation is managed, policed, authorized? That no such terms have been codified evidences a certain disciplinary dispersedness that, if allowed to feed the heady romantic tendencies that sometimes attend histories of sound, could reduce our work to the search for evidence of complex emotions, responses to sound, 'expressed silently in glances, tears, and aurally with the occasional sigh',[1] or for evidence of social relationships long since lost to the vicissitudes and selectivity of the archive. And yet, we would argue, it is precisely the interdisciplinary status of the history of sound that constitutes its greatest strength: it is the ingenuity of against-the-grain reading, the stubborn persistence of critical philosophical attempts to think sound before the phonograph, the intense commitment to methodological inclusivity, that have all lent such energy and insight to work in this area.

If sound studies is, by its very nature, dispersed, then this volume will inevitably want to embrace, celebrate and valorise that dispersal. But we also want to show that the field, despite its disciplinary permissiveness, can be strategically constituted in terms of what is unique to it and thus, by logical extension, how it makes claims for its status *as a discipline*. This is not meant as a regressive move – the discipline's interdisciplinary ground cannot simply be done away with for the purposes of institutional expediency and legibility: quite the contrary, in fact – we would argue that this collection brings to sound studies an intensification and deepening of both the complexities and the productive outcomes of the discipline's work.

Why Europe?

In the context of our current intensely globalized techno-, finance-, media-, ideo- and ethnoscapes, to borrow Arjun Appadurai's taxonomy, this book could be accused of reproducing the kinds of uncritical Eurocentrism that have long since been routinely problematized in the work of Edward Said, Homi Bhabha and many others.[2] In *Provincializing Europe*, for example, Dipesh Chakrabarty famously edges his use of the term 'Europe' by recognizing in its current use 'certain figures of the imagination whose geographical referents remain somewhat indeterminate'.[3] In other words, 'Europe' has come to stand for more than a mere territorial designation and has enabled a range of ideological, political and intellectual *regimina* that enforce a mental habituation to Eurocentrism. As such, Chakrabarty's insights are fully in line with the work of the Subaltern Studies Collective, of which he is a long-standing member: that group's key insight has been that radically disenfranchised groups have historical agency, despite their exclusion from mainstream political discourse (hence the term 'Subaltern', a term taken from Italian Marxist Antonio Gramsci, to denote any radically marginalized individual or group).[4] The key insights of the Subaltern Studies Collective's critique of Eurocentrism are writ large by Chakrabarty in his strategic decentring of Europe, its repositioning as what he terms 'an imaginary figure that remains deeply embedded in clichéd and shorthand forms in some everyday habits of thought that invariably subtend attempts in the social sciences to address questions of political modernity in South Asia'.[5] These 'shorthand forms' that attend thinking (or, rather, *not* thinking) Europe are symptomatic of a strange blind spot in the European (and Eurocentric) imaginary: when it comes to Europe, that is, we (which is to say 'we Europeans') are not always aware that what we are thinking about, and what we are seeking to articulate, is a province, not a centre. Europe, in other words, has come to stand, even in some of the most critical work sensitive to the brutalities of the colonial project, for a certain self-evidence, for a naturalized space that does not have to work to elaborate its own ontology. As Roberto M. Dainotto has argued, the idea of Europe as a self-contained, self-evident and self-sufficient 'centre' emerges only as late as the seventeenth century, especially in the emergence of a growing collective consciousness of Europe's otherness to the rest of the world:

> The growth of an exotic literature from the late seventeenth century is certainly a sign of the times, but it was also the sign of the kind of ideal *place* that Europe was starting to mean … as critical of Europe as these texts could sometimes be, they all proceeded from some shared assumption of what Europe, despite its local differences, actually meant: culture versus nature, society versus kinship, dressed versus naked, cooked versus raw, civilization versus naivety.[6]

Hence Europe as a (relatively) coherent socio-political unit emerges not in a flurry of self-realization, but in a determined self-positioning of its 'civilization' in relation to something else, to an endless taxonomy of outside figures, an eternal litany of enemies, dangers and exotics. Such was the coherence effect of the name *Europe* that, by the late eighteenth century, Rousseau was able to exclaim, with some irony:

> All the powers of Europe constitute, among themselves, some kind of a system that unites through the same religion, through the same set of laws, customs, letters, and commerce, and provides the necessary balance of forces. Add to this: the particular situation of Europe, which is more populated and more united than other continents; the continuous mixing of interests that ties of blood and of commerce, of arts and colonies, have instituted among European monarchs; the multitude of rivers and the variety of their courses, which makes communications easy; the restless mood of its inhabitants which makes them travel incessantly, and brings one in the country of the other; the invention of the printing press, and the common taste in the arts, which has made possible the sharing of scholarship and knowledge; and finally, the multitude and small size of the European States which, interdependent in their

> common need for luxury and in the difference of climates, has always made each people necessary to all others.[7]

Perhaps this is one of the few ways in which Europe, that self-obsessed headland at the Western extremity of the Eurasian continent, can make claims to uniqueness, in its fastidious blindness and deafness to its dependence on its others. And this 'unconscious-ness', perhaps unique in its provinciality, unique, also, in its intense commitment to the question of the divine right of its monarchs, and to the always-already-in-place superiority of its civilizations to those outside its fragile borders, enables also the emergence of Europe as a kind of *quality* or *feeling*.[8] Europe, we might say, came to constitute – even as it tore itself apart in internecine wars over the nature of the sacrament and the power of the church – a key resource in managing the everyday experiential world of its citizens, an *affectively charged* category. Renaissance Humanism, for example, figured the European body as a privileged site of experience, and incubated a quite specific imagination of the web of senses that held that body in place. Like scholasticism before it, Humanism imposed specific affective regimes on the European body, as a feeling sensorium, as auditor and as sounding matter.[9]

In this, then, Europe emerges in the late medieval and early modern moments not simply as a political unit (arguably it had never been quite *that*), but as a set of habituated rituals attaching to a (partially) shared imagination of power, autocracy and citizenship. These rituals are enacted in looking, tasting, touching, smelling and, of course, *listening*, and all constitute an affective regimen enacted on the body for constructing a sense of the always-already-in-place-ness (what in Yiddish is called *doikeyt*) of Europe as both a territory and a way of being *in place*. In this regime, rituals of listening get resolutely attached to processes of placing, of locating the auditory subject in the cultural web of signs and symbols. In this regard, Europe makes a striking (and somewhat strange) contribution to the emergence of what Lacan after Freud would call the modern personality; sound, noise and aurality constitute a key component in that emergence. It is strange, therefore, that the sonic is a component that most historians of experience have tended to ignore or side-line. This book seeks to address the relative scholarly neglect of sound and listening as a resource in constructing the affective worlds of Europe and is keen to demonstrate some of the ways in which historical thinking about sound before the phonographic era can cast new light on the history of ideas.

By focusing on rituals of listening, on the sonic articulation of space, time and power, this book seeks to explore the sonic fabric of Europe as part of a wider exploration of the sonic component of the social. If Europe appears to have fixated very early in its coming to self-consciousness on ritualizing the senses, on stitching the human sensorium into closely managed affective regimes, then understanding the role of sound in that process is crucial. One way in which Europe's fixation on regimes of the sensorium manifests itself is in a very particular construal of the listening subject as always already embedded in discourses of inclusion and exclusion (mirroring European constructions of insiders and outsiders in its emerging colonial outlook). This is particularly intensely manifest in European *pastoral* logic. As we shall see below, in the late Renaissance and early modern European imagination, the pastoral emerges as a privileged site for construing the relationship between sound, noise and silence: the withdrawal to quiet woods, for example, emerges as a code for banishment or withdrawal from court privilege; the hubbub of the city emerges as a symptom of dysfunctional and oppressive modernity (and disease, contagion, crowds, the rabble); silence is a kind of abandonment and noise is a kind of danger. To put it perhaps too succinctly, the pastoral metaphysics of sound proposed a crude binarism in which sound and silence were bound together in an endless dance of immersion in and withdrawal from the numerous sonic modernities already in place, or yet to come.

Paul Alpers has suggested that the European pastoral is 'a form that self-consciously considers questions of scope and limitation'.[10] Limitation as a cultural trope (putting boundaries in place, territorializations, enclosures), we would argue, is intensely bound up with the emerging European early modern fascination for mapping and rationalizing the world. Driven, that is, by an intense sensitivity to limits, these early European pastorals tended to construct intensely bifurcated imaginations of space, territory and sound, seeking to mark clearly the boundaries between 'here' and 'there' as a way of marking what is rendered available to exploitation. Indeed, in a quite unique and localized deployment of the resources of the auditory field, this bifurcation of 'here' and 'there' implicated in the pastoral imagination is also a strategy for organizing the unknown and rendering it knowable.

For Susan A. Stewart, indeed, the pastoral is the site of a profound disconnect between those represented (the silenced other, for example) and the authorial machinery of the poet (the exemplary European subject). In this sense, the pastoral also constitutes an attempt to represent as ideal or beautiful the exile of bodies to deadly silence, to a mute Arcadian scene in which cries of suffering are stifled. Stewart terms this the 'Philoctetes problem': taking the origins of pastoral logic back at least as far as Homer, she notes that Homer recounts in the *Illiad* (and after him, Sophocles in his play *Φιλοκτήτης*, *Philoctetes*,[11] thought to have been completed in 409 BCE) that the hero king Philoctetes is bitten by a water viper and falls gravely ill; his subjects cannot bear the sound of his moans of pain and abandon him to the island of Lemnos.[12] This cruel abandonment is connected, of course, to the *silencing* of suffering subjects, and ensnares that subject (*subjecting* her or him) in a sonic muffling enacted by the pastoral space 'outside' European civilization.

It seems, however, that Alpers and Stewart, despite their apparent disagreement on the terms and usefulness of pastoral logic, are both seeking to articulate a view of pastoral that is avowedly *modern*, one implicated in the specifics of early modern European re-imaginations of sound and power, and one, moreover, consistently in line with the great disenchantment of the world that constituted early European modernity proper. Indeed, as Daniel Chua has suggested, there is an intimate relationship between the intensification of the music–nature alignment and European modernity:

> The grounding of music in the 'natural' is a symptom of modernity. The symptom does not occur tangentially, as if it were a side-effect of the attempt of modernity to cure itself of its failures, but centrally as the very dis-ease that divides the ancient from the modern world. ... the alliance of music and nature lies at the epicentre of an epistemological earthquake.[13]

Pastoral operates alongside the project of European modernity, rather like a demand to tarry at a site of abandonment and disenchantment, to

> peer into the exotic world of suffering and to render it mute, render it calm, render it as a desirous space in which the abandoned hero would not be crying out in pain, but idly reclining under an oak tree, surrounded by the abundance of nature, in still-life communion with it.[14]

Silence and exile are key structural demands of the European modern pastoral and this specifically European logic is mirrored in the emergence of the musical genre of the *pastorale*. In that musical 'form' (probably too reductive a term), stillness (the performative absence of teleology) and spontaneous belonging (the sense of always already being in place) are presented as givens. It is, in particular, a certain (imagined) unbeholdenness (the absence of relation) of the pastoral community to other communities that this musical genre seeks to stage: the form seeks to generate what David Schwarz, drawing on Didier and Rosolato, has termed a 'sonorous envelope';[15] and the baroque form in particular, especially in the pastoral scenes of the French Baroque opera-ballet genre the *pastorale héroïque*, seeks to deliver a structure that can hold what German theorists of the first half of the eighteenth century termed *Aufmerksamkeit* or 'attention' in sufficient measure such that the form is not seen as wanting.[16] In other words, the High Baroque *pastorale* is about delivering the sonic image of a certain kind of abundance in enclosure. There are no outsiders here, no interlopers, no contested elements, and no 'others'. The sonic space of its 'envelope' is completely self-sufficient, as if it were speaking both of Europe *itself* as a self-contained and self-sustaining territory, always in place, never *there* and always *here*, and of some outside non-Europe, never *here*, always *there*. It is precisely this doubled quality of the pastoral, and its investment in silence, that interests us here.

There is, then, a narrative to be sketched in the history of musical pastoral forms, which could be told along the following lines: from, at least, Renaissance court song traditions across Europe – the Italian madrigal, the English and French airs – the High Baroque *pastorale héroïque*, through Gluck (the 'Dance of the blessed spirits', for example), Mozart (in, for example, the opera *Ascanio in Alba*), Beethoven (the 'pastoral' symphony) and the romantics (numerous *Lieder*, *chansons*, operatic scenes etc.) to European musical nationalisms including the so-called 'English pastoral school', each intervention constitutes a re-examination, with the musical resources at hand, of the attention/abundance dialectic.[17] In other words, musical forms do not simply comment on aurality more broadly (on listening rituals, soundscapes, sound production and management) but actively intervene in and shape (and are in turn shaped by) the logics of aurality that attend the culture in which those musical forms are incubated. There may be some cause, then, to be cautious about the splitting off of historical sound studies from the cultural history of music.

If European pastoral forms, beholden to the logic of enclosure we outlined above, play out a set of European Arcadian fantasies intimately bound up with the emergence of the modern European subject, then the aural fields we are interested in exploring in this volume are also fields of power, sites of social ordering, domains of *subjection*. And in these new domains of aurality, agency is redistributed into new and complex sound ecologies, in which subjects struggle to maintain their self-cohesion and the sonic processes of subjection become ever more intense. The early modern moment is sonically a period of rationalization, of rural/urban bifurcation and a period of far-reaching change: the old sonic worlds of the court and church are uprooted with remarkable efficiency and the new sonic order announces other new orders ahead of their arrival. Just as modernity is seeking to detach citizens from the moorings of the old certainties of church and monarch, so new forms of listening, audition and aurality emerge, manifest in the heady Arcadian mannerisms of European cultural forms at the onset of the colonial period. That colonization and pastoral should emerge simultaneously is no happenstance; pastoral is a symptom of a coherent process, described by Dainotto as, 'the surfacing of structures and paradigms' that incubated the idea of Europe and in which aurality constitutes another resource ripe for colonization.[18] As R. Murray Schafer famously put it, 'The territorial conquest of space by sound is the expression of visual rather than aural thinking. Sound is then used to demark property like a fence or wall'.[19]

The Disciplinary Turn

Sound studies is a now well established, if still relatively small field of research, although its origins reach back a long way (much further, arguably, than Pierre Schaeffer's *Traité des objets musicaux* (1966) and R. Murray Schafer's *The Soundscape: Our Sonic Environment and the Tuning of the World* (1977, repr. 1994)). It is only relatively recently (in the last two decades), however, that it has begun to assert itself across the humanities as part of a broader set of resources available to the historian of the senses. Sound studies as a philosophical and methodological orientation (with origins in, for example, acoustic ecology or the French acousmatic tradition) belongs to a wider turn towards thinking culture through the senses. Michael Bull and Les Back's volume *The Auditory Culture Reader*, for example, collects existing work from a range of scholars and theoretical approaches from an overwhelmingly empirical perspective. The ethnographic orientation of that collection means, of course, that historical approaches are very much in the minority (5 out of the 28 contributions), but the resonances of that collection's insights with the work of contributors in this volume should not be understated: to a certain extent, the demands of historical sound studies are always in some sense ethnographic, although for many in the social sciences, our use of the term here might seem somewhat strained. As an undoubtedly key resource in the field, Bull and Back's collection does not make an insubstantial contribution, therefore, to historical approaches to sound and audition. The demands of historical sound studies are often such that new and circuitous methodological routes demand a taking leave of presentist sound studies and that is precisely what our contributors have done in this volume. Like Bull and Back's collection, much of the recent work in sound studies is dominated by a presentist orientation, and understandably so. And yet, as we intimated above, many of these methodological and conceptual resources have been taken up by historians of sound: in this, the work of John Picker, Bruce Smith, Veit Erlmann,[20] Mark M. Smith,[21] Penelope Gouk and James Obelkevich are noteworthy.[22]

There are a number of key texts that have been implicitly drawn on by contributors to this volume. The classic texts in the field include Pierre Schaeffer's *Traité des objets musicaux* (1966) and R. Murray Schafer's *The Soundscape* (1977) already mentioned above, and Jacques Attali's *Noise: The Political Economy of Music* (1985). There is no full translation of Schaeffer's text in English. Drawing on French phenomenology in particular, Schaeffer's work is dedicated to constructing a theory of the sound object (*objet sonore*). Much of what Schaeffer has to say is important for the so-called acousmatic tradition, a tradition that has insisted on the autonomy of the sonic realm. Brian Kane has pointed out that there is considerable distance between the 'historical' meaning of Schaeffer's term 'sound object' and its current usage, a distance often deliberately eclipsed or ignored. In particular, Kane argues, drawing on earlier work by Carlo Palombini, Schaeffer's notion of the sound object is avowedly ahistorical, shot through with 'mythical' thinking and dangerously close to a kind of essentialism unhelpful to the contemporary scholar of sound. This is particularly evident in Schaeffer's understanding of technology, which, according to Kane, is tinged with mysticism:

> ... Schaeffer presents a picture of ahistorical, existential man discovering himself within a teleological horizon. What modern technology reveals for Schaeffer, aside from uncritically optimistic potential, is little more than an abstract glimpse into an ancient 'originary experience'.[23]

Kane's critique is well made and his point, revolving in particular around the peculiar absence of close readings of Schaeffer's original *Thesis*, is that 'readers' of Schaeffer have tended to impute to him the kinds of qualities they find useful for themselves, without looking closely at the intellectual assumptions underpinning such notions as the autonomy of technology, the *epoché* and the sound object. As Kane puts it, 'the philosophical context within which the sound object was originally theorized is quite generally neglected'.[24] Clearly, then, this kind of ahistorical tendency in Schaeffer's phenomenology causes problems for historians of sound and there are good reasons, at least for this collection, to see Schaeffer as historically limited to a specific European philosophical moment.

R. Murray Schafer's *The Soundscape* is a key text in soundscape studies, and emphasizes the abstractable spatial dimensions of sound (often, according to his critics, neglecting human agency). One particularly problematic aspect of Schafer's work is its tendency to attach to an uncritical sonic ruralism. As Richard Oddie puts it:

> … acoustic ecology and environmental phenomenology have both tended to emphasize the positive qualities of wilderness spaces and rural environments, while representing urban environments as manifestations of the negative qualities associated with industrialization, mechanization, and modernity. Schafer, for example, places particular emphasis on the ways in which urban environments encourage a loss of connection with the non-human world, silencing non-human voices beneath the din of modern technology.[25]

Thus, Schafer's acoustic ecology is somewhat out of step with the critical tone of the contributions in this volume (and others like it). Many of the contributors to this collection nonetheless draw on this body of work (sometimes called 'acoustic ecology'), but do so from a historicist orientation, mindful of the need for human agency and historical specificity in theorizations of sound.

The title of Jacques Attali's monograph *Noise* [*Bruits*] is deceptive since, although it offers some insights for the study of sound more broadly (and, indeed, some contributors in this volume draw on Attali's work), it is nonetheless primarily focused on music and music making. Steven Shaviro has recently critiqued Attali's work for its tendency to a 'universal history':

> Attali, like so many European thinkers, consigns everything that lies outside Europe and its Near Eastern antecedents to a vague and undifferentiated 'primitive' category, as if there were no differences worth noting among them, and nothing that any of these other cultures could offer that was different from the European lineage.[26]

One could object, of course, that, as we demonstrate below, there is something different, something almost glacial, about sound as historical material, and that its tendency to local specificity is far less marked than, for example, visually mediated communication. And, perhaps, it is precisely here that the ground of Attali's 'universal history' lies. And yet, the crudeness of Attali's Eurocentrism, his consistent exiling of non-Europe to the margins, and his almost unrelenting denigration of the East throughout *Noise*, are all good reason to take Shaviro's critique seriously.

In the last 20 years, landmark monographs in sound studies have appeared that have transformed the field, widened its interdisciplinary reach and intensified its pluralism. Readers will be familiar with Douglas Kahn's *Noise, Water, Meat* (1999), Jean-François Augoyard and Henry Torgue's *The Sonic Experience: A Guide to Everyday Sounds* (2005), Steve Goodman's *Sonic Warfare* (2009) and Michael Bull's *Sounding Out the City* (2000) and his later *Sound Moves* (2007). Anahid Kassabian's timely and finely-wrought *Ubiquitous Listening: Affect, Attention, and Distributed Subjectivity* (2013) constitutes one of the most accessible and yet challenging recent interventions in the field, interrogating some of the ways, as she puts it, we 'listen to the music we hear everyday'.[27] Kassabian's insights about what she terms 'distributed subjectivity' open up a range of radically destabilizing trajectories in and out of the discipline of sound studies and are useful precisely for their refusal to remain limited to disciplinary territories or preconceived notions of musical (and sonic) value. Michael Bull's work (*Sounding Out the City*, 2000, and *Sound Moves*, 2007) stands out in the literature for its commitment to an ethnographic focus on contemporary listening practices (the earlier book deals with uses of the Sony Walkman, the later with uses of the iPod). As befits this kind of ethnographic work, Bull's work makes extensive recourse to fieldwork and interviews. Again, Bull's work is presentist in its emphasis, and, whilst an absolute key contribution to the field, it leaves plenty of material for this volume to explore. Indeed, contributors to this volume have both drawn on an historicized articulation of his methodology, and, to a certain extent, inevitably departed from his methodological orientation in search of an analytical approach better suited to the specificities of aurality in the pre-phonographic era.

Recent historically oriented monograph contributions to the field include Alain Corbin's *Village Bells: Sound and Meaning in the Nineteenth-Century French Countryside* (1998), Bruce R. Smith's *The Acoustic World Early Modern England: Attending to the O-Factor* (1999), John M. Picker's *Victorian Soundscapes* (2003) and Jonathan Sterne's *The Audible Past: Cultural Origins of Sound Reproduction* (2003). Corbin's well-known account of French rural soundscapes shows how bells and bell-ringing reflected the social, political and ideological makeup of rural life. His approach, born of the French *Annales* tradition, emphasizes the rhythm of the day as articulated by bell-ringing and places the everyday rituals of the countryside at the heart of his study. It is a profound and unique contribution to the field, and one which a number of the contributors in this collection draw upon. Smith's work attempts to recreate the soundworlds of Shakespearean England, considering the physicality of sound and its meanings by drawing on a wide range of sources and case studies. In particular he applies his findings to the world of Elizabethan and Jacobean theatre. Smith's wide-ranging account offers valuable methodologies for exploring the soundscapes of the past, and, indeed, a number of contributors to this collection cite Smith's work. John M. Picker's *Victorian Soundscapes*, like the work of Smith, develops out of literary studies, and, like Smith, Picker draws on an array of materials and case studies to give a detailed account of the texture and character of the Victorian soundscape. The work is characterized in particular by its close reading of a wide range of primary sources and by its careful and painstaking reconstruction of Victorian attitudes to what he calls 'close listening'. Sterne's *The Audible Past* is a detailed account of North America's encounter with sound reproduction technologies from the invention of the phonograph in 1877. This careful and thought-provoking volume is perhaps one of the most comprehensive of the recent achievements in the field of historical sound studies. Its avowedly historical and historicist orientation blends cultural studies, the history of communication technology and cultural history to deliver a comprehensive and textured account of the North American soundscape in the twentieth century. All four historically oriented monographs deal, to varying degrees, with thinking about the acoustic, the sonic and noise and sound more generally, as a historiographically focused question, whilst concentrating also on specific, historically determined and geographically located case studies. In this sense they represent a significant ground on which much work contained within the current volume is built.

The Scope of this Volume

Cultural Histories of Noise, Sound and Listening in Europe, 1300–1918 presents a range of historical case studies on the sounding worlds of the European past and includes: considerations of sound, music and religious experience in the later Middle Ages; interrelationships between sound, religious and civic festivities, poetics and urban soundscapes in the early modern period; the history of the body and perception during the *Ancien Régime*; the sounds of the city in the nineteenth century; and sound and colonial rule at the *Fin de Siècle*. These individual histories also range in geographical orientation to include considerations not only of Britain and France, the countries to date most considered in European historical sound studies in Anglophone scholarship, but also Bosnia-Herzegovina, British Colonial India, Germany, Italy and Portugal.

This volume is divided into three thematic sections: 'Historicizing Aurality'; 'Sound Politics'; and 'Urban Soundscapes of Europe'. 'Historicizing Aurality', consisting of four historical case studies, is concerned with the methodological questions that inevitably emerge when trying to construct a historiography of sound: what, these chapters by Lisa Colton, Veit Erlmann, Emily Laurance and Alexander Bonus all ask, is it possible to know about the sonic past? How, that is, is the sounding past made available to us in numerous and varied written primary source materials? And, how were these sounding worlds interpreted as meaningful by the individuals and communities who experienced them? It is striking that all these chapters bear witness to the subterranean connections between radically different sonic cultures and epochs; in essence, these chapters, especially when *read together*, suggest that what Steven Shaviro terms different 'media regimes' operate according to quite specific local historical logics, and that those logics are often out of historical synch with each other, developing and changing at different speeds according to different models of causality and agency.[28] In the case of historical approaches to sound, these differences are both more evident (in that sound operates differently to the materials of other media regimes) and yet more redolently attached to the slower continuities in European experience: the historical analysis of sound often shows that sound regimes change at what seems like a glacial pace, despite important local specificities, and the connections between apparently radically discontinuous sound regimes are often quite striking. In this regard, the first section demonstrates how historical sound studies, especially in a European context, are concerned always, at some level, with relationships of continuity and change across the *longue durée*.

The second section, 'Sound Politics', consists of three chapters, all concerned with the potential of sound to articulate, intervene in and shape the political sphere. These three case studies by Hedy Law, Jim Sykes and Risto Pekka Pennanen explore the political (in the broadest sense) by scrutinizing some of the ways in which sound and noise were used as resources in forming and maintaining group and individual identities. Sonically mediated identity politics, as a set of political, epistemological and ontological debates about the relationship among individuals, groups and larger social units, brings into focus some of the key disciplinary questions that face anyone seeking to understand how social groups make claims to the political field and how sound enables or disables any such claims.

The third, largest and final section, 'Urban Soundscapes of Europe', consists of five chapters, each of which focuses on determining what the European urban imagination might constitute, sonically. In other words, these chapters, by implicitly challenging the ruralism of R. Murray Schafer and acoustic ecology, set out to make claims for the uniqueness and intensity of the urban soundscape. What, these chapters by Helen Coffey, Daniele V. Filippi, Adam Hansen, Olivier Balaÿ and João Silva all ask, is it possible to experience of the rapidly changing (and irreversibly transformed) urban soundscapes of Europe? What challenges does the sonic archaeology of European urban spaces pose and how can we respond methodologically to those challenges? In particular, the urban, as a key site for the formation of European modernity, is a privileged space when it comes to trying to understand the relationship between sound and power as symptomatic of the new modern order in Europe after the Reformation. What, these chapters all ask, does it mean to call a sound 'urban', and how does that designation help us understand the historical processes that organize and make sense of sound?

Whilst the grouping of chapters into the three sections reflects higher-level thematic interconnections between them, the fluid nature of sound as a medium and its susceptibility to multiple temporal, cross-cultural and cross-locational connectivities is such that shared historical narratives of sound can be traced across the collection as a whole. The long history of the significance of sound as a semantic marker in European communal rituals – religious and civic – can be plotted, for instance, from Helen Coffey's exploration of the role of musicians in the towns of Maximilian I's Germany, through Daniele Filippi's consideration of some of the ways in which heavenly and hellish soundscapes were evoked in urban religious festivals in early modern Italian towns and cities, Risto Pekka Pennanen's analysis of the uses of cannon salvos and bells in the articulation of hegemonic colonial power in Habsburg Bosnia-Herzegovina to Jim Sykes's chapter on uses of drumming in religious ritual in British colonial Ceylon. Similarly, in addition to the chapters collected in Section III, the chapters by Emily Laurance, Hedy Law and Risto Pekka Pennanen also deal with the ways in which particular urban centres could be understood as sonically marked spaces. Indeed, one urban centre that appears across this collection, both explicitly and implicitly as a cultural trope, is Paris, from the eighteenth-century case studies by Veit Erlmann and Hedy Law, both centred on Paris, and the study by Emily Laurance of nineteenth-century street cry collecting in Paris, to the haunting of Lisbon and Lyon by Paris-style

Hausmannization and its aural consequences in the nineteenth century in the chapters by João Silva and Olivier Balaÿ.

David Harvey has famously identified Paris as the 'capital of modernity' and it is not surprising that Paris should figure so prominently here.[29] Indeed, Paris is a key site for the rehearsal of new sonic regimes and stands for a certain attitude to urban modernity. Moreover, the rise of Paris, we might say, as capital of Europe in the nineteenth century, points again to the coherence effect that 'Europe' as an idea, as a set of affective *regimina*, enables. This book is focused precisely on seeking to try to understand how this coherence effect is enacted and enabled in sound, and how Europe as a special place, comes to be, as Chakrabarty puts it, *provincial*.

Notes

1 Leon Botstein, 'Toward a History of Listening', *The Musical Quarterly*, 82/3–4, Special Issue: 'Music as Heard' (Autumn–Winter, 1998): 427–31, at p. 429.
2 See in particular: Arjun Appadurai, *Modernity at Large: Cultural Dimensions of Globalization* (Minneapolis: University of Minnesota Press, 1996); Edward Said, *Orientalism* (New York and London: Routledge and Kegan Paul, 1978); Homi Bhabha, *Location of Culture* (New York and London: Routledge, 1994); Frantz Fanon, *Les Damnés de la Terre* (Paris: Éditions Maspero, 1961) trans. Constance Farrington as *The Wretched of the Earth* (New York: Grove Press, 1963).
3 Dipesh Chakrabarty, *Provincializing Europe: Postcolonial Thought and Historical Difference* (Princeton, NJ: Princeton University Press, 2000), p. 27.
4 See, in particular, Notebook 25 from the later (in English unpublished) *Prison Notebooks* 'On the Margins of History: History of Subaltern Groups'. See Marcus Green, 'Gramsci Cannot Speak: Presentations and Interpretations of Gramsci's Concept of the Subaltern', *Rethinking Marxism*, 14/3 (Fall 2002): 1–23.
5 Chakrabarty, *Europe*, p. ii.
6 Roberto M. Dainotto, *Europe (In Theory)* (Durham, NC: Duke University Press, 2007), p. 47.
7 Jean Jacques Rousseau, *Oeuvres complètes*, vol. 3 (Paris: Gallimard, 1964), p. 567; trans. and quoted in Dainotto, *Europe*, p. 136.
8 For a suggestive reading of the idea of the European unconscious see Fanon, *Wretched*, p. 267ff.
9 For more on this aspect of European Humanism, see in particular Edward Said, *Humanism and Democratic Criticism* (New York: Columbia University Press, 2004), especially p. 28 ff.
10 Paul Alpers, '"The Philoctetes Problem" and the Poetics of Pastoral', *Representations*, 86, Special Issue: 'Music, Rhythm, Language' (Spring, 2004): 4–19.
11 It is known also that Aeschylus and Euripides also wrote versions of the story but theirs have not survived.
12 This argument is rehearsed more fully in Ian Biddle, 'Visitors or the Political Ontology of Noise', *Radical Musicology*, 4 (2009), http://www.radical-musicology.org.uk: pars 1–22: 13. For more on the 'critique' of pastoral, see Susan A. Stewart, *Poetry and the Fate of the Senses* (Chicago: Chicago University Press, 2002), p. 102.
13 Daniel Chua, 'Vincenzo Galilei, Modernity and the Division of Nature', in Suzannah Clark and Alexander Rehding (eds), *Music Theory and Natural Order from the Renaissance to the Early Twentieth Century* (Cambridge: Cambridge University Press, 2001), pp. 17–29, at p. 17.
14 Ian Biddle, 'Visitors': par 13.

15 David Schwarz, *Listening Subjects: Music, Psychoanalysis, Culture* (Duke University Press, 1997), p. 7.
16 In this regard, Johann Christoph Gottsched's famous neo-Aristotelian essay on tragedy, 'Von Tragödien oder Trauerspielen', chapter 10 of his *Versuch einer kritischen Dichtkunst*, is exemplary. Available in the *Ausgewählte Werke*, vol. 2/2 (Berlin and New York: Walter de Gruyter, [1737] 1973), pp. 309–35.
17 See, for instance, Giuseppe Gerbino, *Music and the Myth of Arcadia in Renaissance Italy* (Cambridge: Cambridge University Press, 2014); Jeanice Brooks, *Courtly Song in Late Sixteenth-Century France* (Chicago and London: University of Chicago Press, 2000); Kirsten Gibson, '"So to the Wood went I": Politicizing the Greenwood in Two Songs by John Dowland', *Journal of the Royal Musical Association*, 132 (2007): 221–51.
18 Dainotto, *Europe*, p. 4.
19 R. Murray Schafer, 'Acoustic space', in David Seamon and Robert Mugerauer (eds), *Dwelling, Place and Environment: Towards a Phenomenology of Person and World* (Dordrecht, Netherlands: Martinus Nijhoff, 1985), p. 92.
20 Veit Erlmann's *Hearing Cultures: Essays on Sound, Listening and Modernity* (Oxford and New York: Berg, 2004) was developed out of the 2002 Wenner Gren symposium on hearing cultures, and comprises 10 essays taking historical and/or ethnographic approaches to hearing, mostly, but not exclusively, focused on twentieth-century case studies. The collection is particularly useful in laying out some of the key critical issues attending sound studies, from questions relating to the anthropology of sound; the role of acoustic ecology in shaping sound studies as a discipline; sound as a historical problem in early modern England; American modernism; sound and modernity in Egypt; the critical questioning of the notions of proximity and distance in sound studies; and sound in early motion pictures.
21 Mark M. Smith's *Hearing History: A Reader* (Athens: University of Georgia Press, 2004) takes a broader historical view than Erlmann's collection. Bringing together key work on aurality already in circulation by a range of historians and theorists from the previous decade (together with a handful of earlier pieces), Smith's collection maps the emerging field of historical sound studies as it stood at the beginning of the twenty-first century.
22 See, for example: John M. Picker, *Victorian Soundscapes* (Oxford: Oxford University Press, 2003); Bruce Smith, *The Acoustic World of Early Modern England: Attending to the O-Factor* (Chicago: University of Chicago Press, 1999); Veit Erlmann, *Reason and Resonance: A History of Modern Aurality* (New York: Zone Books, 2014); Mark M. Smith, *Sensory History* (New York: Berg, 2007) and *Listening to Nineteenth-century America* (Chapel Hill, NC: University of North Carolina Press, 2001); Penelope Gouk, *Music, Science, and Natural Magic in Seventeenth-Century England* (New Haven, CT. and London: Yale University Press, 1999).
23 Brian Kane, 'L'Objet Sonore Maintenant: Pierre Schaeffer, Sound Objects and the Phenomenological Reduction', *Organised Sound*, 12/1 (April 2007): 15–24, at p. 22.
24 Kane, 'L'Objet Sonore': 15.
25 Richard Oddie, 'Other Voices: Acoustic Ecology and Urban Soundscapes', in Ingrid Leman-Stefanovic and Stephen B. Scharper (eds), *The Natural City: Re-envisioning the Built Environment* (Toronto: University of Toronto Press, 2012), pp. 161–73, at p. 162.
26 Steven Shaviro, 'Attali's Noise', *Pinocchio Theory*, (15 April 2005) http://www.shaviro.com/Blog (accessed 30 July 2014).
27 Anahid Kassabian, *Ubiquitous Listening: Affect, Attention, and Distributed Subjectivity* (Berkeley and Los Angeles: California University Press, 2013), p. 1.
28 Steven Shaviro, *Post Cinematic Affect* (Ropley, Hants: Zero Books, 2010), p. 67.
29 David Harvey, *Paris, Capital of Modernity* (London and New York: Routledge, 2003).

Section I

Historicizing Aurality

Introduction

Ian Biddle and Kirsten Gibson

Ranging from the late Middle Ages to the twentieth century, the chapters in this section draw on a varied range of materials – religious writings, early modern scientific writings and aesthetic debate, a nineteenth-century collection of transcriptions of street cries and the work of nineteenth-century experimental psychologists – to consider perceptions of sound, real and imagined, as a historically contingent semantic agent. While the historical span of these chapters is broad and the case studies in their individual nature diverse, what unites them is the underlying questions they pose: what has sound *meant* to subjects past? And how have they used sound to mediate, shape or represent their understanding of the world around them? The chapters presented here do not, of course, attempt to answer such questions systematically or exhaustively for the epoch or context under scrutiny as a whole. Rather, the focus here is placed on the small scale, on individual experiences and mediations of sound. In all of the chapters in this section the aim of the authors has been to interrogate how, for certain individuals living in discrete historical and geographical circumstances, sound carried meaning that shaped the ways in which they perceived and understood their inner worlds and the world around them.

Chapter 1 by Lisa Colton, drawing on the religious writings of Richard Rolle, Margery Kempe, Walter Hilton and John Wycliffe, for instance, explores how sound was understood by individuals living on the fringes of official religious doctrine. Colton's chapter attempts not to recapture the musical sounds of a specific piece, nor to summarize the different musical sounds familiar to members of medieval society in their daily lives, but is, rather, concerned with how the individuals whose writings are analysed here perceived musical sound, and the ways in which it was understood by them as meaningful. Colton's study is, in many ways, a series of 'microhistories'.[1] What is significant and striking about Colton's chapter is her focus on authors who stood outside the religious 'mainstream', who were 'less insistent that we should all experience music in an approved fashion', and who, in the case of Rolle and Kempe in particular, presented their writing in deeply personal terms. Through this focus on the experiences and perceptions of religiously non-orthodox individuals, Colton's chapter consciously veers away from the well-trodden path of mainstream musicology of the period, primarily concerning itself with the writings of those engaged in more orthodox religious practices and beliefs, not in order to further examine beliefs about sound in this period in terms of official debate and dominant belief, but to uncover the personal experiences and perceptions of individuals living in that particular historical reality.

Early modern France provides the backdrop for Chapter 2 by Veit Erlmann. His study focuses on Claude Perrault's exploration of sound and hearing in *Du bruit* (1680) in which Perrault proposed the notion that the body was 'animated' throughout – that every nerve, fibre and muscle was equipped with its own reasoning faculty – thus challenging the Cartesian concept of the body as a machine separate from the soul. Erlmann's chapter highlights the significance of Perrault's work, long overlooked by historians, in the history of otology. He goes on to consider Perrault's, somewhat contradictory, writings on early French opera, in which Perrault developed a musical aesthetics based on subjective relativism and aural pleasure in which listening was described not as being based on the mechanism of sympathetic resonance but on culture, particularly mediated through the concept of 'taste'. Erlmann's study suggests a subtle retelling of the narrative of the early modern history of otology, positioning Perrault's work as a significant precursor to later tendencies to assign unprecedented embodied primacy to subjective auditory experience and, reflecting the breadth of Perrault's work, touches on a wide-ranging set of not always concordant sources and debates – scientific and aesthetic – to illuminate our view of early modern French understandings of hearing and aural pleasure.

Also focused on Paris, Chapter 3 by Emily Laurance takes as its central focus Georges Kastner's *Les Voix de Paris* (1857). Drawing on the popular *tableau* format, Kastner's *Voix de Paris* is devoted to evoking Parisian urban experience in terms of aurality, particularly from the perspective of Parisian street vendor cries (the figure of the street crier, of course, inhabits other chapters in this collection). Kastner's 'aural *tableau*' of Paris includes a general discussion of 'the cry', linking it with the French national character, a history of the street vendor cry from the Middle Ages to his current day and almost 400 musical transcriptions of then-current street cries, organized by categories of the goods that they were used to sell. Kastner, a trained musician, deployed the cries he had collected in a three-movement cantata, *Les Cris de Paris*, affirming that one of the central aims of his collecting and transcribing work was to generate and use these 'raw' musical materials in his compositional practice, particularly in relation to the development of a nationalist musical style. Like the work in previous chapters in this section, Laurance's concerns in this chapter are not to attempt to recapture, revive or analyse these captured sounds from nineteenth-century Paris, nor to musically analyse the re-workings of these transcriptions in Kastner's composition. Rather, she examines the more implicit underlying aims of Kastner's work, and that of his contemporaries, in terms of what it can tell us about the ways in which sound, and ideas about sound, became a significant medium for negotiating the broader issues of national identity in post-revolutionary France, class identities, social order and, ultimately, the refashioning of French history in order to attempt to reclaim France's pre-Revolutionary past for a modern bourgeois society.

The final chapter in this section, by Alexander Bonus, turns our attention to the nineteenth-century laboratories of experimental psychologists. It is in this context that new concepts of precision-oriented, mechanically regulated musical time emerged as a positive ideal, one that led, in the twentieth century, to the ubiquity of the metronome in the training and practice regimes of classical musicians, and, perhaps more significantly, pervasive understandings of 'good' and 'bad' musical rhythm in the twentieth century. Bonus considers, in particular, the work of Wilhelm Max Wundt, who founded the first experimental psychology laboratory at the University of Leipzig in 1879. Among the assembly of clockwork instruments employed in his research into the variables of human perception and action was a Maelzel metronome. Wundt's research involved the exposure of subjects to repetitive mechanical sounds in order to examine how long the mind was able to hold a string of basic ideas or 'impressions'; the metronome became *the* essential control for these experiments. Wundt's experiments on consciousness had far-reaching consequences that were felt well beyond the confines of the scientific laboratory. Indeed, as Bonus demonstrates, they helped shape and define modern concepts of 'rhythm'. Most fundamentally, Wundt's work radically shifted the concept of musical rhythm from a subjective, internal pulse reference to an objective, unerringly precise phenomenon independent of human agency, and this concept, and by extension its inherently implied musical values of 'good' and 'bad' rhythm, was taken up at the *fin de siècle* by other psychologists who turned their attentions to assessing musical performance against the yardstick of metronomic accuracy. Some of these psychologists, such as Carl Emil Seashore, extended the influences of these scientific ideals to a wider public by disseminating their work in important music publications, and by the 1920s such ideals were becoming pervasive among both amateur and professional music practitioners. Like Chapter 2, Bonus's chapter eloquently draws to our attention the mutually interdependent and constantly shifting relationships between scientific and aesthetic judgement in our mediation of understandings and perceptions of sound in the modern world.

Despite the historical and thematic diversity of these case studies, and their focus on individual experiences, perceptions and mediations of sound, some important broad themes emerge across the wide historical span these chapters cover: the interdependent – though often conflicted – relationships between religious, scientific and aesthetic thought in the ways in which subjects have understood sound as meaningful; the ever-present place of sound as a bridge between the material and the immaterial, the bodily and the sensual, on the one hand, and the spiritual or psychological, on the other; the endurance of sound as a significant marker of space, place and identity; and the semantic agency of sound in the articulation of both hegemonic and counter-hegemonic ideals. The persistence of these ideas about, and engagements with, sound over the *longue durée* is indicative that sonic history is, in fact, as much about what is not historical as what is: it is as much about trying to understand the slower-moving subterranean processes by which we come to listen, as much as it is about trying to understand local networks for enacting that longer and older set of inclinations. The individual chapters here deal with detailed historical specificity, with microhistories, and, in some cases, with mapping individual narratives onto the broader historical contexts in which they lived, but seen collectively, side-by-side, they begin also to reveal longer, slow-moving narratives of how sound has been understood to meaningfully shape and reflect human experience. These themes, as we shall see, will be woven throughout this collection as a whole, and it is the relationship between sound, identity, power and politics that will be the focus of the following section.

Note

1 Giovanni Levi, 'On Microhistory', in Peter Burke (ed.), *New Perspectives on Historical Writing*, Second Edition (Cambridge: Polity Press, 2001), pp. 97–119.

1 'Sowndys and melodiis'

Perceptions of Sound and Music in Late Medieval England

Lisa Colton

The perception of noise and music by individuals, and the meanings that they attributed to sounds heard and imagined, are revealed in various texts from the later Middle Ages. Many writers incorporated references to 'heavenly' or 'hellish' sounds into their religious writings, often as a metaphor for the divine or as representative of more earthly, morally corrupting delights. In spite of moves by the Cistercians and other religious groups to regulate plainchant, and to eradicate unauthorized or 'lascivious' melodies, new musical styles developed within the Christian church that explored ever more ambitious textures, rhythms, harmonies and performance practices.[1] Outside of regular, institutional worship, the piety of individuals was also informed by sound in ways that have attracted less attention, especially when such authors did not also compose music.

This chapter will draw on texts by English mystics and religious commentators whose writings were considered unorthodox in order to examine how sound was understood by writers living on the fringes of official religious doctrine. Focusing on ideas generated by writers whose spiritual practices were criticized as heretical allows the exploration of sound as it was perceived away from the 'mainstream' approved practices of the Church. Examples will include the spiritual song of Richard Rolle (d. 1349), sounds experienced by Margery Kempe (*c.*1373–after 1438) and the description of performance found in the reforming tracts of John Wycliffe (d. 1384) and his contemporaries. The devotional texts of these individuals were informed by the teachings of the Christian church, but were, paradoxically, also positioned against the prevailing discourse of their age. Many authors drew freely on ideas or even textual phrases from much older theological writings, but employed them within their own distinctive style of writing.[2] In order to explore especially personal perceptions of musical sound, by people who were less insistent that we should all experience music in an approved fashion, the examples avoid texts such as treatises on plainchant, harmony or the modes. My question is not 'what did music sound like', but 'how did *these* writers perceive sound, and what did it mean to them'?[3] The selected writers were not trying to tell their audiences about sound: they were using musical references to explore wider questions of the nature of faith, God or prayer. The link between hearing and the other senses, and between the parts of the body that sensed sound, smell, taste and the visual, will recur in these examples, and for that reason this chapter will start with texts that concern song as a corporeal experience.

Music in the Spiritual Body

Musical sound was a fundamental part of everyday life in the Middle Ages; it was neither restricted to particular times of the day, nor to the architectural confines of particular buildings. The boundaries between sacred and secular forms of song were also fluid, and song was therefore worthy of careful scrutiny by theologians, who feared its morally corrupting properties. In a notable example, the Franciscan Bishop of Kilkenny, Richard Ledrede, compiled the Red Book of Ossory (*c.*1320) for the vicars choral at St Canice's Cathedral, appropriating vernacular songs for various liturgical feast days by the substitution of spiritual words.[4] For Ledrede, the songs could fulfil a sacred purpose if their original lyrics were erased and replaced by devotional content, fit to fill the mouths of his clerics. In a specific reference to the physical experience of singing secular poetry, Ledrede claimed that his new texts had a role in moralizing clerics from within their own bodies, 'lest the throats and mouths [of the brethren] sanctified by God should be defiled by theatrical and worldly songs'.[5] This practice of contrafactum (in which melodies were furnished with replacement texts) created alternative forms of song for new purposes, and was a common way of creating music, sometimes changing meaning subtly and sometimes offering a more wholesale change in a different language and with a subject switched from courtly love to admiration of the Virgin Mary, for example. The sacred and the secular were rarely successfully separated from one another; furthermore, some musicians seem to have delighted in the use of religious and decidedly secular texts and melodies in combination.[6]

An example of such practices in England can be seen in the motet *Herodis in pretorio / Herodis in atrio / Hey hure lure* in a mid-to-late fourteenth-century collection of sacred vocal music from Durham Cathedral (Durham Cathedral Library, MS C. I.20, f. 1^{r}), in which two upper texts in Latin, relating to Herod and the murder of the Holy Innocents, are placed above a tenor whose French lyrics, though problematic to translate, seem to carry a potentially salacious meaning: 'Hey, you hairy, strapping fellow, hanging out. Place me downwards, line me up [with you]'.[7] Frank Harrison has suggested that the fact that the tenor lyrics are written in red ink (rather than black or brown) might indicate that they were 'not for use when the motet was performed in choir', but the visual effect is to draw attention to the lyrics, and the scribe could more easily have covered his source by leaving the tenor tune un-texted.[8] A modern commentator can only begin to speculate on the complex implications of sacred/profane references in this motet, not least because its performance took place in the homosocial musical culture of Durham Cathedral Priory.[9]

But how was music imagined or experienced in relation to the medieval body and its senses? Sources of evidence that have been relatively neglected by musicologists (though they are well known to scholars of medieval literature and theology) are the writings of the English mystical tradition and texts written by critics of the established church in the late fourteenth and fifteenth centuries. The English mystics of the later Middle Ages were not, on the one hand, a 'school' as such, in that they did not follow a common theology, or base their writings explicitly on one another's in a coherent or consistent manner. On the other hand, they can be seen to have shared common concerns with the experience of God in relation to their bodies and to the world around them, and they also reflect, in diverse ways, an awareness of previous and contemporary mystical writings from England and continental Europe. Some mystical writers were associated with monastic communities, or were trained within the Church. Walter Hilton (1340–96), for example, was the head of the Augustinian Priory at Thurgarton, Nottinghamshire. Others, like Margery Kempe, were lay people, hermits or anchorites, who sought authentication of their deeply personal experiences within the wider Christian church. In this way, all such writers were both part of their communities (broadly defined) and positioned in opposition to aspects of theological doctrine or to the expectations of society for religious thinkers. In the writings of the English mystics, it is possible to explore both prevailing understandings of faith and human experience and, perhaps more usefully in this study, intensely personal reactions to, and reflections on, sound and the spiritual body.

Richard Rolle: Spiritual Song

The mystical experiences of Richard Rolle were inextricably linked with his sensory perception, the *canor*, *fervor* and *dulcor* – melody, warmth, sweetness – that were frequently described in his various writings, notably in his most famous work the *Incendium Amoris*, or 'Fire of Love', which still survives in over 40 copies from the fourteenth and fifteenth centuries. As Nicholas Watson has noted, Rolle's sensory experiences were usually described in a ranked order, from sight of heaven through touch (*fervor*), smell or taste (*dulcor*) to sound and song (*canor*).[10] Rolle experienced *fervor* and *dulcor* in his chapel nine months before his receipt of the gift of *canor*, spiritual song, a time lapse that related his musical inspiration to the pregnant female body. His gift of song was described as a reflection of the music of heaven, in Rolle's typically alliterative style, and was evidently a deeply physical experience:

> While I was sitting in the same chapel and saying the night-psalms before supper as best I could, I heard as it were a ringing of psalmody, or rather of singing, above me. And when I was stretched out in prayer and with all my desire towards heavenly things, I do not know how but I soon felt a symphony of song within myself and caught up from heaven the most delicious harmony, which remained with me in my mind. For my thought was forthwith changed into a tuneful song; and I had as it were melodies in my meditation, and I also gave out the same song in my prayers and psalmody.[11]

Rolle's writings reveal that he avoided communal singing and listening to the organ in church (*Incendium Amoris*, ch. 33). He also advocated a manner of singing that seems similar to meditative practices found in many religions:

> [The lover of the Godhead] will be sweet and godly and full of song … Then it is that there happens to such a lover what I have never found in any learned writing or have heard expounded, namely that this song will spring to his lips,

> and he will sing his prayers in a spiritual symphony of celestial sweetness. It will make him slow of speech … and what once occupied him not more than an hour he will now find difficult to complete in half a day. And while this is happening he will sit alone, mixing as little as possible with those who sing psalms, and deliberately not singing with the rest.
>
> (*Incendium Amoris*, ch. 32)[12]

Spiritual song, Rolle says, 'is not an affair of those outward cadences which are used in church and elsewhere; nor does it blend much with those audible sounds that are made by the human voice and heard by physical ears; but among angel melodies it has its own acceptable harmony' (*Incendium Amoris*, ch. 33).[13] It is traditional in the literature to understand Rolle's descriptions of spiritual song as internalized – and essentially silent – forms of music, as a form of sound that had 'no physical expression'.[14] To do so neglects the crossover between spiritual song and the descriptive terms Rolle used to explain it, rooted as they were in his knowledge and experience of the liturgy; for example, we are told in the *Incendium Amoris* that on the eve of Assumption day, one year, Rolle attended church to pray, and that the next day he sang at Matins and gave a sermon at Mass. The *Incendium Amoris* makes it clear that the true lover of God will not be able to stand hearing psalmody unless his own inner song is attuned to it (*Incendium Amoris*, ch. 33). Rolle's accommodation as an anchorite by the Cistercian nuns of Hampole brought him into contact with the monastic liturgy, personal piety and musical practices of enclosed women, including anchoress Margaret Kirkby.[15]

From writings such as the *Incendium Amoris*, it is evident that Rolle found the bodily experience of song more powerful and meaningful than any textual elements conveyed in the liturgy; in a sense, this was a reversal of standard ideals in which the music was expected to enhance the text. In addition, his alliterative style – in English and in Latin – was an important example of a prevailing manner of writing in later Medieval England, especially in the midlands and the north, one that arguably foregrounded the sonic properties of words over their semantic meanings.[16] This style can be seen (and heard) in Rolle's work, including glosses and commentaries on devotional texts, such as his *Super Psalmum Vicesimum*, the *Super Canticum Canticorum* and his *English Psalter*, as well as prose and poetry that was informed by religious themes and experiences.

An anonymous contemporary of Rolle, working in a similar tradition, produced an alliterative poem on the penitential psalm *Miserere mei, Deus* (Psalm 50 (51)), and gives a useful viewpoint of the relationship between the body and that which it perceives through sound in private devotions. This writer described his hearing as 'my ears within', a phrase which playfully elided the unseen, hidden physiological 'ear' with the aural (and therefore not visually perceptible) experience of sound.

> *Auditui meo dabis* [*gaudium et leticiam et exultabunt ossa humilitata*]
> To myn herynge myn eris within,
> Lorde, sere solace sall þu sende,
> With ioy and blysse þat neuer sall blyn,
> And laste in welthe withowtten ende.[17]

Margery Kempe and Walter Hilton: Sounds of Good and Evil

The Boke of Margery Kempe is unique in its portrayal, in a literary style usually reserved for hagiography, of a lay female English mystic, Margery Kempe.[18] Kempe's story paints a particularly colourful image of her world through its inclusion of her pilgrimages to Italy, the Holy Land and Santiago de Compostela, her efforts to live a chaste life in middle age (despite having borne 14 children) and her notorious 'cryings' that overwhelmed her on many occasions.[19] Margery's senses were a fundamental part of her identity as a woman and as a visionary. The *Boke of Margery Kempe*, which appears to have been written down by a priest in the late 1430s, reveals details of Margery's life and attitudes in frequently startling frankness. It is often described as an 'autobiography', though the term is anachronistic and misleading: Margery's book is properly understood under its own terminology, a 'treatise', which 'teaches its readers strategies for managing the emotional, and ultimately spiritual, damage brought on by feelings of uncertainty, unworthiness and despair'.[20]

The outpourings of tears and crying that characterized Margery's mystical experiences, her 'plentyvows teerys of contricyon day be day' (Preface: 42–3) have formed the focus of a great many scholarly discussions, but the music and sounds that feature at key points in her book have received little attention.[21] Margery's tears were, after all, not only plenteous but also 'boystows', loud and violent (for example, I, 3: 394; I, 17: 1268). The reaction from those around her was not only because of her physical abundance of tears, but also their disruptive sound, which was an irritation to many. On one occasion, 'sche cryed so lowde that it myth ben herd al abowte the chirche and owte of the chirche' (I, 57: 4687–8). Margery's tears were contextualized for her reader by examples from other holy figures who wept. Her scribe drew on texts such as Jacques de Vitry's *Vita Maria Oigniacensis*, in which Marie is forced to leave a church when her tears distract the priest from his duties, and Bridget's *Liber Revelationem Celestium* as part of the authorization of her spirituality.[22]

Apart from the sound of her own crying, Margery's other religious experiences were often also steeped in sensory perception. Her first encounter with heavenly melody is found in Book I:

> On a nygth, as this creatur [i.e. Margery] lay in hir bedde wyth hir husbond, sche herd a sownd of melodye so swet and delectable, hir thowt, as sche had ben in paradyse. And therwyth sche styrt owt of hir bedde and seyd:
>
> 'Alas, that evyr I dede synne, it is ful mery in hevyn!'
>
> Thys melody was so swete that it passyd alle the melodye that evyr myght be herd in this world wythowtyn ony comparyson, and caused this creatur, whan sche herd ony myrth or melodye / aftyrward, for to have ful plentyvows and habundawnt teerys of hy devocyon, wyth greet sobbyngys and syhyngys aftyr the blysse of heven, not dredyng the schamys and the spytys of the wretchyd world. And evyr aftyr this drawt, sche had in hir mende the myrth and the

> melodye that was in heven, so mech that sche cowd not wyl restreyn hyrself fro the spekyng therof. For wher sche was in ony cumpanye, sche wold sey oftyntyme:
>
> 'It is ful mery in hevyn!'
>
> (I, 3: 324–39)

It was as a result of this experience of heavenly melody that Margery withdrew from sexual relations with her husband, emphasizing her wish to live a chaste life. In one of Margery's moments of sorrow over her lack of virginity, Christ comforted her by saying: 'so schalt thu dawnsyn in hevyn wyth other holy maydens and virgynes' (I, 22: 1684–5). Significantly, texts from this period commonly associate the sweetness of songs of heaven with the chaste former living of those residing there. The anonymous early thirteenth-century text *Hali Meiðhad*, for example, discussed the hierarchy of status between virginity, marriage and widowhood, and stated that it is virgins who have the sweetest song of these groups in heaven.[23] As such, references to Margery's hearing of sweet heavenly sounds reflected her intention and destiny to be counted as one of the virgins of heaven.

Margery's focus upon her senses reaches a climax in the description of her mystical marriage to Jesus, in which she experiences touch, smell and sound:

> sche felt many gret comfortys, bothe gostly [spiritual] comfortys and bodily comfortys. Sumtyme sche felt swet smellys wyth hir nose; it wer swettar, hir thowt, than evyr was ony swet erdly thing that sche smellyd beforn … Sumtyme sche herd wyth hir bodily erys sweche sowndys and melodiis that sche myth not wel heryn what a man seyd to hir in that tyme, less he spoke the lowder. The sowndys and melodiis had sche herd nyhand every day the terme of xxv yere whan this boke was wretyn, and specialy whan sche was in devowt prayer, also many tymes whil sche was at Rome and in Inglond bothe. (I, 35: 2862–73)

How might Margery's perception of the sound of her tears and of her mystical sounds been influenced by those around her? We know that she consulted many religious thinkers of the period, including mystics such as Julian of Norwich and 'many worshepful clerkys, archebysshopys, bysshoppys, doctowrs of dyvynyte and bachelers also' (Preface: 67–8), and that she was read many devotional texts, including writings by Walter Hilton – perhaps his *Scale of Perfection* – and the *Incendium Amoris* by Rolle mentioned above. The mystical experience of sweet smells in addition to hearing the voice of God is something that Julian of Norwich, who Margery met, expressed in her *Shewings*, extending the reference, sensually, to sight and taste: 'we endlesly ben al had in God, hym verily seand and fulsumly feland, hym gostly heryng, and hym delectably smellyng and hym swetely swelowyng' (ch. 43: 1540–42).[24] Such bodily experiences were considered potentially heretical, leading some commentators to warn against taking any unexpected 'experience' of God as authentic. Walter Hilton's *Scale of Perfection*, for example, was keen to note that certain feelings and sensory experiences might have evil rather than godly origins:

> By this that I have seid myght thu sumwhat undirstonde that visiones or revelaciouns of ony maner spirite, bodili apperynge or in ymagynynge, slepand or wakand, or ellis ony othere feelinge in the bodili wittes maad as it were goosteli; either in sownynge of the eere, or saverynge in the mouth, or smellynge in the nose, or ellis ony felable heete as it were fier glowand and warmand the breest, or ony othere partie of the bodi, or onythinge that mai be feelyd bi bodili wit, though it be never so comfortable and lykande, aren not verili contemplacion; ne thei aren but symple and secundarie though thei be good, in regard of goostli vertues and in goosteli knowynge and loovyng of God.
>
> (I, 10: 200–207)[25]

For Hilton, it was not that sound could not be a sign of the divine, but that it should not automatically be trusted to be so, and that heavenly sounds were worth nothing without loving Christ (I, 47: 1352–8). Furthermore, it seems that Hilton wished to distinguish between the significance of good and evil sounds because of the very power that sung prayer had in strengthening resolve against sin:

> alle othere praieres, as the Pater Noster or the Ave Maria or ympnys or psalmes or other devoute seyynges of Holi Chirche aren turnyd as it were into gostli mirthe and swete songe, bi the which thei [Christians] aren comfortid and strengthed agens alle synnes and mykil relevyd of bodili dishese. Of this degree speketh Seynt Poul thus: *Nolite inebriari vino sed impleamini spiritu sancto, loquentes vobismetipsis in ympnis et psalmis, et canticis spiritualibus, cantantes et psallentes in cordibus vestris domino* [Ephesians 5:18–19]. Be not drunken with wyn, but be ye fulfilled of the Holi Goost, seiande to youresilf in ympnes and psalmes and goostli songes, syngynge and phalmynge [psalming] in youre hertes to oure Lorde.
>
> (I, 7: 134–42)

The loss of aural perception was also significant to Margery Kempe. On one occasion she was tempted to adulterous lechery; she and the man went together to the service of Evensong, but she was unable to hear it, or to say her Pater noster (I, 4: 444). Then, while meditating on the suffering of Christ at the Passion, Margery heard 'so hedows a melodye that sche mygth not ber it', a sound that literally floored her, hoping that the hideous tune would desist (I, 17: 1242). An opposition of good and bad sounds was also used to emphasize the righteousness of Margery's campaign to refrain from sex with her husband; during this time, Margery was listening to the celebration of Mass, 'hering hir messe', when 'sche herd a gret noyse and a dredful' (I, 9: 655–6), whereupon a piece of masonry and a wooden roof beam fell down from the church tower and struck her, but by a miracle she survived without injury. The heavenly sounds of the Mass as setting for the drama of this episode, which is signified by terrible sounds, was a clever juxtaposition in which Margery was vindicated by her miraculous escape. The sounds of Mass, and of the church and its services more generally, would have been comfortably familiar to all readers of Margery's account. The impact of the moment was heightened by its occurrence within a setting more commonly experienced as – one imagines – gentle spoken delivery of daily liturgical custom through the architectural space, occasional ringing of bells, or the performance of simple plainchant, with its mixture of recitation and melody. Indeed, the drama would have been less effective within the narrative if placed against a more special sonic event, such as a polyphonic carol, a rhythmically complex motet or a three-part setting of the Credo, such as existed in Margery's day and which someone of her social status and experience of pilgrimage would have likely encountered.

Singers and Singing in Late Medieval England

Strange sounds – strange in terms of the infrequency of their incorporation into services outside of the highest status religious institutions – had begun to infiltrate worship across England more fully by the early fifteenth century. Aside from the regular performance of chant in the liturgy, choristers in the later Middle Ages were increasingly elaborating the basic elements of religious song with additional parts in harmony, often in virtuosic display of theoretical, rhythmic and vocal techniques. The fear of effeminizing pure, clerical identity through polyphonic singing was reflected in England as early as 1217, when a statute forbade Cistercians from performing three- and four-part singing in the abbeys of Dore and Tintern.[26] Later Cistercian statutes forbade singing too loudly (1258), the use of new songs (1302) and the practices of singing with syncopated rhythms and with hocketing (1320).[27]

John Wycliffe and the Lollards also spoke out against the use of polyphonic song in worship, noting the increasingly familiar presence of a small group of trained singers performing vain songs at the expense of the common chanting from a choral community.[28] At the same time, they were well aware of the importance of the senses in proper devotion, as can be seen from Wycliffe's text *On the Five Outer Wits*:

> It is necessary to know the five wits, the outer and the inner, and how to spend them in good use, and in the loving of God. The first is the sight of the eye; the second the hearing of the ear; the third the taste of the mouth; the forth the smelling of the nose; the fifth the handling or touching of members. Keep so thy sight, that thou see nothing that is not permitted to be seen, or may harm thy soul. And keep thy hearing so that thou hear no evil speech, or thing that is neither honest nor profitable. And keep so thy taste, that thou swallow no more than is needed, or ministers to thy bodily sustenance. And keep so thy smelling, that it make thee not to eat over much, neither delight not over much in smelling.[29]

Not only what was listened to, as here, but also the manner of listening and performance were of concern to Wycliffe. The tract *On Feigned Contemplative Life*, once attributed to Wycliffe but likely by an unknown contemporary, draws on Augustine's warning, saying, 'As long as the song liketh me more than doth the sentence that is sung, so often I confess that I sin grievously', and objects to occasions:

> When they are forty or fifty in a choir, three or four proud vicious fellows shall so play the most devout service, that no man shall hear the sentence, and all others shall be dumb, and look on as fools, while strumpets and thieves praise Sir Jack or Hobb, and William the proud clerk, saying, How small they play their notes, and that they serve well God and holy church, while they despise God to his face, and hinder other Christian men of their devotion.[30]

But it was not only non-specialist singers who objected to display. In one piece by the French composer Mayshuet, preserved complete in the Old Hall Manuscript, the complex polyphonic genre of the motet was used to protest at the vanity of singers adding elaborate songs to the liturgy.[31] In this motet, *Are post libamina / Nunc surgunt* (the texts are heard simultaneously in performance), the poet rails against various sensory pleasures, the enjoyment of which is linked to sinfulness:

> Triplum text
>
> At the altar after the oblation we will sing odes and songs with praise and good intentions. It is best of all that we praise with careful movement of voice, as reason orders: it may be accompanied by string instruments. This way of singing is worth more than ignorance, blind assumption and complicated vulgarity. Singers, the great number of whose skills fall short, seek vainglory. Sing willingly, not worthlessly, to the glory of God, so that in eternity you may be led into the kingdom of heaven. We should give forth sweet harmonies, highly lyrical, with joyful tone. Our sound should never be hastened or anticipated, but always restrained by listening with attention. The active, distinguished Frenchman composed this song on French melodies; but after he revised it with the Latin language, it more often became sweet to the English, replacing Deo Gracias.
>
> Duplum text
>
> Now there arise in the people mercenary men; they change the best gold into lead and sweetly fragrant flowers into distorted smells. These men are saying, nor am I deceived, 'Most singers, when they see in the middle of a church some great person, seek their best song, most pleasing to themselves'. Such singers multiply the short notes and they reflect upon the song not for the love of God, I believe, but of the great person. You hypocrites, you have never paid attention to the Holy Gospel, in which you have read the true word

> of God speaking about these matters; verily you are told – you have received your reward.[32]

The composer of this motet appears to enjoy the irony of placing criticism of singers and their vanity within a five-part musical texture, a number of parts that did not become commonplace in England until the latter part of the fifteenth century. The effect of this motet is one of deliberate aural hypocrisy: what you hear is representative of what should not be heard.

Conclusions

The experience of musical sound was a highly personal affair, related not to the formal requirements of the Christian liturgy but to individual spiritual identity. To Margery Kempe, melody could range from the most ethereal tunes to those of pure, excruciating dissonance. Her metaphorical allusions to music placed sound within the body of the textual protagonist, but the reader experienced the imagined realities of heaven and hell through Margery's 'ears' and in combination with their own perception of sound. Rolle's spiritual song slowed down the temporal experience of liturgical chant to such an extent that the text of such melodies was almost eradicated, leaving behind lengthy inward melodies that formed his understanding of Christ.

The impression given here is quite different in some cases to that found in the writings of more central figures found within musicological discourse. When Augustine and other church scholars lamented singing, it was primarily for elements such as the lascivious qualities of the church modes and the dangerously livening effect of melodies that might incite dancing more than devotion. The Lollards seem also to have disapproved of the vanity of performance, and, in particular, clerks making a visual as well as aural spectacle of themselves. In an extension of this position, Rolle's spiritual song was critical of even the established texts of liturgical celebrations, preferring to focus on melodic properties as reflective of the Divine. Orifices such as the mouth, throat, nose and ears were all vulnerable to being defiled by the performance of inappropriate songs; however, the writers discussed here focused not on actual songs but on particular types of sonic experience. Sound rather than text was the issue. Within the church, Bishop Ledrede was happy to write sacred texts to fit the melodies of popular amorous songs, but by contrast, Rolle argued that 'it is the tune that makes the song, not the words that are chanted' (*Incendium Amoris*, ch. 32).[33] Music theorists and others who formed part of the mainstream of musical discussion in the Middle Ages sometimes referred to music as if distanced from the body that produced and perceived it; by comparison, for Rolle, Wycliffe, Hilton and Kempe, spiritual identity was a bodily experience rooted in sound and its multi-sensual perception.

Notes

1 See Chrysogonus Waddell, 'The Origin and Early Evolution of the Cistercian Antiphonary: Reflections on Two Cistercian Chant Reforms', in M. Basil Pennington

(ed.), *The Cistercian Spirit: A Symposium in Memory of Thomas Merton* (Spencer, MA: Cistercian Publications, 1970), pp. 190–223 and Bruce W. Holsinger, *Music, Body, and Desire in Medieval Culture: Hildegard of Bingen to Chaucer* (Stanford, CA: Stanford University Press, 2001), pp. 112–13.

2 On ancient and early medieval texts that relate music and its experience to the body, see Holsinger, *Music, Body, and Desire*, pp. 27–83.

3 For a discussion of how the notated music of the period might have been realized in performance, including a detailed examination of possibilities for ornamentation and improvisation, see Timothy J. McGee, *The Sound of Medieval Song: Ornamentation and Vocal Style According to the Treatises* (Oxford: Clarendon Press, 1998).

4 Edmund Colledge (ed.), *The Latin Poems of Richard Ledrede, C.F.M., Bishop of Ossory, 1317–1360* (Toronto: Pontif. Inst. of Mediaeval Studies, 1974).

5 Cited in John Stevens and Dennis Libby, 'Carol', *Grove Music Online*, in *Oxford Music Online*, http://www.grove.com (accessed 7 April 2010).

6 On the interplay of sacred and secular references in medieval vocal music, with particular reference to motets, see Sylvia Huot, *Allegorical Play in the Old French Motet* (Stanford, CA: Stanford University Press, 1997) and Lisa Colton, 'The Articulation of Virginity in the Medieval *Chanson de nonne*', *Journal of the Royal Musical Association*, 133/2 (November, 2008): 159–88. On late medieval English examples see John Caldwell, 'Relations Between Liturgical and Vernacular Music in Medieval England', in Susan Rankin and David Hiley (eds), *Music in the Medieval English Liturgy* (Oxford: Clarendon Press, 1993), pp. 285–93.

7 Translation by Peter M. Lefferts in Frank Ll. Harrison (ed.), *Motets of English Provenance, Polyphonic Music of the Fourteenth Century* vol. 15 (Monaco: L'Oiseau-Lyre, 1980), pp. 197–8. Lefferts gives the final line of the tenor ('Mettez moi iuse, accoler moy') as 'Stitch my gizzard, give me a hung!' but I would read 'Place me downwards, line me up [with you]', based on the Anglo-Norman words 'juse' (down, downward) and a version of the modern French verb 'accoler' (to place side by side) (see http://www.anglo-norman.net).

8 Harrison, *Motets of English Provenance*, p. 164.

9 For an examination of the relationship between homosocial musical cultures and homoeroticism, see Holsinger, *Music, Body, and Desire*, pp. 87–187, especially pp. 87–8.

10 Nicholas Watson, *Richard Rolle and the Invention of Authority* (Cambridge: Cambridge University Press, 1991), p. 66.

11 'Dum enim in eadem capella sederem, et in nocte ante cenam psalmos prout potui decantarem, quasi tinnitum psallencium uel pocius canencium supra me ascultaui. Cumque celestibus eciam orando toto desiderio intenderem, nescio quomodo mox in me concentum canorum sensi, et delectabilissimam armoniam celicus excepi, mecum manentem in mente. Nam cogitacio mea continuo in carmen canorum commutabatur, et quasi odas habui meditando, et eciam oracionibus ipsis et psalmodia eundem sonum edidi'. Richard Rolle, *Incendium Amoris*, ch. 15; cited in Margaret Deansley (ed.), *The Incendium Amoris of Richard Rolle of Hampole* (Manchester: Manchester University Press, 1915). A critical edition of the Latin text can also be found at http://www.ccel.org/r/rolle/incendium/incendium.htm.Trans. Watson, *Richard Rolle and the Invention of Authority*, pp. 69–70.

12 Trans. Clifton Wolters (ed.), *Rolle: The Fire of Love* (Harmondsworth: Penguin, 1972), p. 146.

13 Ibid., p. 47.

14 The citation is from Wolters, *Rolle: The Fire of Love*, p. 27, but the interpretation is typical.

15 On liturgical practices in nunneries, see Ann Bagnall-Yardley, *Performing Piety: Musical Culture in Medieval English Nunneries* (New York: Palgrave Macmillan, 2006).

16 On alliterative verse by Rolle and his contemporaries, see Susanna Greer Fein, '*Haue Mercy of Me* (Psalm 51): An Unedited Alliterative Poem from the London Thornton Manuscript', *Modern Philology*, 86/3 (1989): 223–41, at p. 224.

17 Translation of the Latin line of the Psalm: 'To my hearing thou shalt give joy and gladness: and the bones that have been humbled shall rejoice'. Poem cited from Fein, '*Haue Mercy of Me*', where the full poem is edited on pp. 236–41, and this is verse IX, lines 97–100.

18 Katherine Lewis notes the rarity of 'Middle English hagiography relating to the cult of "real" saints who actually lived in later medieval England', and that the majority of these accounts deal with 'high-status men' or with famous clerics and monks; see Katherine Lewis, 'History, Historiography and Re-writing the Past', in Sarah Salih (ed.), *A Companion to Middle English Hagiography* (Woodbridge: D. S. Brewer, 2006), pp. 122–40, at p. 124.

19 Anthony Goodman, *Margery Kempe and Her World* (London: Longman, 2002).

20 Rebecca Krug, 'Margery Kempe', in Larry Scanlon (ed.), *The Cambridge Companion to Medieval English Literature* (Cambridge: Cambridge University Press, 2009), pp. 217–28, at p. 218.

21 All citations from this text – which, following a short preface, is divided into two 'books' – are taken from Barry Windeatt (ed.), *The Book of Margery Kempe* (London: Longman: 2000), and will be identified by book and chapter, and by the corresponding line number in the modern edition (for example, I, 3: 42). The same system will be used for the writings of Walter Hilton and Julian of Norwich.

22 For discussion of the influence of other writings on the Book of Margery Kempe see Karma Lochrie, *Margery Kempe and Translations of the Flesh* (Philadelphia: University of Pennsylvania Press, 1991).

23 This is found in the text *Hali Meiðhad*, as discussed in Bella Millett and Jocelyn Wogan-Brown, *Medieval English Prose for Women* (Oxford: Clarendon Press, 1990), p. xviii.

24 Georgia Ronan Crampton (ed.), *The Shewings of Julian of Norwich* (Kalamazoo, MI: Medieval Institute Publications, 1994).

25 Walter Hilton, *Scale of Perfection, Book 1* (ed.) Thomas H. Bestul (Kalamazoo, MI: Medieval Institute Publications, 2000), also available at http://www.lib.rochester.edu/CAMELOT/teams/hilfr1.htm.

26 Mary Berry and Franklyn Gellnick, 'Cistercian Monks', *Grove Music Online*, http://www.grove.com (accessed 9 April 2010).

27 David H. Williams, *The Cistercians in the Early Middle Ages* (Trowbridge: Cromwell Press, 1998), p. 232.

28 Bruce Holsinger, 'The Vision of Music in a Lollard Florilegium: *Cantus* in the *Rosarium Theologie* (Cambridge, Gonville and Caius College MS 354/581)', *Plainsong and Medieval Music*, 8/2 (1999): 95–106.

29 John Wycliffe, *On the Five Outer Wits*, in Robert Vaughan (ed.), *Tracts and Treatises of John de Wycliffe, D.D. with Selections and Translations from his Manuscripts and Latin Works* (London: Blackburn and Pardon, 1845), ch. 7, p. 117, translated into modern English by Lisa Colton.

30 Anon, *Impedimenta Evangelizantum* (*Of Feigned Contemplative Life*); translated and cited in Vaughan (ed.), *Tracts and Treatises of John de Wycliffe*, ch.16, pp. 49–52, at p. 51.

31 The sources of this motet are London, British Library, Additional 57950 (Old Hall Manuscript), fols 111^{v}–112^{r} (4/2) and Oxford, Bodleian Library, Don. b. 32 (Triplum fragment only). Andrew Wathey has identified the composer with Matteo de Sancto Johanne, who was in England in 1369; 'The Peace of 1360–1369 and Anglo-French Musical Relations', *Early Music History*, 9 (1990): 129–74, at pp. 149–50.

32 Edition in Andrew Hughes and Margaret Bent (eds), *The Old Hall Manuscript*, 3 vols in 4, *Corpus Mensurabilis Musicae* 46 ([Rome]: American Institute of Musicology, Hänssler-Verlag, 1969–73), vol. 1/2, pp. 419–23 (edition), vol. 3, pp. 43–4 (texts).

33 Trans. Wolters, *Rolle: The Fire of Love*, p. 145.

2 The Physiologist at the Opera

Claude Perrault's *Du bruit* (1680) and the Politics of Pleasure in the *Ancien Régime*

Veit Erlmann

When it comes to seventeenth-century French culture and politics, the Louvre is always a good starting point – a much less noticed part of the Louvre than the *Mona Lisa* or the spectacular glass pyramid in the courtyard, that is: an elegant row of double columns at the eastern end called the 'colonnade'. When the Louvre was built, architecture did not know such rows of double columns, and in the eyes of many seventeenth-century beholders, the slender structures may well have 'flirted with precariousness'.[1] The evenly spaced regularity that characterizes classicist architecture is here disrupted and accentuated at the same time by a fragile arrangement that allows the spaces between the individual columns to contract and expand, as if a rippling movement had pushed them against each other.

The colonnade is said to be the work of Claude Perrault (1613–88), one of France's most prominent intellectuals during the reign of Louis XIV. Brother of the celebrated author of fairytales Charles Perrault and of the pioneer of modern hydrology, Pierre Perrault, Claude Perrault designed the Observatory, translated the work of the ancient Roman architect Vitruvius into French, and authored the widely read essay, *Ordonnance des cinq espèces de colonnes selon la méthode des anciens* (Ordonnance of the Five Kinds of Columns after the Method of the Ancients, 1683), considered to be one of the founding texts of modern architecture.[2] But Perrault was also one of the last all-round scholars, whose activities covered a stupendous range of scientific and artistic subjects beyond architecture. A physician by training, he conducted extensive zoological studies, 'compiled' the first volume of the pioneering *Mémoires pour servir à l'histoire naturelle des animaux* (1671/1676), and became a founding member of the Académie Royale des Sciences. And as an astute cultural observer, he also weighed in on the infamous 'Querelle des Anciens et des Modernes', one of the modern era's first culture wars, which was sparked by the premiere of Jean-Baptiste Lully's opera *Alceste* in 1674 and pitted 'moderns' such as the Perrault brothers against 'ancients' such as Racine and Nicolas Boileau.

Perrault's real passion, however, was the ear. It was in a book entitled *Du bruit* (Of Sound) that he first outlined what may well be his most influential contribution to the epistemology of life: the 'animist' conception that the animal body is not a machine, as Descartes had taught, but a self-generating organism.[3] Tucked away in Perrault's multi-volume *Essais de physique* (1680), and having long been neglected by art historians, cultural historians and musicologists alike, *Du bruit* is more than one of many early modern treatises in otology.[4] It is a key document of the late seventeenth-century avant-garde and of considerable significance to the history of modern aurality.

Compressing, Decompressing

We owe a delightfully warm portrait of Claude Perrault and his brothers to the eminent evolutionary biologist and historian of science Stephen Jay Gould. By crossing the boundaries between letters and science, Gould suggests in *The Hedgehog, the Fox, and the Magister's Pox* (his last book, written before his untimely death), the Perrault brothers had become models for the 'modern liberty to move on'.[5] Even though posterity denies Claude Perrault the status of near sanctity that it accorded to Newton, Harvey or Descartes, the architect-zoologist-physician might be regarded as a representative of the 'vernacular' modern: an everyday kind of epistemology that, as Gould puts it, 'often surpasses the stellar as a source of insight about the main thrust of a movement'.[6]

Perhaps Gould has a point. Claude Perrault's *Du bruit* undoubtedly provides us with a fascinating opportunity to examine the meaning of 'liberty' during a period that, in many ways, may be said to have invented the idea of the 'modern', but that also witnessed unprecedented levels of bloodshed, famine and intolerance. Above all, however, the strange intermingling of architecture, opera and otology pervading Perrault's work invites us to revisit some of the more expedient constructions of seventeenth-century culture. The idea, for instance, that the era witnessed the emergence of a new nexus between music, the ear and a triumphant modern form of subjectivity is firmly entrenched in current musical writing. And as even a cursory survey of the standard literature on early modern music and aesthetics shows, this new subject, 'Man', is usually – and for the most part, without much further elaboration – equated with the Cartesian ego, representation and absolute power.[7]

There are a few exceptions though. In what is one of the most thought-provoking discussions of early modern opera to date, Gary Tomlinson, for instance, has brought new life to a debate that, if it did not ignore the broader philosophical implications of early modern opera altogether, has tended to subsume seventeenth-century musical aesthetics under the precepts of *Affektenlehre* or 'Doctrine of Affections'.[8] Working from a Foucauldian perspective, Tomlinson detects in the vocality pioneered in Lully's operas a new type of metaphysics. The recitative, perhaps Lully's most momentous accomplishment, he argues, not only became a source of expressive power, it also created new 'conditions of subjectivity'.[9] Faced with the disintegration of the Renaissance *episteme*, Lully struggled to regain some measure of semiotic predictability by means of musical stock formulas, standardized personae, and a whole arsenal of dramatic conventions, signs, similes and chimeras. The fruit of his labours, the *tragédie en musique*, conjoined music and words in order to achieve a new unity of mind and body constituted through the fixity of representation or, as Tomlinson puts it, 'habituation'.

A similar account of Cartesianism's relationship with early modern aurality is offered by Georgia Cowart in *The Triumph of Pleasure: Louis XIV and the Politics of Spectacle.*[10] Focusing on the opera-ballet and comedy-ballet of the latter half of the seventeenth century, Cowart is concerned with the pivotal – and increasingly counterhegemonic – role that pleasure was beginning to play in the spectacles staged by and for Louis XIV. While during the early part of the monarch's long reign elite audiences were expected to identify with and submit to the heroic images of absolute power projected by the court ballet, beginning in mid-century, strains of resistance began to emerge. Audiences insisted on indulging in the sonic and visual delights afforded by these spectacles on terms of their own choosing, preferring a more egalitarian, everyday kind of libertinism over the martial manifestations of royal might. In short, as Cowart compellingly shows, early modern spectacle could be said to be one of the terrains on which Gould's 'modern liberty' began to assert itself.

Neither Tomlinson nor Cowart questions the larger Cartesian framework however. Thus, while Tomlinson's adumbration of the Lullian recitative as 'habituation' indeed may open up a space for envisaging a new, albeit somewhat passive, type of listener, the aural sovereignty achieved through such 'habituation' remains firmly anchored in Descartes's mind–body dualism. Cowart, similarly, sees the weakening of the quasi-mystical correspondence between king and state power during Louis XIV's final years and the potentially unruly kind of aurality increasingly being espoused by upper-class audiences as mediated by the rise of Cartesian mechanism.[11] And as in Tomlinson's account, the new alliance of pleasure and liberty for her manifests itself first and foremost within the space of representation: in plots and allegorical figures, but much less so at the level of bodily sensation.

However, it is precisely here, in the concept of an aurality operating outside the confines of the Cartesian mind–body dualism, that Perrault's unique contribution to the pursuit of 'modern liberty' lies. Contrary to the notion that the groundswell of anti-aristocratic sentiment and the rise of alternative, 'proto-Enlightenment' visions of leadership and society were fuelled by the growing acceptance and aesthetic relevance of Cartesianism, Perrault's position was a more ambiguous one.[12] The pursuit of 'modern liberty' for the architect-zoologist-physician Perrault began at the interstices of reason and resonance. It began in the cochlea, in a new concept of hearing.

There is a glitch however. Historians of otology differ as to whom is due the honour of being the true pioneer of the resonance theory of hearing. Is it Claude Perrault, as the eighteenth-century inventor of iatromechanist medicine Herman Boerhaave held, or was it rather Perrault's young friend, Joseph-Guichard Duverney, a professor of anatomy at the Jardin du Roi and, like Perrault, a member of the Académie and a rising star on the French medical scene? Three years after *Du bruit*, Duverney had published *Traité de l'organe de l'ouïe*, a work that historians generally consider to be one of the milestones of modern otology and the first major statement of cochlear selective resonance.[13] The *lamina spiralis*, Duverney argues, is

> wider at the beginning of its first revolution than at the end of its last, where it finishes in a point, and its other parts diminishing proportionately in width, we may suppose that since the wider parts may be vibrated without the others

> participating in that vibration they are capable only of slower undulations which consequently correspond to the low notes; and that on the contrary when its narrow parts are struck their undulations are faster and hence correspond to high notes, in the same way as the wider parts of a steel spring make slower undulations and correspond to low notes and the narrower parts make more frequent and faster undulations and hence correspond to high notes; so that, finally, according to the different vibrations in the spiral lamina, the spirits of the nerve, which spread through its substance, receive different impressions which in the brain represent the different appearances of pitches.[14]

To be sure, modern science now locates pitch perception in the opposite direction of what Duverney claimed: the perception of low pitches occurs at the apex of the cochlea and that of high pitches near its basal end. Still, Duverney's skills as an anatomist (as an anatomist of the inner ear, at any rate) surpassed those of Perrault, as a comparison between Perrault's and Duverney's plates shows. Unlike Perrault's rather crude depiction of the cochlea (Figure 2.1) Duverney clearly depicted the 'spiral canal' (AA) and the 'spiral membrane' (BB) next to it (Figure 2.2), where 1, 2 and 3 mark the bony lamina and 4, 5 and 6 the basilar membrane.

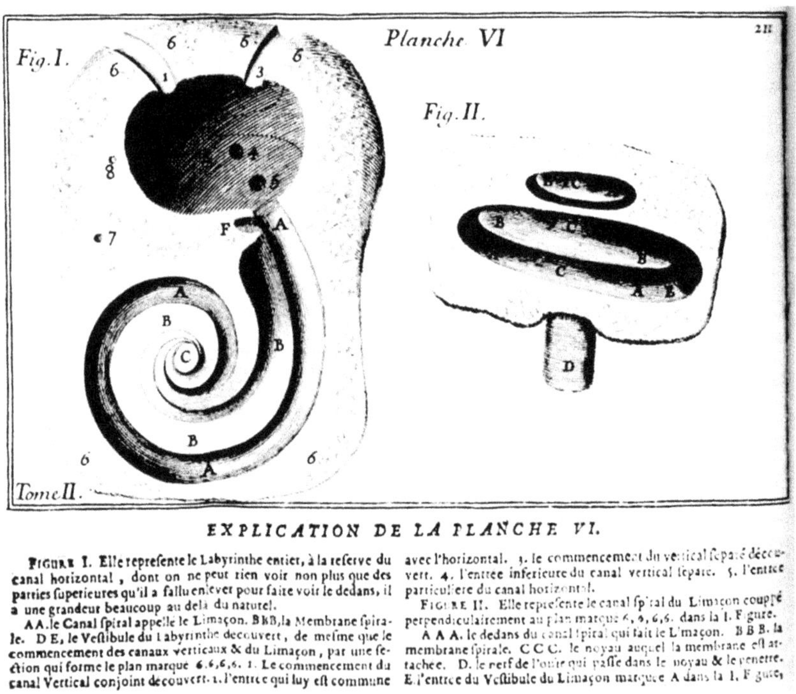

Figure 2.1 Claude Perrault, *Essais de physique*, Figure 6. Figure I. shows a cross-section of the labyrinth, with AA being the cochlea, BBB the lamina spiralis, and DE the vestibule. Figure II. shows the same part perpendicular to the axis 6 6 6 6 on left. D represents the auditory nerve.

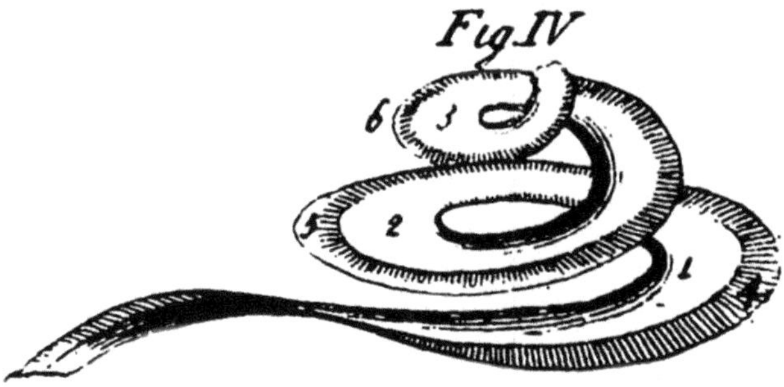

Figure 2.2 Joseph-Guichard Duverney, *Traité de l'organe de l'ouïe*, Figure IV. A close-up of the lamina spiralis.

Yet Perrault's lacunae should not deceive us. They obscure the extraordinary significance that his broader 'animist' concept has for the physiology of hearing, seventeenth-century musical aesthetics and the politics of pleasure in the *Ancien Régime*. The notion that every nerve, fibre and muscle was equipped with its own reasoning faculty not only constituted a frontal assault on Descartes's mechanistic concept of the body as a soulless machine. It also had enormous consequences for the ear, granting subjective, sensory experience an unprecedented primacy. Perrault's animism points toward a new type of subjectivity – to modern liberty. Indeed, *Du bruit* is the first work to highlight the central role of hearing in the making of that off-Cartesian, as one might call it, subjectivity.

How, then, did Perrault's argument in *Du bruit* unfold? The work is divided into three parts. The first two parts contain a lengthy, if rudimentary, summary of the velocity of sound, echo, vibration, sympathetic resonance and other acoustic phenomena, but they also show, more importantly, Perrault embracing the corpuscular theory of air motion according to which sound is carried by subtle, elastic particles of air moving toward the ear. There then follows a third part on the ear that is further subdivided into four chapters. The first of these chapters presents a copiously illustrated, if rather unreliable, discussion of the anatomy of the ear, while the remaining three chapters treat the comparative physiology of hearing. It is these latter chapters that form the most speculative, albeit most interesting, part of the book, because they provided Perrault with the framework within which to promulgate his broader critique of the Cartesian mechanism.

The basis of this critique was the idea that solid bodies possess two essential qualities: hardness and spring, or in present-day usage, cohesion and elasticity. These were said to be as much due to internal factors such as the material and shape of the object as the result of external factors such as the pressure exercised by the surrounding air. Consequently, it was thought to be the natural propensity of bodies to resist separation and to tend toward cohesion, a feature Perrault, using an ancient Greek term, labelled *péristaltique*. In a like vein, human bodies were held to exist always in a state of tension. Muscles and membranes are naturally flexed, and the action of the animal spirits 'destined for movement does not consist in operating contractions … but, quite to the contrary, to produce relaxation in the muscles opposed to those contracting'.[15]

Although this concept of muscular motion was novel, the underlying physics was not. It was essentially a mechanistic concept, and as such, like all seventeenth-century physics, was modelled on the Cartesian machine metaphor, the predominant paradigm of European scientific thought. But Perrault's notion of peristaltic movement differed from Cartesianism in one crucial respect. For Descartes, movement could occur only if an external force acts upon inherently inert matter. As the famous illustration in *L'homme* had shown, for Descartes, the role of the animal spirits was to convey the sensation they had transmitted to the brain back to the body organ that experienced the sensation and thus cause it to move. If a fire were about to burn a foot, the animal spirits would transmit the feeling of heat back to the muscle so that it would contract and remove the foot from the source of danger. Similarly, the animal spirits would fill up the muscles attached to the eye, shortening them, and thus cause the eye to move left or right. For Perrault, quite the opposite was the case. Instead of a chain reaction in which each part of the body machine transmits its movement to the next part, the peristaltic interaction between cohesion and elasticity was something that Perrault considered to be dispersed throughout the entire body machine.

This revision of Descartes's notion of inert matter inevitably posed a dilemma for Perrault's attempt to understand auditory perception. Though the *Essais de physique* are full of machine analogies – the eye is compared to a telescope, the lungs of an ostrich are depicted next to a pair of bellows, and so on – sensation for Perrault ultimately could be only another form of *péristaltique*. But how to demonstrate the way the tension between the antagonistic forces of cohesion and elasticity plays itself out in the ear? Once the vibrating air gets past the eardrum and meets with rigid, bony structures, what happens to peristaltic action? *Du bruit* is remarkably lapidary here. Since particles of air, Perrault wrote, have elasticity and therefore, like all matter, tend to return to their original position – thus displacing other particles that had since taken their place in a continuous series of 'impulsions' finally reaching the ear – it followed that the organ of hearing was agitated in 'about the same manner'.[16]

The problem with this statement is that there simply is no air in the inner ear. And so, by holding on to the age-old idea of the air implantatus, Perrault failed to apply the full potential of the notion of peristaltic action to hearing. The ear, to be sure, no longer quite functioned on the model of Descartes's rumbling pulleys and wheels, but at the same time, it had not fully acquired the plasticity of living matter that Perrault assumed to operate throughout the rest of the body, either. As a consequence, Perrault's physiology could only fall back on analogies.

It is instructive to follow Perrault's argument more closely, because although later rejected by Duverney, it remained extremely influential until well into the late eighteenth century, throwing into sharp relief the predicament facing the late seventeenth-century and eighteenth-century physiology of hearing. Basically, what Perrault argued for was a qualitative link between the physics of sounding bodies and the ear, the result being an anatomy in which only those components mattered that maintained a specific physiological relation to the object perceived. This sort of reasoning drew on a rather simplistic analogy between the ear and the eye. 'I am looking for parts in the organ of hearing that have some analogy to parts in other senses, in order to see if by analogy they might have the same function', Perrault wrote.[17] What the Greeks and Romans used to call 'imago' (image) and the Scholastics misinterpreted as species, he went on to say, might in fact be a useful way to conceptualize the analogy between seeing and hearing and to argue that sensory impressions are caused by identical processes, be it in the eye or in the ear:[18] as light rays are bundled by the retina, the labyrinth focuses the vibrating air. In addition, all sense organs are ultimately attached to the nervous system through nerves with the same substance and an almost similar delicacy.

If each sense organ, then, had something in common with the other sense organs, it followed that the different sensory impressions received by these organs had to be caused by some parts that were unique to this organ. The highly polished, watery surface of the retina, for instance, shares with the optic nerve some of its fluidity so as to ease the passage of the mirror images. Likewise, the consistency of the basilar membrane rests somewhere between the softness of the auditory nerve and the hardness of the cochlea. Because it made sense to assume, said Perrault, 'that dryness is one of the conditions most required for bodies to be able to produce sound, so the same quality must also be present in the organ sensing the movement that caused the sound'.[19] The connection was so close that he even attributed the fact that the auditory nerve is exceptionally harder in lions and sheep to these animals' 'very dry temperament'. And, in a similar vein, certain pathologies such as poor hearing had to be caused by the moist winds of southern France, which diminished the dryness of the membrane.[20]

Obviously, all of this was not only pure Aristotle (in the sense that perception was thought to be mediated by 'qualities'), it also fell well short of Perrault's broader peristaltic claims. Having reached the limits of mechanical explanation, Perrault therefore set out in the third chapter to advance an ambitious, if somewhat circuitous, theory of auditory perception and, ultimately, an entirely different metaphysics of the subject. Although still conceived in the long shadow of Descartes, significant parts of this theory also resonated more closely with late seventeenth-century cultural practice than with Cartesian dogma.

One of the major points separating Perrault's views from Descartes's reflexological theory concerned the role of the soul. Whereas Descartes had denied that the soul has any impact on the working of the body, Perrault believed that the relationship between the mind and the body, and more particularly between the mind and the sensory organs, could not be explained without assuming the agency of some kind of 'reasoning' force in every part of the animal organism. He arrived at this conclusion by way of a complex argument that led him to query the central role Descartes had accorded the brain in judging sense impressions. How can the brain, Perrault wondered, handle two contrary sense impressions at the same time without getting confused? How can the rays of light that meet the eye in a straight line reach the brain undisturbed when the optic nerve itself is bent? And, finally, since nerve and sense organ must form a homogeneous whole, as they do in the retina, wouldn't the opacity of the optic nerve obstruct the delicate vibrations being received by the transparent humours of the retina and thus hinder them from reaching the brain? The solution to these quandaries, according to Perrault, could lie only in the rejection of Descartes's wax-seal theory of sensation. Or put differently, instead of figuring the body as a robot, a 'new system' would have to assume the existence of 'interior senses' that function independently of the body organs reserved for exterior sensation.[21]

The reason for this hypothesis was that the soul, by virtue of it being linked to all body parts, did not need the brain to ponder (*contempler*) these impressions. It could perfectly scrutinize them in each organ, where they were present in the moment of sensation. The role of the brain, in this view, was thus a more limited one – to prepare the animal spirits necessary for the exterior senses to be properly disposed for perception. Likewise, memory was not located in the brain, as Descartes's theory of traces would have it, but in every part of the body. In short, in contrast to a wound-up clock, the Perrauldian body consisted of fibres that were animated by a living, superior force lodged within it. The human body, if it was a machine at all, was capable of 'doing something different from what it is made for'.[22] Living matter had the innate ability to adapt.

Perrault had little difficulty in finding evidence for this stunning reversal of Cartesian mechanism. Having dissected countless animals from the royal menagerie – one infected specimen of which, a giraffe, is said to have caused his death – he was convinced that the operation of such a 'reasoning' faculty could be easily observed in a horse, for instance. If Cartesian dogma was correct in imagining the movements of a horse to be only the quasi-automatic effect of a series of successive sense impressions devoid of any autonomous reasoning, why was the animal able to recombine these impressions in the reverse order and thus use the same path on its way home? Likewise, how was it possible that a viper whose head had been cut off not only continued to move, but also found the way back to its favourite hiding spot, if not through the memory stored in its sentient organ of touch?

At this point in the argument, oddly, Perrault denied the difference between voluntary action guided by the soul and those movements that occur without the active participation of the soul, such as the beating of the heart, or movements that even defy the soul, such as illness. Instead, he argued for two types of thought: a 'distinct' way of thinking, unique to humans and meant for 'the things that we attend to with care', and a 'confused' way of thinking, shared by humans and beasts and designed 'for the things which habit has made so simple that they no longer require express and exact thinking'. One example of the former was the regularity and vital importance of the cardiac contractions, which simply persuade the soul not to interrupt them. Ailments, by contrast, result from the soul's occasional failure to pay sufficient attention to the body's interior functioning.

Distinct or confused, the larger point Perrault was pressing here was of course that the functioning of the eye and ear would remain incomprehensible without 'assuming a judgment and reasoning no machine would ever be able to attain, no matter how subtle and well-constructed one may imagine it to be'.[23] Because there are two types of thought, the senses also are governed by two kinds of judgment that help them to correct errors to which they might otherwise be prone. The sense of touch, for instance, is much more easily deceived and consequently more in need of 'distinct' judgment than sight and hearing, where 'habitual and confused [judgment] are almost always adequate'.[24] Thus, the ear as a rule has no difficulty in distinguishing a male from a female voice, just as it is seldom deceived by ventriloquist trickery. Some exceptions do, of course, exist, such as when the ears are cupped by the hand and a noise is heard like that of a waterfall, or when a piece of iron attached to a string that in turn is wound around a finger stopping an ear produces a bell-like sound. On the whole, however, Perrault proclaimed, the ear is capable of an admirable 'certitude and discernment'.[25]

Here *Du bruit* ends, leaving the present-day reader startled and frustrated at the same time. One is startled because animism, one of the key paradigms of seventeenth-century natural science, whose significance extended well into the eighteenth century, is premiered in a text that, contrary to the taken-for-granted hierarchy of the five senses, takes not vision as its testing ground, but hearing. At the same time, the dramatic shift from Cartesian dogma to a more flexible mind–body relationship or vital materialism, as one might call it, seems mired in contradictions. Perrault, it appears, was unable to reconcile the innovative thrust of his broader theory with his views of auditory sensation articulated toward the end of *Du bruit*.

It was Duverney's achievement to close this gap between animism and otology, at least in part. In the third section of his *Traité*, Duverney offers the first scientifically plausible theory of *tintement* or tinnitus. This, he writes, consists of the 'perception of a sound that does not exist, or of a sound that exists inside'.[26] The reason for this sensation does not lie, as the ancients believed, in some form or other of mental deficiency, but in the nature of the auditory nerve. It makes no difference whether its fibres are affected (*ébranler*) on the side of the ear or on the side of the brain. 'The result will always be the same sensation'.[27] The *tintements*, then, are 'veritable sounds', that are perceived 'such as they are, yet the ear is unable to relate them to an exterior object'.[28]

Duverney, unlike his mentor, was indifferent toward philosophy. Yet although Duverney's discussion of tinnitus was superior to Perrault's concept of cavity resonance both texts support the same off-Cartesian model of subjectivity. Like Descartes, Duverney and Perrault do not call into doubt the existence of an exterior 'reality'. The idea of reality is still inextricably linked to the object world. But Duverney and Perrault differ from Descartes in that a sound that is 'veritable' and that 'does not exist' at the same time is less a logical impossibility than the admission of a parallel interior, subjective reality. In this reality sensations of things exist that do not of necessity fall under the master philosopher's verdict of being illusions of the body.

Folie de l'Écoute

In 1672, Perrault had attended a performance of the 'pastorale' *Les peines et les plaisirs de l'amour* (1672) by Robert Cambert, one of the pioneers of French opera. The event inspired him to draft a text titled 'Sçavoir si la musique à plusieurs parties a esté connüe et mise en usage par les Anciens' (Whether Part Music Was Known to and Used by the Ancients).[29] Meant to serve as an introduction to another, much more ambitious, essay titled 'De la musique des anciens' (Of the Music of the Ancients) that he later appended to the *Essais de physique* – and which we will explore further below – 'Sçavoir' was published only as recently as 1914, without it having been possible to determine why Perrault decided to withhold its publication. One reason might be that its key propositions had become superseded during the period leading up to *Du bruit* by the much sharper criticism Perrault advanced in his Vitruvius translation and, most crucially, as part of his embroilment in the Querelle des Anciens et des Modernes.[30]

Another reason for Perrault's decision not to publish his early text might be the same flirtation with 'precariousness' that had shaped his colonnade project earlier. While, on the surface, 'Sçavoir' foregrounds the contrasting 'ancient' and 'modern' positions on the emerging musical theatre, the listening stance that Perrault advocates in this text is much less clearly circumscribed, bordering perilously close on a concept of subjectivity that diverges significantly from Cartesianism and, by the same token, might be seen as undermining the increasingly repressive politics defining Louis XIV's late period.

What, then, is 'Sçavoir' about? Perrault recounts a dispute he claims to have overheard between two spectators at the start of the performance of Cambert's 'pastorale'. At first, the disputants, Paleologue (Expert on Things Ancient) and Philalethe (Lover of Truth), quarrel over the value of the novel type of spectacle. Philalethe, who had been present at the work's premiere, is utterly ravished. Showing no interest in the author and composer of the piece, much less in finding out about the designer of the stage machinery, his only concern is 'to know how someone was able to produce such surprising things' – 'de scauoir comment on pouuoit faire reüssir des choses si surprenantes'. Paleologue, meanwhile, declares that the only thing he is amazed about is how people who have seen Italian opera in Venice can 'admire so little'.[31] But as the curtain rises and the violins begin to intone the overture, something unexpected happens. Unlike the raucous spectators around them, who continue with their conversations and care only about the names of the female singers, the two antagonists suddenly find themselves transfixed by the interplay between different voices and the accompaniment of instruments, the 'symphonie'. Even the obstinate Paleologue becomes more 'docile', listening with 'less contempt' than he did at the start of their conversation:

> He even entered, in a way, into its 'feelings' [*il entra mesme en quelque façon dans ses sentimens*] when the piece was finished and they remained in their place a little longer while the crowd dispersed. He [Paleologue] drew his [Philalete's] attention to all the things that made the piece that had been played so marvellous [*merueilleuse*] and so much superior to the spectacles of the ancient in terms of intellect [*esprit*] and ingenuity [*inuention*].

The marvellous, Paleologue goes on to elaborate, consists mainly in a 'shrewd economy' (*œconomie iudicieuse*):

> in the clever way in which the composer introduced ballet episodes and machinery that are 'necessary' to the topic at hand; the variety of characters represented by the different roles that provide an opportunity to broach both happy and sad subjects; the use of two familiar genres of song like the drinking song and the love song; the way the plot of the fable, with its intrigues, is made comprehensible by the concatenation [*enchainement*] of songs that ordinarily must also contain propositions of a general nature; the abstention from the recitative, which is devoid of charm [*grace*] when sung; the technique of having sing together several characters who are in conversation; and, finally, to have all this go on for three hours without being boring and notwithstanding the inconvenience all music, no matter how good, becomes subject to when it becomes insufferable because of its long duration.[32]

At first glance, the passage reads like a laundry list of all the things early modern audiences found to be either remarkable or objectionable in the marvellous effects, *le merveilleux*, so central to the new form of spectacle. But the passage is also interesting because it wrestles with some of the potentially more troubling implications of the marvellous for the nascent project of an off-Cartesian theory of hearing and subjectivity. At stake in this brief text is nothing less than the irrational side of the 'animated' body.

One phrase, in particular, should give us pause: 'he even entered in a way into its feelings', 'il entra mesme en quelque façon dans ses sentimens'. It stands in marked contrast to the key tenets of Cartesian representation and to the aesthetics of *Affektenlehre*, both of which do not call for immersion and identification, but distance and mirroring, and as such it merits more detailed discussion. For instance, it is not entirely clear whether *ses* refers to Philalethe's feelings, which Paleologue had earlier sought to resist, or whether, by contrast, the possessive noun points to the remainder of the sentence, 'lorrsque la piece estant acheuée ils demeurent encore quelque temps à leur place pendant que la foule se retiroit', 'when the piece was finished and they remained in their place a little longer while the crowd dispersed', and, hence, to the piece (*la piece*) itself.

Equally, if not more ambiguous is the word *sentimens*. What is the space of emotion in early modern opera? What exactly was the meaning of sentiment during the middle of the seventeenth century, and how did it relate to what, according to classical dogma, constituted man's mental faculties: memory, imagination and reason? And finally, what did Perrault mean by the phrase 'he entered' (*il entra*)? To understand this remark, it is helpful to take a brief look at the plot of *Les peines*. The prologue has the Muses order Apollo to follow the rule of the steely Jupiter and to abandon love. Apollo, however, refusing to obey Jupiter, casts his lot with the Graces, asserting love's supreme power to transform pain into pleasure. This course of action, in turn, enrages the Muses, who denigrate

Cupid and castigate the fruits of love as pains, only to see the ultimate defeat of the Muses and the triumph of love.

As would be expected of a pastorale, the *sentimens* that the disbelieving Paleologue eventually 'enters into', leads to the ultimate triumph of love, of pleasure over pain. But as Cowart rightly points out, there is a subplot that adds an altogether more defiant tone to *Les peines*. In aligning the Muses with Jupiter's (read, Louis XIV's) ruthless rule, the work invokes, much like Moliere's and Lully's *Le bourgeois gentilhomme* produced two years earlier, the 'metaphor of the transformation of pain into pleasure as the means by which a utopia of spectacle could serve as the model for a new society'.[33]

Equally helpful in understanding Paleologue's immersive moment might be to revisit some of the older scholarship of the 'marvellous' and contrast it with more recent attempts to elucidate early modern spectacle in terms of its perceived parallels with postmodern or 'Neo-Baroque' forms of affective immersion, virtuality and the madness of sensory perception.[34] In the literature on French early modern opera, the marvellous has largely been framed in terms of its relationship to the precepts of 'classical' theatre. While in the spoken tragedy – Racine's *Phèdre* (1677), for instance – the supersensible realm is banned from the stage, opera thrives on presenting a world filled with supernatural characters, magic moments and scene changes, all made possible by intricate machinery and – music. In fact, in the eyes of some contemporaneous commentators, music's role in the lyrical theatre was essential to the privileged place of the spectacular. In the *tragédie en musique*, Jean-François Marmontel wrote, 'everything is false, but everything is in agreement; and it is this agreement that constitutes truth. Music gives charm to the marvellous, the marvellous in turn gives credibility [*vraisemblance*] to music'.[35]

Yet spoken tragedy and opera are also united by a more hidden link, beyond the obvious opposition between what is not shown and, hence, exterior to the text, in the tragedy, and what is recovered and integrated into the text, in the case of opera. There is, in fact, a 'familiar strangeness' between the two genres. As Catherine Kintzler, the foremost authority on French scenic aesthetics of the classical era, writes: 'Opera treats the exteriority of the theatre as though it was not exterior'. It does so 'by not only integrating this exteriority into the plot, but above all by bringing the poetic rules of the classical drama to bear on it, by treating this exteriority as a theatrical element in its own right'.[36]

Exteriority is not merely exhibited, then, but truly repatriated and interiorized at the heart of the theatre. In this way, exteriority narrows the gulf between opera and theatre by bringing, in a perverted manner, what 'should distance opera from theatre in principle … closer to it'.[37] The key to this coexistence of the two genres within the same aesthetic frame and thus also the guarantee for the effect of the marvellous to succeed is of course, as Marmontel was well aware, verisimilitude, *vraisemblance*. The supernatural events must be plausible. They must refer to a quasi-nature, one that if it is not accessible to the senses of ordinary mortals, at least might be possible if only one had access to it. Thus, when a supernatural being performs a magic act, this act becomes plausible only when it is part of some sort of intermediary chain of then 'known' natural causes: the appearance or disappearance of a God can be accompanied only by an earthquake, a thunderstorm or a meteor; the murder of Sangaride in Lully's opera *Atys* (1676) occurs in a moment of dementia brought about by the fury Alecton, and so forth.

Verisimilitude also depended on a subtle interplay between illusion and reality, distance and immersion. On the one hand, the assumption that the realm of the supersensible in the *tragédie en musique* is plausible depends on a separation of this world from the ordinary world of the spectator. Nature as the ultimate reference point of poetic representation is present, but only as a nature whose probability is mediated by the marvellous. The typical audience response corresponding to this distanced stance, typified in part by Philalethe's interest in the 'surprising' effects of Cambert's pastorale, is thus one in which the object of wonder and the concomitant desire for rational mastery of this object are less about the marvellous itself than about the ambiguous relationship between the rational and the irrational: 'How did they do this?'

Conversely, the principle of verisimilitude functions only when the suspicion that the communication between separate realms is the result of artifice is at least temporarily suspended. Only when the process of mediation between the ordinary and the marvellous is somehow naturalized can early modern opera work its magic. Contemporaries of the *tragédie en musique* commented at length on verisimilitude's ability to produce an enchanting effect on the audience, but as a rule failed to dwell on the specific musical qualities of this *enchantement*. Yet it is precisely here, in the physical linkages between sound and the body, that for a present-day observer such as Kintzler, music's genuine role and indeed necessity for the lyrical theatre resides. More than a mere mode of expression proper to mythical characters such as Orpheus or the Sirens, music constitutes the very element in and through which deities, demons and inspired poets act. But, she cautions, the way music imparts this sense of the ordinary extraordinary to the audience is not by some mysterious spiritual or psychological force, but by virtue of it being a 'natural (and, hence, probable) sounding box for a number of passions or situations being dealt with in the work'. Music finds itself in mechanical conformity with those passions as a result of the 'sympathetic transmission of vibrations'.[38] In short, the naturalization of the marvellous, for Kintzler, is a question of resonance.

A similar interaction of verisimilitude and the marvellous underpins opera's instrumental sounds. Such music, Kintzler suggests, must be understood as an excerpt of nature or as an approximation of nature, not as nature appears in its teeming variety, but as it truly is. It is through this principle of truth to nature (itself a key concept of early modern science), by concentrating on what is essential about the sounds of nature – the 'rolling' of thunder rendered in a drum roll, the flash of lightning imitated in a single stroke of the tympani, the running brook made audible in the rapid runs of the woodwinds, and so on – that music transforms the ordinary and physical into the true. At the same time, however, this passage from noise to musical sound, this poetization of noise not only fictionalizes reality, it is itself fundamentally fictional, drawing the audience into the simultaneously real and fantastic realm of the sounds heard in opera.[39]

Music, of course, is not a natural language, and what legitimizes music's presence on the classical stage is therefore its insertion into the play of rhetorical tropes or, as Kintzler calls it more generally, the 'hysterization' of music.[40] Music's ability to make the inaudible audible depends on the familiarity with a code, such as when a composer wishes to render the phrase 'trembling with cold', he does so by adding a tremolo to the tone accompanying the word 'cold'.

Kintzler's adumbration of the *tragédie en musique* as a genre that oscillated between a pathetization and a hysterization of music is a powerful one, no doubt. Nonetheless, there are several problems with her interpretation. There is, first of all, Kintzler's terminological confusion. The use of the term 'sounding box' (*caisse de résonance*) as a metaphor for music all told (rather than for specific acoustic aspects of it) is fanciful, at best. Another, perhaps more damaging problem is of a methodological nature. Many of the sources Kintzler adduces in support of her argument are culled from the writings of Abbé Dubos and Jean-Philippe Rameau, that is, from the mature period of the *tragédie en musique*. This move is problematic for the following reason: as is well known, Rameau's theory of harmony, expounded with great persuasion in his *Génération harmonique* (1737), was due in no small part to his discovery of the work of Joseph Sauveur (1653–1716) on harmonic partials published in 1701 and to his acquaintance with the scientist Jean-Jacques Dortous de Mairan (1678–1771).[41] As early as 1722, Rameau had allowed himself to be instructed in acoustical matters by Mairan, finding especially fertile ground for his theory of harmony in Mairan's concept of the *corps sonore*, or sounding body that transmits its vibrations to the perceiving ear, and in the concomitant idea that the ear perceives pitch on the basis of the sympathetic vibrations of the basilar membrane discovered by Duverney.[42] Harmony, Rameau confidently begins his treatise, 'is a natural effect, the cause of which resides in the air agitated by the percussion of each individual *corps sonore*'.[43]

Of course, neither the famous 'Proposition 3' of *Génération harmonique*, in which Rameau acknowledges his debt to Mairan, nor Fontenelle's 'Extrait des Registres de l'Academie Roïale des Sciences' that he proudly appended to his work, mention Duverney's theory of hearing. Still, it is clear that Rameau's work relied on a much more sophisticated knowledge of acoustics and the physiology of hearing than the one attained during the period that saw Cambert's pastorals and Lully's first *tragédies en musique*. And so to apply this theory retroactively to the naturalization of the marvellous through resonance during the formative period of the *tragédie en musique* half a century earlier is simply bad history.

Returning to 'Sçavoir', Perrault's objective was of course, first and foremost, to grant 'modern' harmony superiority over ancient Greek monophonic music and thus also to ground nascent French opera's legitimacy vis-à-vis the classical drama in the claim that not only had there to be a space for dealing with *le merveilleux*, but that opera was that space. Yet at the same time, the marvellous exerted a more ambiguous fascination on Perrault, as is exemplified by Paleologue's magic transformation as he 'enters' into the *sentimens* of the opera. Harmony's effects, Perrault wants to suggest, similar to those of the monophonic music of ancient Greece, are located at an emotional level, to be sure. But modern harmony differs from the simple cause–effect relationship underpinning ancient music, as he saw it, because it is predicated on a richer sense of multiple criss-crossing correlations between subject and object.

And thus the affective communion with music's *sentimens* generated by harmony does not so much depend on the stable interchange between two vibrating systems (music and the ears) guaranteed by the laws of nature, as on a coming to sound – or, better still, a remaining in sound. The ideal listening stance that Perrault seems to impute to his two opera lovers is not one based on a fixed vantage point vis-à-vis an object mechanically transmitted via sympathetic resonance. Nor, in fact, are these listeners asked, as later audiences during the early eighteenth century were, to engage with the theatrical figures emotionally, through mimetic mirroring. Rather, Philalethe's and Paleologue's preoccupation with the marvellous can be described as involving what one might call, borrowing a phrase from Kaja Silverman's work on male subjectivity, which itself draws on Max Scheler's *Wesen und Formen der Sympathie* (The Nature and Forms of Sympathy) (1923), 'exteriorizing' identification.[44] These spectators' theatrical experiences always teetered on the brink of ecstasy, even madness, involving a delirious form of listening, a veritable *folie de l'écoute* similar to the visual frenzy or *folie du voir* that Christine Buci-Glucksmann has located at the heart of 'Baroque' aesthetics.[45]

One cannot but note the marked contrast between the instability of this identificatory position, this auditory frenzy, and the subordination of the senses to the mastered clarity of Cartesian representation generally held to be the sole basis of early modern arts. Instead of the containment of the object of representation within an optical frame or a system of stable signifiers, early modern *folie de l'écoute* can be said to liberate its object from any frame and, in a sense, even erase the object altogether. It is in this latter sense that Buci-Glucksmann adumbrates early modern opera as akin to '"Baroque" poetry's phonic multiplication of the signifier and as ruins of language, lacking any hermeneutic dimension'.[46] And of course, this is also what Perrault's Paleologue has in mind when he discredits the sung recitative as being 'devoid of charm'.It thus makes sense to rethink the status of the perceiving subject. Early modern opera, one might say, instead of excluding the perceiving subject from the space of representation and reducing him or her to the role of a stable, if largely passive observer, stages the act of seeing and hearing itself. Listening to the *bruit agréable* of Cambert's pastorale here comes to be understood in the larger sense of the ear becoming absorbed by its own potential for the infinite expansion of the audible – as the possibility of gaining untold access to an unheard, yet eminently representable world.[47]

The idea of the *folie de l'écoute* erasing the object of audition and the concomitant concept of a listening devoid of a subject are not meant to denote some form of personal idiosyncrasy or pathology. Rather such objectless and subjectless perception is part of a broader logic of perception typical of the early modern period, one that Gilles Deleuze, in *The Fold: Leibniz and the Baroque* (1993), has termed 'fold-in'.[48] The affective content of early modern opera, Deleuze contends, is not the result of a binary correspondence between text and music, which would be rather arbitrary and, one suspects, more in keeping with the Cartesian roots of *Affektenlehre*. Instead, the text is enveloped by harmony, as though the melody of the voice is inflected by the harmonies so as to render the text's affective dimension.[49]

There are several ways to read Deleuze's cryptic depiction of early modern opera as a fold-in. One of these would be to see in the operatic voice a tool for multiplying and ultimately destroying in an act of sheer folly the sphere of meaning and replacing it with pure sound. Another reading would proceed more from the standpoint of the listener and see in the Deleuzian fold-in of music and text the attempt to oppose a more off-Cartesian notion of the subject to that of Descartes's own, all too confident, subject predicated on stable patterns of signification. In this view, it would have been Deleuze's goal to destabilize listening subjects such as Perrault's Paleologue and Philalethe by containing them in a potentially endless series of folds without a content or frame, thereby thwarting any claims to substance or coherence. Subjects such as these, obviously, do not so much perceive *from* a perspective as they 'come to' or 'remain in' one.[50] In short, theirs is a relationship to the world that is not determined by semiotic stability, but is mediated through a more resonant type of knowledge, or what Deleuze calls 'point of view'.[51]

It is tempting to draw a parallel between Deleuze's 'point of view' and what film sound scholar Rick Altman, in a by now canonic collection on film sound, has termed 'point-of-audition', or POA, for short.[52] From a mere technical standpoint, POA sound is akin to the 'point-of-view' shots ubiquitous in Hollywood cinema. Like these, it is the dominant model of the Hollywood soundtrack during the 1930s and a major form of subject placement. For instance, when in Howard Hawks's *Only Angels Have Wings* (1939) Noah Beery's plane takes off into the night, the soundtrack for the next shot of Cary Grant and Jean Arthur is that of the sound of the (unseen) airplane motor fading into the distance. POA sound, in other words, here is used as a type of sound that would be heard from a point within the film's diegesis by a specific character. As such, it is a common way of positioning the spectator in a specific relationship to the plot on screen.

But POA sound is also highly fictional:

> Point-of-audition sound always carries signs of its own fictional audition. As such, point-of-audition sound always has the effect of luring the listener into the diegesis not at the point of enunciation of the sound, but at the point of its audition. Point-of-audition sound thus relates us to the narrative not as external auditors, identified with the camera and its position … nor as participant in the dialogue (the standard situation of the 'intelligible' approach), but as internal auditor.[53]

Of course, POA sound in film is primarily produced by the montage of different patterns of volume and reverb. Yet even though in the context of early modern opera listening was not subject to film's more advanced technique of manipulation, Perrault's model of listening predicated on 'exteriorizing' identification bears a striking similarity to what Altman calls 'perfect interpellation'. He writes: 'We are asked not to hear, but to identify with someone who will hear for us'.

Point-of-audition sound 'inserts us into the narrative at the very intersection of two spaces which the image alone is incapable of linking, thus giving us the sensation of controlling the relationship between those spaces'.[54]

POA sound thus enables a widening of the subject's sphere of imagination. But at the same time, this act of hubris threatens the same subject with the possibility of its own dissolution. Losing themselves in the phantom space of *sentimens*, Perrault's spectators do not restrict their auditory experience to the sounds and the affects represented by them under the rules of *Affektenlehre*. Their hearing takes place entirely outside the dichotomy of the subject and the phenomenal world, or better still, parallel to it.

One understands why listening to the marvellous might take opera audiences at mid-century down a slippery path. And one sees the cause of the uncertainty into which Perrault releases the reader at the end of *Du bruit* and, especially, why he needs to invoke the ear's quasi-inbuilt 'certainty' of judgment.

A Matter of Taste

Sometime after having completed 'Sçavoir,' Perrault appended to *Du bruit* the afore-mentioned postlude entitled 'De la musique des anciens', in which he distanced himself even further from the claims and cultural models of the 'ancients'. But in so doing, Perrault not only accentuated the relativist stance of 'Sçavoir', he also placed it on a more solid theoretical footing by grounding it in the concept of good taste. Originally conceived as a response to Giovanni Battista Doni's account of ancient music *De Praestantia musicae veteris* (1647), 'De la musique des anciens' is a blistering attack on the idolization, current throughout the seventeenth century, of ancient Greek music. This idolization, Perrault argued, was simply based on a misreading of the ancient sources. The word *symphonia*, for instance, did not mean that ancient Greek music was the sophisticated polyphonic music modern audiences took it to be. Quite the contrary, it was monophonic and monotonous, little more than a 'rhythmic grumble', a 'noise fit for the infancy of the world'.[55] To understand the linkage between Perrault's anti-Hellenic criticism and his otological pursuits, we have to retrace our steps and return to the Louvre and Perrault the architect.

In 1673, Perrault had translated into French Vitruvius' *De architectura, libri decem* (Ten Books of Architecture).[56] Rediscovered during the Renaissance, Vitruvius for two centuries had been the undisputed authority on architecture, aligning in Pythagorean fashion the proportions of buildings with the ratios governing musical harmony. Although by the late seventeenth century such precepts were no longer taken at face value, the idea that musical harmony might in and of itself lend solidity and aesthetic unity to edifices still carried considerable, if not practical ideological weight, as is evident in the concurrent publication of *Architecture harmonique* (1679) by none other than the *maître de musique* at the Sainte-Chapelle in Paris, René Ouvrard and *Cours d'architecture* (1675–83) by the director of the Académie Royale d'Architecture and one of Perrault's fiercest critics, François Blondel.[57] Using the intervallic proportions obtained on the monochord as his yardstick, Ouvrard had been dreaming up buildings that would function as vast resonators solely by dint of the harmonic proportions underlying their construction. Blondel, for his part, insisted that the ancient architect's proportions were in agreement with the laws of nature and, like music, were grounded in what he, following Vitruvius, called *eurythmie*.

Perrault's extensive footnotes to Vitruvius' text leave little doubt that he considered his Roman predecessor's (and by extension, Ouvrard's and Blondel's) views as hopelessly outdated. It may thus well be argued that by the time he wrote 'De la musique des anciens', Perrault had already fixed his attention on the full-blown assault on Grecophile architecture he would soon launch in *Ordonnance*. The music essay, then, was but an opening salvo in the culture wars of the late seventeenth century, a mere prelude to the broader project of challenging not just the hegemony of ancient Greek architecture, but that of antiquity and the foes of 'modern liberty' all told. If, Perrault reasoned, it could be shown that the ancients were 'a little ignorant in precisely the science [music] in which they took the most pride', not only would the exalted place of the ancients in the history of music become untenable, the very usefulness of musical ratios for determining the sizes and proportions of buildings would also become open to doubt.

It was precisely this musical argument that served as the jumping-off point for Perrault's new theory of architecture. The latitude ancient architects had taken in determining the dimensions of buildings, he stated in the opening paragraphs of *Ordonnance*, show 'just how ill founded is the opinion of people who believe that correct architectural proportions were as certain and invariable as the proportions that give musical harmony its beauty and appeal – proportions that do not depend on us but that nature has established with absolutely immutable precision and that cannot be changed without immediately offending even the least sensitive ear'.[58]

A proper comparison between music and architecture thus demanded consideration of more than harmonic proportions. Individual and national differences would come into play, as would, perhaps more significantly, epistemological factors. For all their structural homologies, the eye and the ear do not perceive in quite the same manner. Whereas the eye is able to convey to the mind the effect of proportional relationships through knowledge, the ear cannot provide such knowledge.[59] Still, Perrault mused, why is it that even though the eye is capable of producing positive knowledge, we often appreciate proper proportions without actually knowing their reason? Are such preferences grounded in something objective, such as musical harmonies, or are they merely founded in custom? Antedating some of the propositions set forth in Charles Perrault's (ultimately more influential) *Parallèle des anciens et des modernes* (1688), Claude Perrault supposed two kinds of beauty to inform our judgment: a beauty based on convincing reasons and an arbitrary sort of beauty. While the former he claimed to be easily apprehended by everyone, arbitrary beauty he saw as entirely based on external associations, such as the patterns of speech or certain forms of aristocratic *couture* that, although unattractive in themselves, were deemed beautiful because of the 'worthiness and patronage of people at court'. Thus, it is not reason that forms the basis for aesthetic judgment, but power relationships.[60]

The same holds true for architecture which displays both positive beauty and beauty that is only arbitrary. Yet far from denying the existence of classical proportions, Perrault considered that these sorts of 'positive' beauty belong to the realm of *bon sens*, or common sense proper. Arbitrary beauty, by contrast, is the object of what he called *bon goût*, good taste. Thus, although it takes only common sense to realize that a large edifice built of precisely cut blocks of marble is more beautiful than a small stone house 'where nothing is plumb, level or square', only the architect – or, more broadly, the *honnête homme*, the man of distinction – has the knowledge of arbitrary beauty required for *bon goût*.[61]

With this notion of good taste, Perrault plainly put himself at variance with the prevailing opinion that the beautiful has its origin in the natural proportions inherent in musical harmony and in architecture and that, hence, in Nicolas Boileau's famous dictum, 'rien n'est beau que le vrai', 'nothing is beautiful but the truth'. But the anti-Vitruvian, anti-conservative criticism was not the only point behind the issue of good taste. Its significance went beyond architecture, broaching a new sense of art as an autonomous sphere. Nowhere does this become clearer than in one of the key propositions set forth in *Ordonnance*, which directly links Perrault's aesthetics to his scientific research on hearing and sight. This was the idea, fundamental to Vitruvian dogma, that in order to allow the truth of harmonious proportions to appear, the architect has to take into account the optical distortions of dimensions that arise when statues on top of buildings are viewed from a distance and, consequently, have to be slightly tilted forward. Perrault systematically refuted this theory, arguing – as he had in fact shown in his *Du bruit* – that hearing and sight are 'not as susceptible to surprise and deceit as people claim'.[62] Aesthetic pleasure, in other words, is acquired pleasure, or, as Descartes had taught, it is the result of habituation and reflex.

How, then, did Perrault frame the specifically musical component of his argument? The point of departure, once again, was the perennial question of whether or not the music of the ancients was superior to that of Perrault's day. Like Descartes 50 years earlier, Perrault was prepared to grant ancient Greek music exceptional powers. And like the celebrated philosopher, he believed the penchant of Greek musicians for what he called 'excessive transport' to reside in the fact that 'all they put their effort into was to touch the heart and the senses, which is quite a lot easier than to satisfy the intellect [*esprit*]'. Small wonder, then, that the ancients 'did not endeavour to look for another genre of perfection, since it would have been to the detriment of their goal of pleasing the multitude, which is usually more easily moved by the senses and the heart than by the intellect'.[63]

And so it was with 'modern' music:

> Those who do not possess an intellect capable of being touched by what is admirable in the diversity and beautifully arranged consonances of part music [*musique à plusieurs parties*], but instead only have ears to judge the clarity and intonation of the voice and a heart to love the gaiety of a cadence or the plaintive sweetness of a sad mode, will prefer a beautiful voice or a sweet flute solo to a music in parts whose beauty remains unknown to them. A musically

informed mind, by contrast, will be ravished with admiration upon hearing an excellent counterpoint, poorly sung though it may be.[64]

The association of taste with social status highlights the precariousness of Perrault's project of grounding aesthetic relativism in anatomical 'animism'. Having repudiated Pythagorean musical dogma no less than Cartesian physiology, Perrault struggled to anchor the compatibility of sense and reason within the larger framework of absolutism's 'modernizing' agenda. The simultaneously resonating and 'reasoning' cochlea, while potentially enlarging the subject's sphere of liberty, also threatened to undermine the established order. Taste, always a good way to end an argument, bridged these conflicting strands of late seventeenth-century point of audition. But even as Perrault's work straddled several aesthetic, political and scientific projects without resolving the contradictions inherent in either of these projects, it is precisely the everyday, 'vernacular' character of his modernity that lends his theory of hearing a compelling force. Even in hearing, it appears, to 'move on', one sometimes has to go back.

Notes

1 Antoine Picon, 'The Freestanding Column in Eighteenth-Century Religious Architecture', in Lorraine Daston (ed.), *Things That Talk: Object Lessons from Art and Science* (New York: Zone Books, 2004), pp. 67–100, at p. 75.

2 The best account by far of Claude Perrault's life and work is Antoine Picon, *Claude Perrault, 1613–1688, ou la curiosité d'un classique* (Paris: Picard Editeur, 1988). An older, but still useful, portrait is Wolfgang Herrmann, *The Theory of Claude Perrault* (London: A. Zwemmer, 1973). For a modern English translation of the ordonnance essay, see Claude Perrault, *Ordonnance for the Five Kinds of Columns after the Methods of the Ancients*, trans. Indra Kagis McEwen (Santa Monica, CA: Getty Center for the History of Art and the Humanities, 1993). For more on the Perrault family, see André Hallays, *Les Perrault* (Paris: Librairie Académie Perrin, 1926). On Perrault's zoological work, see Anita Guerrini, 'The "Virtual Menagerie": The *Histoire des animaux* Project', *Configurations*, 14/1–2 (2006): 29–42. The best discussion of Perrault's role in the architecture of the Louvre is Michael Petzet, *Claude Perrault und die Architektur des Sonnenkönigs* (Munich: Deutscher Kunstverlag München Berlin, 2000).

3 Claude Perrault, *Essais de physique; ou, Receuil de plusieurs traitez touchant les choses naturelles*, 4 vols (Paris: Jean Baptiste Coignard, 1680). I am translating *bruit* as 'sound', because Perrault intended the concept to encompass more than just noise. A reprint of parts of *Du bruit* appeared as Claude Perrault, *Du bruit et de la musique des anciens: Préface de François Lesure* (Geneva: Éditions Minkoff, 2003). This edition is based on the Leiden edition of Perrault's *Œuvres diverses de physique et de mécanique*, vol. 2 (1721) and, regrettably, omits the third part on the anatomy of the ear. In what follows, I am quoting from the first edition of the *Essais de physique* throughout. All translations given in this chapter are my own, unless otherwise stated.

4 Exceptions to this neglect are, among others, Alistair C. Crombie, *Styles of Scientific Thinking in the European Tradition: The History of Argument and Explanation Especially in the Mathematical and Biomedical Sciences and Arts*, 3 vols (London: Gerald Duckworth & Company, 1995), vol. 2, pp. 1158–63; and André Charrak, *Raison et perception: Fonder l'harmonie au XVIIIe siècle* (Paris: Librairie philosophique J. Vrin, 2001), pp. 230–59.

5 Stephen Jay Gould, *The Hedgehog, the Fox, and the Magister's Pox: Mending the Gap between Science and the Humanities* (New York: Knopf, 2003), p. 81. The Perrault brothers are eulogized at pp. 75–81.
6 Ibid., p. 22.
7 Richard Taruskin, *The Oxford History of Western Music: The Earliest Notations to the Sixteenth Century*, 6 vols (Oxford: Oxford University Press, 2005), vol. 1, p. 797.
8 Gary Tomlinson, *Metaphysical Song: An Essay on Opera* (Princeton, NJ: Princeton University Press, 1999), pp. 40–61.
9 Ibid., p. 48.
10 Georgia Cowart, *The Triumph of Pleasure: Louis XIV and the Politics of Spectacle* (Chicago: University of Chicago Press, 2009).
11 Cowart, *Triumph of Pleasure*, p. 162.
12 Cowart, *Triumph of Pleasure*, p. 172.
13 Joseph-Guichard Duverney, *Traité de l'organe de l'ouïe. Nouvelle édition* (Leiden: Joh. Arn. Langerak, 1731), first printed 1683. For a reprint of the English translation of 1737, see *A Treatise of the Organ of Hearing* (1737; New York: AMS Press, 1973).
14 Duverney, *Traité de l'organe de l'ouïe*, pp. 96–8; *A Treatise*, pp. 83–4.
15 Perrault, *Essais de physique*, vol. 1, p. 162.
16 Perrault, *Essais de physique*, vol. 2, pp. 333–4.
17 Perrault, *Essais de physique*, vol. 2, pp. 217–18.
18 Ibid., p. 219.
19 Perrault, *Essais de physique*, vol. 2, pp. 248–9.
20 Ibid., p. 251.
21 Perrault, *Essais de physique*, vol. 2, pp. 265–6.
22 Ibid., p. 284. For a more detailed discussion of the ups and downs of Cartesian mechanism in the second half of the seventeenth century, see Jacques Roger, *The Life Sciences in Eighteenth-Century French Thought* (Stanford, CA: Stanford University Press, 1997), pp. 133–204.
23 Perrault, *Essais de physique*, vol. 2, p. 312.
24 Ibid., p. 313.
25 Ibid., p. 331.
26 Duverney, *Traité de l'organe de l'ouïe*, pp. 160–61.
27 Ibid., p. 170.
28 Ibid., pp. 167–8.
29 Reprinted as Claude Perrault, 'Préface manuscrite du Traité de la musique de Claude Perrault (Bibl. Nat. manuscr. fr. 25,350)', in Claude Perrault, *Du bruit et de la musique des anciens*, pp. 576–91.
30 While it is true that Claude Perrault played a prominent, albeit often overlooked part in the dispute, he was not alone in his defence of Lully's innovations. Charles Perrault soon joined the fray with the brief, if ultimately more influential tract, *Critique de l'opéra*, followed by Pierre Perrault's rejoinder *Critique des deux tragédies*, written in response to a sharp rebuttal of the opera penned by Racine. The texts in question are available in two collections: William Brooks, Buford Norman and Jeanne Morgan Zarucchi (eds), *Alceste, suivi de La Querelle d'Alceste: Anciens et Modernes avant 1680* (Paris: Droz, 1994) and François Lesure (ed.), *Textes sur Lully et l'opéra français* (Geneva: Editions Minkoff, 1987).
31 Perrault, 'Préface manuscrite', p. 577.
32 Perrault, 'Préface manuscrite', pp. 586–7.
33 Cowart, *The Triumph of Pleasure*, p. 121.
34 On the parallels between early modern spectacle and postmodern aesthetics, see, for instance, Angela Ndalianis, *Neo-Baroque Aesthetics and Contemporary Entertainment* (Cambridge, MA: MIT Press, 2004).
35 Jean-François Marmontel, 'Opéra', *Œuvres complètes*, 19 vols (Paris: Verdière, 1818–20), vol. 14, p. 409.

36 Catherine Kintzler, *Théâtre et opéra à l'âge classique: Une familière étrangeté* (Paris: Fayard, 2004), pp. 10–11.
37 Ibid., p. 11.
38 Catherine Kintzler, *Poétique de l'opéra français de Corneille à Rousseau* (1991; Paris: Minerve, 2006), pp. 301–2.
39 Kintzler, *Poétique de l'opéra français*, pp. 303–5.
40 Ibid., pp. 303–8.
41 Jean-Philippe Rameau, *Génération harmonique, ou traité de musique théorique et pratique* (1737), in Jean-Philippe Rameau, *Complete Theoretical Writings*, ed. Erwin R. Jacobi (n.p.: American Institute of Musicology, 1968).
42 Jean-Jacques Dortous de Mairan, 'Sur la propagation du son dans les différens tons qui le modifient', in *Histoire de l'Académie royale des sciences avec les mémoires de mathématique et de physique pour la même année tirés des registres de cette académie–Année 1737* (Paris: L'imprimerie Royale, 1737), pp. 1–58. The *corps sonore* and its discovery by Rameau are expertly discussed in Thomas Christensen, *Rameau and Musical Thought in the Musical Enlightenment* (Cambridge: Cambridge University Press, 1993), pp. 133–68.
43 Rameau, *Génération harmonique*, p. 15. Rameau refers his readers, incorrectly, only to the 'Mémoires de l'Académie des Sciences de l'année 1720. pag. 11', which do not contain any of Mairan's work.
44 On identification in early eighteenth-century opera see Downing A. Thomas, *Aesthetics of Opera in the Ancien Régime, 1647–1785* (Cambridge: Cambridge University Press, 2002), pp. 201–64. On 'exteriorizing' identification, see Kaja Silverman, *Male Subjectivity at the Margins* (New York: Routledge, 1992), p. 264; Max Scheler, *Wesen und Formen der Sympathie* (Bonn: F. Cohen, 1923).
45 Christine Buci-Glucksmann, *La folie du voir: Une esthétique du virtuel* (Paris: Galilée, 2002).
46 Ibid., p. 112.
47 Buci-Glucksmann, *La folie du voir*, p. 93.
48 Gilles Deleuze, *The Fold: Leibniz and the Baroque* (Minneapolis: University of Minnesota Press, 1993), pp. 132–7.
49 Ibid., p. 136.
50 Deleuze, *The Fold*, pp. 19–20.
51 Ibid., p. 22. For a fascinating discussion of 'point of view' in 'Baroque' art see Mieke Bal, *Quoting Caravaggio: Contemporary Art, Preposterous History* (Chicago: University of Chicago Press, 1999), pp. 1–44.
52 Rick Altman (ed.), *Sound Theory/Sound Practice* (New York: Routledge, 1992).
53 Ibid., p. 60.
54 Altman (ed.), *Sound Theory*, pp. 60–61.
55 Perrault, *Essais de physique*, vol. 1, p. 353.
56 Claude Perrault, *Les dix livres d'architecture de Vitruve* (1673; Paris: Bibliothèque de l'Image, 1995).
57 René Ouvrard, *Architecture harmonique, ou Application de la doctrine des proportions de la musique à l'architecture* (Paris, 1677); François Blondel, *Cours d'architecture* (Paris: Lambert Roulland, 1675), pp. 756–60.
58 Perrault, *Ordonnance*, p. 48.
59 Perrault, *Ordonnance*, p. 49.
60 Ibid., p. 52.
61 Perrault, *Ordonnance*, p. 54.
62 Ibid., p. 160.
63 Perrault, *Essais de physique*, vol. 2, p. 388.
64 Ibid., p. 390.

3 Georges Kastner's *Les Voix de Paris* (1857)

A Study in Musical *Flânerie*

Emily Laurance

Louis-Sébastien Mercier's pre-revolutionary work *Tableau de Paris* (1781–88) is one of the most famous examples of literary non-fiction devoted to the description of Paris. Mercier's use of the highly theatrical but static term *tableau* implied that his was a comprehensive account of his subject, encompassing a thorough survey of urban variety. Taken together, the observations contained in Mercier's *tableau* could presumably sum up Parisian identity and character at a single glance. In his book *Paris and the Nineteenth Century* Christopher Prendergast claims that the *tableau*, or popular urban guidebook, represents a typically nineteenth-century way of thinking about and understanding Paris. He calls the *tableau* 'one of the founding genres of a certain discourse of "Paris", [which became] increasingly popular throughout the nineteenth century'.[1] According to Prendergast, Mercier's work, one of the inaugurators of the genre, 'offered something intellectually stronger than mere notation; it effectively invented a form of urban anthropology'.[2]

The *tableau* depicted Paris as a source of aesthetic discovery; simultaneously, it sought to render the city coherent via the description and classification of a bewildering variety of social types. 'The issue of the identity of the city, and of identities within it, constituted a veritable nineteenth-century obsession. In the later part of the century, Maxime du Camp wrote that never before had the city been so minutely described, monitored, surveyed, classified, generally "cleaned up" taxonomically as well as practically'.[3] The *tableau* provided an aesthetic or a scientific distance from its subject, in either case assuming a perspective of mastery over the urban landscape. This perspective was only possible, however, because the *tableau* tended to minimize aspects of urban life that were pathological, alienating or brutal. Instead, the *tableau* asserted that the city was fundamentally comprehensible. Aesthetic pleasure and systematic knowledge of urban life could be achieved by the true connoisseur, at whom the *tableau* was directed. In some *tableaux* this connoisseur was identified as a *flâneur*, a literary type made famous by writers such as Charles Baudelaire and Henry James. James's definition held that 'the prime requisite of an expert *flâneur* [is] the simple, sensuous, confident relish of pleasure'.[4]

> That, of course, is the view of the comfortably furnished, travelling gentleman, a view amply confirmed and provided for in a welter of nineteenth-century guidebooks and journalistic writing, themselves often constructed in the form of a chapter-by-chapter itinerary through the city.[5]

This privileged perspective was the domain of educated, often wealthy men.

James's version of *flânerie* tended to emphasize its experiential, bon vivant nature. As with the *tableau* generally, much of the discourse surrounding the figure of the *flâneur* revolves around his rapturous aesthetic celebration of urban variety (in all its chaos). Coupled with this trope of urban variety is an emphasis on the city's kaleidoscopic and constant change. But the *flâneur* also related to his urban environment through the taxonomic practices of social science. There is a fundamental tension between the knowledge gained through experiential living and the control asserted through systematic categorization, both of which characterize the *flâneur*'s relationship to his urban environment. These dual perspectives reveal an ambivalent attitude towards the realities of urban life: on the one hand, the *flâneur* is charged to be open to the vagaries of experience, including the random experiences available on city streets. On the other, these experiences have to be brought to order and to conform to a systematic understanding of reality. In either case the *flâneur*'s encounter with the city is one of discovery – in scientific research and analysis, in aesthetic contemplation or in some combination of the two.

The *tableau* was such a pervasive interpretive framework of nineteenth-century Paris that there were versions devoted specifically to *aural* urban experience. Very often these chronicles of the Parisian soundscape were specifically concerned with the music used by street vendors to assist them in their trade. The Parisian street vendor was a frequent denizen of popular art, music and literature since the Middle Ages, but the nineteenth century saw an intensified fascination with the phenomenon. Street cries provided in microcosm the requisite elements of dynamic variety and randomness, as well as supplying a nostalgic connection to the city's pre-revolutionary past. Prendergast, in fact, refers to the development of a whole 'literature of picturesque cacophony'.[6] An example of a *tableau* specifically documenting the *sounds* of the city is Victor Fournel's 1887 *Les Cris de Paris*. Fournel's version of 'picturesque cacophony' was mainly a nostalgic one, coloured by a sense of loss related in part to the immense physical changes Paris had undergone due to the policies of Baron Haussmann and the subsequent disappearance of much of the old city. There is a sense of purposeful antiquarianism in his work, a real desire to capture not just the contemporary taxonomy of street criers but their historical antecedents as well:

> Old Paris, from one end to another of his walls, was nothing but one ceaseless symphony, which married together all tones of the provocative voices of the travelling merchants. Far from increasing with time, the number and the variety of these street cries has decreased. It is easy to understand it. Formerly, before the discovery of printing, before the invention of the gazettes and leaflets, the means of publicity were singularly restricted. Few people could read. For advertisements or announcements, it was necessary to resort to the human voice. Then everything was shouted through the streets.[7]

As far as the work is about music, it is the work of an amateur, in the sense of a non-professional aficionado, and most of the citations of street cries are from literary or other printed sources; very seldom does Fournel turn his focus on the individual musical formula per se. He does have an acute appreciation for the sensory experience of their musical aggregate, however. Just as nineteenth-century tableaux described Paris in terms of shifting patterns of light, Fournel described the city's shifting layers of ambient sound, hearing their intermingling as a 'raucous yet agreeable "concert" ... signify[ing] a world of everyday transactions at once "naturally" differentiated and "harmoniously" orchestrated'.[8]

The most impressive and comprehensive attempt at an aural *tableau*, however, is found in Georges Kastner's treatise on Parisian street cries, aptly titled *Les Voix de Paris*. Kastner himself fits much of the essential profile of the *flâneur*: classically trained as a composer, he was freed from the necessity of earning his living due to an advantageous marriage.[9] Thus, in addition to composing, he was able to spend much of his time as an independent man of letters, writing several theoretical treatises along the way. In *Les Voix de Paris* his broad musical theoretical interests led him to consider street vendor cries from several disciplinary perspectives, thus capturing the twin identity of the *flâneur* as scientist and aesthetic connoisseur. Kastner's multiple roles of scientist, connoisseur and artist result in a hybrid narrative tone of voice. On the one hand, Kastner affects a dry detachment in his documentary work, as when he dispassionately provides the mundane rationale for his topic's existence:

> Wandering merchants have no better advertisement than the cry: no other means of publicity could offer them the same advantages. Without it how could they ever get the attention of households living on the highest floors of tall Parisian houses?[10]

But like Fournel, Kastner also luxuriates in the endless variety of street vendor cries, resulting in a celebratory perspective, which glories in the very confusion of his subject:

> The voices of Paris, which, in the manner of a basso continuo, accompany the rumble of carriages from morning to night, are innumerable and defy the observer's patience. Infinitely varied, reproduced by a thousand echoes, they form a sort of perpetual polymorphic canon, the last note of which even the most able musician cannot hope to capture.[11]

In these florid descriptions of street life the constant noise of the city is transformed into a *Grande Symphonie de Ville*. Kastner's simultaneous analytic and panegyric perspectives on street cries set up an opposition between an abstract sense of social order (in the form of traditional occupations, hierarchical relationships and formulaic melodies) and experiential disorder, as when individual cries, each intelligible on its own, combine to create constant unintelligible noise. Kastner's book strives to create a reassuring sense of the stable and continuous social structures that underlie this chaos, and to render, through acts of understanding and classification, the chaos intelligible again.

Kastner claims that his work has two main goals, both of which express the tension inherent in the role of the *flâneur*: '… the musician must ascertain two things about the popular cry', he writes, 'the various forms of that cry, and its proper application',[12] by which he means the direct study of an object and the evaluation of its potential as artistic raw material. Kastner's first goal, establishing 'the various forms of that cry', takes up the bulk of his prose. He begins by outlining the nature of cries (of pain, surprise or fear, for example) in general, establishing their status as a universal expression of human feeling. He then documents specific examples derived from French historical and literary sources, presenting them as emblematic of French character. This includes the history of Parisian street vendor cries from the Middle Ages down to his own day, chronicling the habits of every subgroup of street vendors in the city. Kastner includes every musical reference to street vendor cries he can find, including examples from Clément Jannequin, Philippe Gaubert and Joseph Mainzer, to name just a few. As Kastner finally turns his attention to the contemporary scene, his methods conform to what we would now term fieldwork. The quality of the documentation is striking, as well as the attention to detail and overall scope. Kastner supplies almost 400 of his own musical transcriptions of then-current Parisian street vendor cries, organized by type of goods sold into a systematic glossary and reproduced on high quality plates (see Figure 3.1). Kastner's high degree of musical precision suggests that he is truly attempting to give his readers 'the various forms of that cry'.

Neither history nor ethnography was Kastner's primary goal, however. Rather, he was ultimately more concerned with what he called the 'proper application' of the street cry. In his introduction to *Shadows in the Field*, Timothy Cooley remarks: 'collecting data for goals quite external to the field experience itself is strikingly common in the history of ethnographic enquiry'. He connects this to 'the science paradigm of the modern era', in which 'music was an objectively observable fact to be collected in the field and manipulated in the laboratory'.[13] Kastner's treatise neatly fits into this discussion: while *Les Voix* supplies us with a large repository of musically notated street vendor cries, which Kastner seems to have collected himself, this was not his principal aim. Instead Kastner, having established street cries as representatives of French national identity, now writes to argue for the *musical* worth of Parisian street cries and further, for their consideration as musical source material for original large-scale works expressing French national themes. Ultimately, therefore, *Les Voix* is a composition treatise above all else. This is, incidentally, true in some senses of virtually all of Kastner's prose works, some of which include monographs on the music of the French military, the history of music associated with sirens and a general treatise on orchestration, the work for which he is best known. These titles indicate not only a general pedagogical interest in composition, but also a tendency to champion mundane and/or marginalized musical repertoires and genres. It is in his role as composer that he supplies his *coup de grâce*: a '*Grande Symphonie Humoristique*' called *Les Cris de Paris*, a three-movement cantata representing an orchestrated version of Parisian street life, complete with full orchestra, soloists, choruses and a separate *banda* composed of 24 of the newly-invented saxhorns. In this appendix,

Figure 3.1 Georges Kastner: *Les Voix de Paris*. Cris Notés, Série V, Planche XII, 30–53: Marchands et Marchandes de Légumes.

Source: Courtesy of The Music Library, University of California, Berkeley.

he offers his own fictionalized musical reinterpretation of the Parisian streetscape. His extensive inclusion of vendor cries as sonic raw material infuses the work with an immediacy and experiential discovery sought by the true *flâneur*.

Underlying the outwardly celebratory tone of Kastner's work, however, is the same tension that typifies many other *tableaux* devoted to the subject of Paris. As with his literary counterparts, Kastner is aesthetically, rhetorically and socially removed from his subject, revealing his uneasiness with its less ordered, less predictable and less polished sides. This is evident in the several ways Kastner insists on making his unruly subject conform to rigid systems throughout the various stages of his work. First, when Kastner takes the role of the historian, his main aim is to mould his material to fit his own theories of French national character. Similarly, in his discussion of contemporary street *crieurs*, he takes very little account of them as social, modern individuals, but instead uses them mainly as symbolic expressions of French national identity. Finally, Kastner's evaluations of his collected street cries depend entirely on how well they reinforce musical theoretical systems, the tastes of the musical establishment and the aims of musical nationalism. I will examine each one of these issues in turn.

Many of the historical concerns evident in Kastner's work reflect the larger concerns of contemporary professional historians, some of the most prominent of whom include Jules Michelet, Augustin Thierry and Prosper de Barante. After the Revolution and the Napoleonic Wars, this group of scholars expressed a crisis of confidence in received historical tradition, in particular a loss of faith in the legacy of the Enlightenment. In *Beyond the Enlightenment* Charles Rearick writes that 'it was not until after tumultuous years of revolution and war that doubts about individual reason and criticism became a dominant theme in French thought. By 1815 in the eyes of many, and not only conservatives, the inadequacy of rationalism had been demonstrated by the Revolution itself: could a philosophy be sound if it led to such destruction?'[14] Turning inward, these young historians sought to discover in history the key to a national character, one that transcended tyranny and revolutionary anarchy. But in doing so, they wrestled with some fundamental problems about their nation's past: how does one, when defining French national character, reclaim France's pre-revolutionary past for a modern bourgeois society? Further, how does one elevate the role of the common people within the narrative of a monarchical past? And, given the disruption of long-term national structures, how does one define French national character?

For many young historians, the key to this new history, and to the delineation of national character, lay in the folklore, customs and oral traditions of ordinary people. But as Rearick states:

> ... it was not ... the dearth of other sources that led the French historian to folklore. It was primarily a new conception of what history would be. Folklore became important to the historian because he had new ideas of what constituted the truth about the past.[15]

Reliance on oral evidence especially characterized the work of Jules Michelet, whose seminal *Histoire de la Révolution Française* was published four years before *Les Voix de Paris*. In that work Michelet relied heavily on oral evidence, seeing in it a truth unavailable from official documentation and a key to French identity:

> When I say oral tradition, I mean *national* tradition, which remained generally scattered in the mouths of the people, which everybody said and repeated, peasants, townsfolk, old men, women, even children; which you can hear if you enter of an evening into a village tavern; which you may gather if, finding on the road a passer-by at rest, you begin to converse with him about the rain, the season, then the high price of victuals, then the times of the Emperor, then the times of the Revolution.[16]

How did the emerging status of folkways and oral tradition in historical enquiry influence Kastner's work? What led Kastner to collect street cries in the first place? Kastner claims in his preface to be interested in street cries primarily as raw materials for musical compositions:

> Upon consideration [he writes] it seemed that there were discoveries to be made in this ocean of sound, and our first preoccupation has been to extract a subject for musical composition.[17]

But Kastner, like Michelet, is also using oral evidence to refashion national history. He is at some pains to demonstrate the musical worth of the street cry, but there are two reasons for this: the first is to justify their use in his own composition, the other is to legitimate the status he claims for them in the sweep of French history.

Kastner's relationship with his subjects is complicated because he viewed Parisian street vendors as both exotic and familiar. Kastner, originally Swiss, was a cosmopolitan man living in a cosmopolitan city. The working class *marchands* that he heard regularly were often French provincials who had migrated to the city. Georgina Born and David Hesmondhalgh, in their introduction to *Western Music and its Others*, might well have been talking about Kastner's own compositional habits when they spoke of 'modernist composers who have drawn on … other musics'. Typically, they claim, these composers are 'not producing that [other] music but drawing upon it in order to enrich their own compositional frame. They are … appropriating and re-presenting it. Crucially, in doing so, they intend not only to evoke that other music, but to create a distance from it and transcend it'.[18] This judgment of Kastner has some rationale, especially because of his marked tendency to view the Parisian *crieur* as an idealized type. But it is complicated by the fact that, while Kastner may experience street cries as 'other', he is also asserting his right to identify with them as a Frenchman. After all, it was an ordinary repertoire he could hear every day.

But just as relevant is Kastner's use of the Parisian street cry tradition to *bridge* distances – between levels and functions of society, between past and present and between varieties of musical experience. Kastner's conviction of the continuity of the French *crieur* led him to write his treatise primarily within a historical frame. Within this frame Kastner, like his contemporaries, sought to demonstrate that 'the popular imagination was at the root of literature, law, religion, historiography and even institutional development':[19]

> Once we were convinced of the almost musical character of the cries of Paris, we were not surprised that the historical character of the subject excited our curiosity. Here at least we had numerous predecessors, but their research lacked order. There remained the task of introducing a dominant theory to the history of the popular cries, to show their relationship to the history of languages, arts and society, and to characterize diverse ages with the discernment of an attentive examination.[20]

Just as Kastner argues in his introduction that the human cry is the origin of both human language and song, (showing clear debts to Rousseau and Herder), the cries of Paris are similarly intertwined with the development of French customs, institutions and character. They provide continuity from the civil society of the *Ancien Régime* through subsequent expressions of revolutionary fervour and ultimately to the soundscape of modern commerce, the characteristic expression of bourgeois public life. While obviously the expression of an individual, the cry was also, perhaps more importantly, a collective expression. Throughout the upheavals of French history the *crieur* represents the love of freedom inherent in the French character:

> Above all the *crieur* was free: whether selling, travelling, or even begging, he varied according to his will formulas designed to elicit consumerism, attention or pity from those passing.[21]

Kastner's historical survey of the Parisian *crieur* begins in the Middle Ages. This is surely determined in part by the extant sources, but it also bears some relation to the historical trends of Kastner's day. Under the influence of Romanticism many French historians 'rediscovered' the Middle Ages, claiming that the era saw the birth of national consciousness and the organic formation of many current social institutions:

> Men of varying political orientations were attempting to reconstruct the unbroken and unifying elements in French history in order to explain both the French Revolution and the movement of contemporary history. Voltaire's debunking of the Middle Ages less than a century earlier was … unacceptable to these historians who were searching for the origins of institutions and, in often biological imagery, the foundations of modern French society.[22]

Kastner explicitly describes the connection of the *crieur* to the formation of public institutions in his chapter on the Middle Ages. 'Perhaps never was the relationship of the popular cry to the political organization and social habits of a city revealed with more evidence than in the Paris of the Middle Ages'.[23] In the Middle Ages there were, alongside the commercial *crieurs*, public officials whose cries were the ordinances and documentation of the time. *Crieurs* were thus integral to all realms of public life: they were even employed as official mourners.

In his chapter on revolutionary cries, Kastner likewise argues for the centrality of the cry. It is, however, in this chapter that he exhibits the least sympathy with his subject. While Michelet saw in the Revolution the nobility of the common person, exemplified in his capacity for sacrifice, Kastner exhibited an uneasiness and ambivalence towards the Revolution which was not uncommon at mid-century. Here the *crieur*, elsewhere the subject of celebration, is dangerously close to inciting a mob:

> Revolutions, as those in Paris know better than anyone, achieve no great things except the conditions to commit excess; the path of despotism as well as that of anarchy always leads to extremity …
>
> In revolutionary times the collective cry exhibits a great variety of application, but with a singular uniformity in its formulas. We could define it in this way: the combination of the words *vive* and *à bas* with the names that designate now one of the idols of popularity, now one of its victims.[24]

Nevertheless, Kastner's continued insistence on the *crieur* as the public spokesmen for all defining periods of French history enables him to create out of the *crieur* a kind of national everyman. In every era it is the *crieur* who expresses the fundamental motivations and concerns of the French people. He is both a leader and a representative, an individual who gives voice to the collective. He is in this respect not dissimilar to the fictional character of Jacques Bonhomme, a rhetorical device first developed by Augustin Thierry and Prosper de Barante – a personalized representation of the experiences of the ordinary Frenchman.[25] Certainly the status of the *crieur* as generic specimen enables him to better represent a grand sweep of historical continuity and national identity.

Because Kastner is primarily interested in the aesthetic qualities of his subject, his discussion of the *crieur* himself is incidental and impersonal; Kastner's interest is mainly in the *crieur* as a representative type and in his musical material rather than on the shape of his individual life. Although Kastner does describe the cry as the expression of an individual, his view of the cry is as an expression of a collective genius, one that is instinctual, but not reflective. The generic nature of the *crieurs* is further reinforced by Kastner's predominant narrative strategy, which more often than not presents his subject through the fictionalized eyes of a historical observer. This takes several forms: in his description of the *crieurs* of the thirteenth century Kastner places his reader imaginatively into the body of the poet Guillaume de Villaneuve, and the anonymous author of

Dits d'un mercier, his two most frequently cited sources for medieval street cry representations. Consciously linking his own work to the tableau genre, Kastner uses Sebastien Mercier in a similar fashion when discussing the street cries of the eighteenth century. Even in Edward Thierry's libretto for Kastner's cantata *Les Cris de Paris* there is always present a middle-class character who functions as an outside observer of the passing urban scene. In the first movement, *Le Matin*, he is called *le dormeur* – a hapless sleep-deprived figure who is serenaded in the early hours by the *marchands* below his window. In the second movement, *Le Jour*, he becomes *le dormeur s'eveillé*; by the time we reach *Le Soir* he is *le promeneur*, wandering the streets and experiencing the urban cacophony directly. As such the character echoes Rousseau's *promeneur*, as well as the literary character of the *flâneur*.

Just as Kastner seeks to bring his subject in line with a comprehensible understanding of French character, Kastner in his capacity as composer has a similar mission, that is, to bring order to otherwise unruly musical elements. In Kastner's introductory remarks regarding the cries of his own day he emphasizes their commonality and the inherently fluid, unfixed nature of their forms. This stresses continuity of practice and collective composition over the authority of the individual *crieur*.

> For the most part, the cries form melodic groups, each of which has its own colour and particular style, and which depends on one or several primitive formulas which, by passing through all kinds of mouths, get altered, modified, and gives birth to numerous variants. The *crieurs*' song is frank, natural, often energetic, and sometimes tender or gracious.[26]

But Kastner unquestionably also values the individual street cry in all of its musical particularities as well. Throughout Kastner's text he returns to the primary evidence of the cries themselves, musically notated and printed in columns on separate plates. The contemporary examples he undoubtedly collected himself, as evidenced by some of the performance practice particulars he cites:

> Also the voice of the honest purveyor of hydrogen in its oxidized state relaxes on a note in his chest register, out of which the syllable *eau* falls. After holding this note several minutes, the water-bearer ends brusquely at the octave or sixth above, using a portamento with both assurance and success.[27]

Les Voix was written to argue for the *musical* worth of Parisian street cries and for their consideration as motives in original large-scale musical works. But while Kastner the composer hears in Parisian street cries wonderful musical potential, they are in their original form far too primitive to be imported wholesale into a piece of music, as they do not fit well into recognized pitch and rhythmic systems:

> Truth be told, not all of our small-time singing vendors are virtuosos. There are some of them who have a tin ear, who sing quite badly, and even *à la turque*,

with all sorts of strange inflections, quartertones being only the beginning. In the course of our wanderings, we have rarely heard beautiful voices.[28]

However, the tradition of each individual cry can go through a developmental process; after achieving its 'perfected form, the cry is no longer an unpleasant noise whose precise intonation escapes all rules of art; it is a vocal manifestation with pitches clearly positioned in the orderly scale of sounds'.[29] Kastner's first task in bringing order to this material, then, is its transcription into familiar and standardized musical terms. The greater the inherent musicality of the cry, Kastner implies, the fewer problems posed by its notation:

> Notwithstanding, we would mention the magnificent baritone voice of an adult chimneysweeper that we heard with rapture, which made us listen to, at regular intervals, the classic *haut en bas*. The child that followed his companion echoed the cry at the interval of a sixth, with perfect accuracy; this double cry in imitation is one of the ones that we heard executed the best.[30]

While Kastner remains faithful to the essential musical characteristics of each cry, the street cries in his cantata are further modified to fit into recognized harmonic systems (see Figure 3.2). And when combined, they are fashioned together into a rhythmic structure that, while exhibiting some surface variety, lines up in all metric particulars. Although Kastner refers to them as 'voix confusées', any remaining sense of disorder is limited to the staggered entry of each individual cry.

While Kastner's street cry transcriptions provide us a glimpse of Kastner's own experience in the field, they remain fundamentally incidental and impersonal. They almost exclusively describe musical traits, almost never straying into the experience of the individual *crieur*. Kastner places value on the features of the musical object, and on the *crieur* as a representative type, but not so much on the one as expressive of the other's experience. For Kastner a cry is the manifestation of a collective genius, one that is instinctual, but not reflective. Kastner was, like Michelet, 'drawn to the impersonal forces of beliefs, customs and laws'.[31] His fascination with the street cry offers us a musical corollary to Michelet's own rehabilitation of myths and legends into French history. For Michelet, '[m]yths and legends possessed value because they were translations of collective truths expressed in imaginative form'.[32] This collective sonic expression daily links those who create it with those who perennially hear it. While Kastner celebrates the commercial enterprise as the impetus for the cries' creation, it is ultimately not via commerce that the *crieur* integrates society, but by the power of sound itself. Sound, continuous, unbounded and inescapable, is one of the more democratically unifying forces in the city. The power of this perpetual sound enables Kastner, through an act of imaginative will, to link the past and present in a similar fashion. At the same time, however, Kastner uses the devices of the cry and of the *crieur* to assert an integrated national identity in an effort to come to terms with France's fractious recent history.

Figure 3.2 Georges Kastner: *Les Cris de Paris*. I: Le Matin.

Figure 3.2 Continued

Figure 3.2 Continued

Figure 3.2 Continued

Figure 3.2 Continued

Figure 3.2 Continued

Figure 3.2 Continued

Figure 3.2 Continued

In his survey of street cry representations Kastner took special care in detailing street cries in musical theatre. But perhaps the most extensive example of street cry use in an opera, Gustave Charpentier's *Louise*, postdates Kastner's work. It is likely that Kastner's treatise influenced Charpentier's opera, and the opera is worth examining here because it demonstrates some of the same attitudes towards its musical subject evident in Kastner. Charpentier's street cries demonstrate a concern with musical authenticity that is comparable to Kastner's. In his book on realism in opera Manfred Kelkel evaluates just how far this tendency goes by comparing Charpentier's use of street cries to Kastner's.[33] The close resemblance of some of Charpentier's cries to those of the earlier treatise might lead one to believe that Charpentier depended on Kastner's work as source material, but while there are many close parallels between the two, there are no exact matches. They are just similar enough, without being identical, to suggest that Charpentier, while probably familiar with Kastner's work, did not depend on it as a direct source. More likely both writers transcribed their sources faithfully from life; the differences between them are due to the distance of a generation, and the fact that the level of variation in the repertoire was relatively high, something to which Kastner himself attests:

> In general, the cries of the food sellers and of the wandering artisans seem to have the oldest and most richly diverse chants; however, in spite of the changes they have undergone while passing through different mouths, they have not wholly lost traces of the original cries, or of the melodic types that have brought forth all of these variants.[34]

It is not so much the use of street cries in *Louise* that bring Kastner's work to mind as the point of view through which he presents them. In Julien, the male lead of *Louise*, Charpentier creates a figure that serves many of the same functions as Kastner's bourgeois observer, for whom the inhabitants of the city streets provide aesthetic gratification. In the fourth scene of Act II Julien, in one of the opera's few soliloquies, thinks aloud about his problems with Louise. Since he has no auditor on stage, Julien's speech feels directed at the actual audience, increasing their identification with him and reinforcing the status that Julien enjoys as dramatic lead. At the end of these musings, Julien asks: 'And who is there to help me?' Immediately following, as though conjured up by these ruminations, a series of street vendors starts to sing. First a chair-mender is heard in the distance, then a ragpicker, an artichoke vendor, a carrot vendor and a bird-seed seller. Like Kastner, Julien responds to this particular version of 'picturesque cacophony' by rhapsodizing about it, seeing in the street cries the raw materials in a poetic construct of the city:

> Julien: (*with enthusiasm*) Ah! the song of Paris, that thrills and enchants my soul! The old refrains of the town when it wakes! Morning sounds – my ear rejoices with the cries of Paris. Voices of the street, are you a song of victory to our love? (*Louise* Act II/2nd tableau/scene iv)[35]

The pre-eminence of Julien's sentimental perspective here is not only the result of his important dramatic status, or even of his comparative class privilege. It is also an artistic one, for it is, according to Prendergast '… a matter of the narrator's privileged capacity – socially as well as aesthetically – for making subjective magic out of his sensory environment'.[36]

Compared to that of Kastner in his cantata *Les Cris de Paris*, Charpentier's musical language in *Louise* is free enough to combine street cries in more musically haphazard ways. The successive harmonies that Charpentier introduces, while not remote, do suggest that each vendor's entry is independent and occupies its own musical space. Additionally, Charpentier's street cries are far more rhythmically differentiated than Kastner's. Ultimately, however, what begins in scene four as seeming disorder gradually coheres to a greater and greater degree. Finally, as harmonies and rhythms coalesce, Julien bursts in, making sense of the experience by his climactic poetic interpretation. As with Kastner, Julien's vision of the street vendors around him is essentially impersonal and generic. Each vendor's music, while particular to him, seems to be subservient to impressions of it held by Julien and by the audience. Through Julien's eyes and ears, we experience their disparate musical motives as part of a larger codified scheme, a pre-existent aesthetic construct about Parisian street life. It is this scheme that forms how we hear each character, rather than what they themselves are singing (we are, after all, not shopping for vegetables). Besides the voyeuristic aesthetic structure in which these street vendor cries exist, their recognized pre-existence outside the opera and their continual literal repetition also contribute to our sense of predetermination. It is almost as if each street vendor is entranced and, zombie-like, programmed to sing only their assigned characteristic music.

Charpentier's depiction of Julien as a *flâneur* demonstrates the staying power this type of figure had right up to the end of the century. It also shows just how potent aural experiences of the city could be in developing versions of the *flâneur*. This is partly because of the nature of sound. Sound, continuous, unbounded and inescapable, is a force tying different parts of the city together, crossing boundaries to reach rich and poor alike. 'Every day', Kastner writes, 'and, so to speak, at the same time, they tap our ears with the regularity of a clock chiming, which pleasantly awakens us at dawn and doesn't stop following us until nightfall'.[37] Kastner's bourgeois observer, like his alter ego the *flâneur*, celebrates the sheer scope of sound in his peregrinations and, in doing so, represents the imaginative link between Kastner's readers/listeners and all the historical and geographical variety of the Parisian soundscape. At the same time, however, he articulates a sentimental distance whose attempts at capturing his subject are ultimately impeded by his own attempts at mastery.

Notes

1 Christopher Prendergast, *Paris and the Nineteenth Century* (Oxford, UK and Cambridge, MA: Blackwell, 1992), p. 184.
2 Ibid., p. 184.
3 Ibid., p. 2.
4 Henry James, *Roderick Hudson* (Boston and New York: Houghton Mifflin Co., 1917), pp. 15–16.
5 Prendergast, *Paris and the Nineteenth Century*, p. 4.
6 Prendergast, *Paris and the Nineteenth Century*, p. 127.
7 'Le vieux Paris, d'un bout à l'autre de son enceinte, n'était qu'une symphonie incessante, où se mariaient sur tous les tons les voix provocatrices des marchands ambulants. Loin d'augmenter avec le temps, le nombre et la variété de ces cris de la rue ont beaucoup décru. Il est facile de le comprendre. Jadis, avant la découverte de l'imprimerie, avant l'invention des gazettes et des prospectus, les moyens de publicité étaient singulièrement restreints. Peu de gens savaient lire. A défaut d'annonces ou d'affiches, il fallait bien recourir la voix humaine. Tout alors se criait par les rues …'

Victor Fournel, *Les Cris de Paris: Types et physionomies d'autrefois* (Paris: Librairie de Firmin-Didot, 1887), pp. 5–6. All translations in this chapter are my own unless otherwise stated.

8 Prendergast, *Paris and the Nineteenth Century*, p. 127.

9 Thomasin La May and Stewart A. Carter, 'Jean-Georges Kastner', *Grove Music Online*, http://www.oxfordmusiconline.com (accessed 13 May 2010).

10 'Pour les marchands nomads il n'est pas de meilleure enseigne que le cri: aucun autre moyen de publicité ne pourrait leur offrir les mêmes avantages. Comment parviendraient-ils sans cela à obtenir l'attention de la ménagère qui habite les étages les plus élevés des hautes maisons de Paris?' Georges Kastner, *Les Voix de Paris* (Paris: G. Brandus, Dufour, 1857), p. 78.

11 'Les voix de Paris, qu'accompagne du matin au soir, en manière de basse continue, le roulement des voitures, sont innombrables et défient la patience de l'observateur. Variées à l'infini, reproduites par mille échos, elles forment une sorte de canon polymorphique perpétuel dont le plus habile musicien perd l'espoir de trouver la dernière note'. Ibid., p. 77.

12 'Ainsi le musicien doit rechercher deux choses dans le cri populaire, les formes variées de ce cri, puis les applications qu'il autorise'. Ibid., p. 21.

13 Timothy J. Cooley, 'Casting Shadows in the Field: An Introduction', in Gregory Barz and Timothy J. Cooley (eds), *Shadows in the Field: New Perspectives for Fieldwork in Ethnomusicology* (New York: Oxford University Press, 1997), pp. 3–19, at p. 5.

14 Charles Rearick, *Beyond the Enlightenment: Historians and Folklore in Nineteenth Century France* (Bloomington: Indiana University Press, 1974), p. 12.

15 Rearick, *Beyond the Enlightenment*, p. 20.

16 Quoted in Paul Thompson, *The Voice of the Past* (Oxford: Oxford University Press, 1978), p. 22.

17 Kastner, *Les Voix de Paris*, p. vi.

18 Georgina Born and David Hesmondhalgh, 'Introduction', in Georgina Born and David Hesmondhalgh (eds), *Western Music and its Others: Difference, Representation, and Appropriation in Music* (Berkeley: University of California Press, 2000), pp. 1–58, at pp. 15–16.

19 Rearick, *Beyond the Enlightenment*, p. 3.

20 Kastner, *Les Voix de Paris*, p. vi.

21 Kastner, *Les Voix de Paris*, p. 35.

22 Stephen A. Kippur, *Jules Michelet: A Study of Mind and Sensibility* (Albany, NY: State University of New York Press, 1981), p. 57.

23 Kastner, *Les Voix de Paris*, p. 23.

24 Kastner, *Les Voix de Paris*, p. 70.

25 Rearick, *Beyond the Enlightenment*, pp. 28–9.

26 'Ils forment, pour la plupart, des groupes mélodiques qui ont chacun sa couleur et son style particulier, et qui dépendent d'une ou de plusieurs formules primitives, lesquelles, en passant par toutes sortes de bouches, se sont altérées, modifiées et ont enfanté des variantes plus ou moins nombreuses. Le chant des *crieries* est franc, naturel, souvent énergique, quelquefois tendre et gracieux'. Kastner, *Les Voix de Paris*, p. 82.

27 'Aussi la voix de l'honnête pourvoyeur d'hydrogène à l'état d'oxyde se prélasse-t-elle sur la note du registre de poitrine où vient tomber la syllabe *eau*. Après avoir filé pendant quelques minutes cette note, le porteur d'eau conclut brusquement à la sixte supérieures, au moyen d'un port de voix exécuté avec autant de hardiesse que de succès'. Ibid., p. 84.

28 'Il faut reconnaître une vérité, c'est que tous nos petits commerçants chanteurs ne sont pas des virtuoses. Il y en a qui ont l'oreille dure, qui chantent fort mal, et même *à la turque*, avec toutes sortes d'inflexions étranges, dont les quarts de ton ne font pas seuls les frais. Dans le cours de nos pérégrinations, nous avons bien rarement entendu de belles voix'. Ibid., p. 83.

29 'Dans cette forme perfectionnée, le cri n'est plus un bruit rude et grossier, dont l'intonation précise échappe aux règles de l'art; c'est une manifestation vocale qui a son rang marqué dans l'échelle régulière des sons'. Kastner, *Les Voix de Paris*, p. 3.

30 'Nous citerons pourtant la magnifique voix de baryton d'un ramoneur adulte que nous écoutions avec ravissement pendant qu'elle nous faisait entendre, à intervalles réguliers, le classique *haut en bas*. L'enfant qui suivait le compagnon répétait ce cri en écho à la sixte, avec une justesse parfait; ce double cri en imitation est un de ceux que nous avons entendu exécuter le mieux'. Ibid., p. 83.

31 Ceri Crossley, *French Historians and Romanticism: Thierry, Guizot, the Saint-Simonians, Quinet, Michelet* (London: Routledge, 1993), p. 190.

32 Ibid., p. 195.

33 Manfred Kelkel, *Naturalisme, vérisme et réalisme dans l'opéra: de 1890 à 1930* (Paris: Libr. philosophique J. Vrin, 1984).

34 'En général, les marchands d'articles de consommation et quelques crieurs parmi les artisans nomades paraissent avoir les mélopées les plus anciennes et les plus richement diversifiées; cependant, malgré les modifications qu'ils ont subies en passant par tant de bouches différentes, on ne perd pas tout à fait la trace des cris originaux, des types mélodiques qui ont enfanté toutes ces variantes'. Kastner, *Les Voix de Paris*, p. 81.

35 'Julien: (*avec enthousiasme*): Ah! chanson de Paris où vibre et palpite mon âme! ... Naïfs et vieux refrains du faubourg qui s'éveille, aube sonore qui réjouit mon oreille! Cris de Paris ... voix de la rue! Êtes-vous le chant de victoire de notre amour triomphant?' Gustave Charpentier, *Louise*, libretto. English translation by Henry Grafton-Chapman. Liner Notes, Sony Classical 46429.

36 Prendergast, *Paris and the Nineteenth Century*, p. 126.

37 '... chaque jour, et pour ainsi dire aux mêmes heures, frappent notre oreille avec la régularité d'une sonnerie d'horloge; qui, bon gré, nous éveillent dés l'aurore, et ne cessent de nous poursuivre jusqu'à l'entrée de la nuit...' Kastner, *Les Voix de Paris*, p. 77.

4 Refashioning Rhythm

Hearing, Acting and Reacting to Metronomic Sound in Experimental Psychology and Beyond, *c.*1875–1920[1]

Alexander Bonus

Any inquiry into the state of Western musical culture today would find very few performers, pedagogues, composers or critics who would admit to being unfamiliar with the tempo machines known collectively as 'metronomes'. For many who consider themselves musically inclined, performing with the 'right' rhythm in the 'proper' tempo is inexorably intertwined with these now-commonplace technologies. The sound emitted from metronomes – once made from clockwork, and now from more precise quartz crystal oscillators and digital circuitry – provides singers, conductors and instrumentalists both young and old with an apparently trustworthy basis for understanding the tempos of nearly every piece in the Western musical canon.[2] The metronome's artificial clicks, they are often told, objectively underpin the metre, pulse and rhythm of works as divergent as Renaissance masses and Suzuki exercises to complex Baroque sonatas and Romantic-era symphonies. It would seem pointless to consider otherwise, given that popular references continue to describe metronomes as innocuous 'devices that measure the time of music'.[3] This seemingly absolute truth, in which the 'time of music' is deemed a scientific, mechanically measurable constant across history, has been largely unchallenged in musical pedagogy or scholarship since the twentieth century. Indeed, some at the vanguard of performance, education and criticism have done much to support the metronomic basis of musical temporality, despite the conceptual anachronism it represents when placed against former epistemologies of time, the history of technology and past performance practices.

In the pursuit for rhythmical accuracy today, musicians therefore seem completely justified in applying a series of continuous mechanical sounds to compositions that pre-date, sometimes by centuries, Johann Maelzel's clockwork-metronome patent from 1815.[4] Pianist Emmanuel Ax, when learning a composition by John Adams, exposed the anachronistic view of musical time held by many modern performers, when he stated in 2005, 'Most classical pianists – myself at the head of the list, don't necessarily have such good rhythm. ... We don't necessarily have a good feel for the inexorable pulse'. Guided by this belief in a mechanically oriented temporality, in which an inexorable pulse regulates all printed note values (as is often heard in many minimalist compositions from the late twentieth century), Ax then intimates, '[when rehearsing Stravinsky's *Pulcinella* reduction with cellist Yo-Yo Ma] I had a hard time just being accurate, not rushing, not slowing down. … Sometimes rhythm takes a backseat. So I practiced with a metronome, which I've been doing ever since'.[5]

Professional musicians such as Ax occasionally admit to this mechanical bias underlying modern musicianship: playing to an objective metronomic reference inculcates performers in the now laudable ability to play compositions, regardless of era, with that most inexpressive of Industrial-driven values – precision. Comments such as Ax's indicate how musical times have changed: metronomes have not always been so highly esteemed in professional musical culture. For instance, Ignacy Jan Paderewski (1860–1941) dismissed musicians who relied upon such machines as 'metronome believers'. As I document elsewhere, the beginnings of the twentieth century witnessed these 'believers' furthering the presumption that both 'good rhythm' and the exact intentions of long-since departed composers could be attained through an ability, in Herbert van Karajan's phrase, to 'sustain the metronome'.[6] Thus it should come as no surprise that the celebrated twentieth-century conductor prided himself on recollecting and repeating mechanical tempi with scientific precision.[7] Similar statements by renowned twentieth-century musicians signalled that a drastic cultural shift had indeed occurred. Many Western performers do strongly believe in automatic metronomes as desirable and necessary regulators of musical temporality, and that these modern machines should be applied to centuries' worth of repertoire.[8]

When considering the anti-metronomic aesthetics expressed by musicians of Paderewski's ilk, the wide-scale acceptance of metronomes in our age should incite serious scholarly inquiry: during an era that purportedly championed the power of individual expression, artistic genius and the creative spirit, when and under what circumstances did the metronome's 'inexorable pulse' become the foundational reference of musical time? Why did the synchronization of musical gestures to a continuous beat-per-minute click-track ever become a meaningful and acceptable performance activity? And, although 'metronome believers' seldom broach such a question, it must ultimately be asked: how could lifeless technological sound ever succeed in becoming so oddly integral to that uniquely human trait known as musicality?

As this brief and limited study suggests, the origins of a metronomic performance practice were not found in the realms of compositional theory, traditional pedagogy or nineteenth-century musical aesthetics. This present history focuses on what I consider to be a pivotal moment in the understanding of musical time, in which precision-oriented, mechanically regulated tempo and rhythm first emerged as positive ideals. These values began in what might seem to be an unlikely home for modern musicality: in the nineteenth-century laboratories of experimental psychologists. As I argue, the self-dubbed 'New Psychologists', working in venues far from contemporaneous music studios or salons, founded a laboratory 'rhythm' exclusively referenced and regulated through invariable mechanical noise.[9] During an age when skilled musicians and educators upheld the primacy of rhythmic intuition and interpretation, these empirical scientists redefined musical performances with mechanical qualities. They were the first truly influential 'metronome believers', as they actively subsumed an individual's embodied sense of time and motion to the machine's beats-per-minute designations.

This chapter analyses these scientists' philosophies and procedures to illustrate how they initially refashioned musical rhythm to reflect the unwavering ticks of metronomes. Evidence suggests that, through novel scientific methods, they founded a performance practice based in mechanical objectivity, precision and accuracy, and that this performance practice eventually became a standard for twentieth-century culture. Their initial valuing of a metronome-guided musicality spawned a uniquely modern aesthetic that is still endorsed today. The present study thus charts a 'rhythm' and 'tempo' born from an effectively forgotten technoscientific tradition, in which hearing, perceiving and acting under continual metronomic sound first represented a modern musical virtue.[10]

Acknowledging Pre-Metronomic Rhythm

The prevalent use of tempo clocks in pedagogy and performance practices today has likely helped to steer the discourse about musical temporality away from historical evidence: continuous metronomic sound represents a drastic aesthetical departure from past creative and expressive musical traditions. The notion that automatic metronomes can define and dictate an absolute musical time reflects a distinct aesthetic paradigm shift, what I consider to be a *metronomic turn* for modern musical culture. An intuitive feeling for the musical pulse ran counter to the inexorable, redundant clicking of clockwork.[11] Prior to the twentieth century, 'good rhythm' and precise metronomic regulation shared little relation to each other, conceptually or practically for so-called 'artistic' performers and educators in the Western musical tradition.

For over a century, many illustrious musicians realized the distinction between mechanical rhythm and vital musical motion. Even Ludwig van Beethoven, an early proponent of Maelzel's tempo scale, had occasion later in life to damn the use of the metronome.[12] It is highly unlikely that Beethoven ever considered the machine a constant tempo regulator for an entire work, a single movement or even an isolated musical phrase. In keeping with the performance practices of his age, Beethoven acknowledged that musicality transcended any initial metronome number or click, explaining, 'sentiment has also its peculiar rhythm'.[13] Such sentimental rhythms were ultimately derived from a seemingly unquantifiable epistemology, and Liszt and others commonly paraphrased what was considered to be Beethoven's final assessment on the mechanical dictation of musical tempo: 'Better no metronome marks at all'.[14]

Apart from tempo and affect words that hinted at a subjective underpinning to musical pulse and metre, instructions on how to perform sentimental rhythms were seldom printed alongside metronome numbers in nineteenth-century music publications. Nevertheless, many skilled musicians over the century seemed to regard the anti-positivistic, anti-metronomic nature of musical temporality as common knowledge. Pianist and conductor Ignaz Moscheles, despite being one of the first 'metronoming' editors of Beethoven's posthumous editions, acknowledged the sheer absurdity of playing any composition to an implied metronome beat, because 'the musical world knows that marking time by a metronome is but a slight guide for performers and conductors'.[15] Composer and conductor Hector Berlioz, who considered the employment of metronomes in certain operatic situations potentially useful, likewise considered playing under constant metronomic regulation to be both physically and conceptually unfathomable, commenting, 'I even doubt whether it would be possible to observe so flat a uniformity during a certain number of bars'.[16]

Comments reaching into the first decades of the twentieth century reiterate the pronounced anti-metronomic aesthetics and practices fundamental to historical musical rhythm.[17] Besides such renowned musicians as Paderewski and Rachmaninoff, less famous pedagogues and performers rejected Maelzel's metronome for imposing an undesirable, a-rhythmical sound upon traditional epistemologies of musical time. In 1886, American educator Albert Ross Parsons reported that if a knowledgeable and expressive performer 'consults absolute metronomic signs at all, [he] does so chiefly with a view to comparing them with his own impressions on the subject'.[18] Parsons warned that if novices, by contrast, played 'heedless of everything save the inexorable tick-tack of the conscienceless machine, then, confessedly, [the metronome] is no aid to practice'.[19] It seems that the clockwork machine's obtrusive 'tick-tack' sound did not sufficiently or accurately explain the vastly more nuanced qualities of musical rhythm for either the most skilled or most fledgling of musicians. Vocal pedagogue Julia E. Crane, in her *Music Teacher's Manual* (1889), perhaps most succinctly summarized the failure of this particular technology as it related to musical interpretation when she warned, 'by accurate [musical] rhythm is not meant metronome accuracy'.[20] American piano prodigy Joseph Hofmann, 16 years after Crane's instruction, reiterated why musical time and precise metronomic sound represented two seemingly antithetical rhythmical qualities. '*Never Play with a Metronome*', Hofmann exclaimed, 'for according to the metronome a really musical rhythm is unrhythmical – and on the other hand, the keeping of absolutely strict time is thoroughly unmusical and deadlike'.[21] Prior to the rise of Modern-Age metronomic aesthetics and the machines that supported them, the values of technological precision seemed wholly incommensurate with the living qualities of musical expression in time.[22] Indeed, most skilled musicians trained in nineteenth-century traditions heard Maelzel's metronome for what it truly was: an artificially contrived time-beater absent of the pulse of life itself.

Perhaps unbeknownst to the majority of nineteenth-century musical artists, the metronome's 'deadlike' rhythm garnered very influential and zealous promoters – well before most composers, players or pedagogues ever found it appropriate to perform under its inexorable click. Indeed, the metronomic performance practice grew and thrived in what some musicians might have considered a non-artistic discipline, one that employed clockwork metronomes with anti-expressive intentions. Within a distinctly contrasting performance culture, perceiving and responding to constant metronomic sound was of paramount importance. Experimental psychologists, with their uniquely technoscientific worldview, first considered hearing the unremitting 'tick-tack' of 'conscienceless' clockwork to be the very foundation of modern 'rhythm'. By furthering their mechanistic ideals of rhythmic action, a new concept took root in twentieth-century performance practices. It was first through the precision-oriented methods and machines of nineteenth-century experimental psychologists that the metronome became the scientific tempo regulator of modern musicianship.

Founding a New Metronomic Science

The origins of this formerly 'anti-musical' aesthetic can be directly traced to the work of Wilhelm Max Wundt (1832–1920), who, in founding the first experimental psychology laboratory at the University of Leipzig in 1879,[23] used clockwork metronomes as continuous rhythm-regulators when researching the variables of human perception and action. The new psychology, Wundt proclaimed, 'has to investigate that which we call internal experience, – i.e. our own sensation and feeling, our thought and volition … Man himself, not as he appears from without, but as he is in his own immediate experience, is the real problem of psychology'.[24] In his Leipzig laboratory Wundt and his students sought to quantify the subjectivities of human nature using an assemblage of automatic machines that referenced a variety of physical and mental acts.[25] Among the laboratory apparatuses essential to their research was a Maelzel metronome, a tool that Beethoven scholar Gustav Nottebohm once called 'soulless clockwork'.[26] Its purpose was to regulate the perceptions and motions of living subjects in order to discover the hidden laws driving all human performances.

Wundt alluded to a 'real problem' that psychological experimentation could hope to answer. His science sought to record, regulate and ultimately 'solve for' the many variables intrinsic to individual thought and activity. The pressing issue – that all people move, react, sense and perceive somewhat differently to each other – was not a concern native to nineteenth-century music performance or composition. This technoscientific 'problem' of individuality, the history of which can only be glossed presently, first became evident in the astronomical observatory, where it was dubbed the 'personal equation'. Using the most precise clocks and viewing equipment, trained astronomical observers almost never registered identical data about the time and location of heavenly phenomena that passed through the crosshairs of their telescopes. Early nineteenth-century astronomers began to realize how variable the process of perception and response could be, and that a great deal of this imprecision could be attributed to the auditory and visual senses of even the most disciplined of scientists.[27] Thus nineteenth-century astronomy's first essential influence upon Wundt's theories, as Edward Wheeler Scripture recounted in his text *Thinking, Feeling, Doing* (1895), was in finding that 'everybody disagreed with everybody else. Moreover, men who disagreed in one way at one time would be likely to disagree differently at another time; so that a man did not even agree with himself. … [Ultimately] each person had a peculiar error of his own'. And this 'peculiar error' of individual performance as measured against exacting clockwork measurement 'was called by the queer name, "personal equation"'.[28]

Astronomy's second contribution to experimental psychology, which made research on the personal equation possible, was the chronographic method, otherwise known as chronography. In the 1850s, the observatory chronograph became the pivotal technology used to 'solve' astronomers' personal equations. It was a large ticking pendulum clock – essentially, an ultra-precise metronome – that had the capacity to electrically record exacting temporal markers on a barrel drum lined with smoked paper. An astronomer now only had to hit a telegraph key to register the exact time of a stellar event, and this procedure drastically reduced the inconsistencies caused by more subjective observational methods. *Stoddart's* noted how the far-superior chronographic method 'relieves the observer of all anxiety regarding the time, the clock-beats, and the making of the record in his note-book, and enable him to give his undivided attention to the observation of the transits, which, under the other [eye-and-ear] method with its attendant difficulties, he was, to a certain extent, prevented from doing'.[29] Thus the chronograph – a state-of-the-art metronomic technology – simplified the most error-prone aspects of astronomical observation, those derived from the individual's unique interpretations and performances in time.[30]

Desiring to discover how human variables applied to a host of other physical activities and mental processes, Wundt's experimental psychology inherited both an interest in the personal-equation problems once vexing astronomers and the chronographic method that recorded and reduced their problematic differences. 'When we learn how long it takes to perceive, to will, to remember, &c.', noted American psychologist James Cattell in 1888, these temporal data 'are in themselves of the same interest to the psychologist, as the distances of the stars to the astronomer, or atomic weights to the chemist'.[31] In his machine-centric laboratory, Wundt practised what he occasionally called 'psychological chronometry', the tabulation and reduction of human subjectivity through controlled, repeatable experimentation.[32] Wundt believed that mechanically precise measurements derived from chronographic apparatuses could eventually lead scientists to the absolute laws regulating human perception, motion, sensation and even the nature of consciousness.[33] 'Now it is possible to measure all these organic changes', former Wundt student E.B. Titchener proclaimed in his popular textbook *A Beginner's Psychology*, 'The observer [subject] may therefore be harnessed to some such system of [chronographic] recording apparatus, and may be subjected to some pleasant or unpleasant stimulus; he reports what he feels, and the experimenter is able to compare the report with the record from the instrument'.[34]

Occasionally it seemed that experimental psychologists' intentions were no less ambitious than the objective reduction of the mind and body to mathematical-mechanical measurements. As their science progressed in both ambition and authority, experimental psychologists trained in Wundt's technoscientific methods seemed intent on quantifying every aspect of human behaviour. Apart from testing the perceptions of sight and sound, as well as physical reactions and tolerances, experimental psychologists eventually sought to define the laws underlying performances in time and rhythm using that trusted technique practised so successfully in astronomical observatories. Only a decade after Wundt's founding of the first experimental psychology laboratory, experimental psychologists began employing the chronographic method for the purpose of ascertaining the personal equation of musicality.

Hearing Technoscientific Rhythm

The accepted cultural practice of performing musical rhythms with a baseline standard of metronomic regularity came about through an unlikely psychological experiment. One of many seemingly intangible mental traits Wundt wished to quantify through chronometric research was the duration of the human attention span.[35] Wundt theorized that by exposing a subject to repetitive mechanical sounds, psychologists could discover how long the mind held a string of basic ideas or impressions. He believed that the data derived from such a procedure would yield a 'range of consciousness': an individual's capacity for perception, memory and recollection within a standard timeframe.[36]

A continuous, automatic sounding-apparatus was deemed an essential control to the experiment, thus it was Maelzel's clockwork metronome, and in following years more precise metronomic apparatuses, that facilitated these consciousness studies. As the father of experimental psychology codified, 'metronome strokes will serve as simple ideas' projecting the 'relative simplicity and uniformity of sound-impressions' for these chronographically inspired experiments.[37] Requiring the most consistent and precise temporal measurements possible, Wundt created his own metronomic testing device – a retooled, electro-magnetic Maelzel metronome, far more accurate and unwavering than any nineteenth-century consumer (or composer) could purchase for musical endeavours (Figure 4.1). It was Wundt

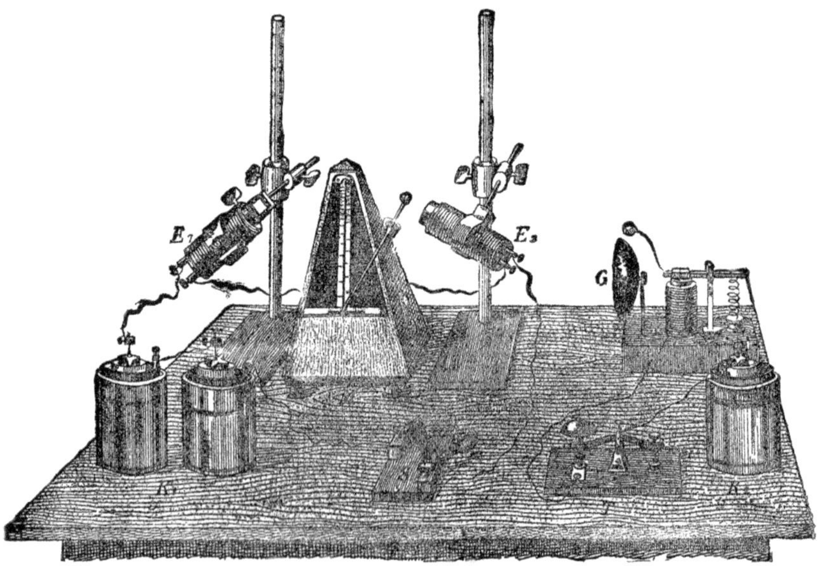

Figure 4.1 Wilhelm Wundt's electro-magnetic metronome apparatus for 'range of consciousness' experiments.[38]

who effectively invented the modern, battery-powered metronome: the vastly more regulative tempo clock that only came into common musical usage during the mid-twentieth century.

Using his state-of-the-art metronome in the role of chronographic regulator, Wundt conceptually and practically invented the modern, scientific ideal of 'rhythm' during investigations into the mental phenomena he considered intrinsic to attention. The tests exposed subjects to the monotonous succession of metronomic sounds, those 'equally accented rhythmic stimuli', at various rates of speed.[39] Upon eliciting subjects' reactions to his metronome's redundant noise, experimental psychologists found that the typical nineteenth-century mind often perceived the incessant and relatively equal metronome ticks as discrete sound-groups, usually in patterns of two or three. American psychologist William James summarized the phenomenon in 1890:

> Our spontaneous [mental] tendency is to break up any monotonously given [i.e. metronomic] series of sounds into some sort of rhythm. We involuntarily accentuate every second, or third, or fourth beat, or we break the series in still more intricate ways. Whenever we thus [mentally] grasp the [metronomic] impressions in rhythmic form, we can identify a longer string of them without confusion.[40]

This discovered mental process by which redundant metronomic sounds transformed into rhythmical patterns became standard scientific knowledge, and experimental psychologists would repeat and revise Wundt's metronomic perception experiments well into the twentieth century.[41] Wundt acolyte Thaddeus Bolton, an early American researcher on the 'psychological chronometry' of musical time, reiterated in 1894 that scientific rhythm was exclusively dictated through mechanized, sonic impressions. 'We come now to the consideration of the nature of the rhythmical group', Bolton stated, 'When the [metronomic] impressions are uniform in length and intensity [i.e. volume], the mind enforces a grouping by giving fictitious values to the impressions, generally with respect to intensity, but sometimes with respect to duration'.[42] Experimental psychologists' 'rhythm' – a mental phenomenon resulting from metronomic sound – was decidedly unlike the experiential, subjective rhythm understood and actualized by skilled musicians of the century, or earlier centuries for that matter. In *A Text-book of Experimental Psychology* (1909), Charles Samuel Myers summarized this radically new conception of Western musical time, in which 'the simplest material for rhythm consists of a series of identical, regularly repeated, and equally accented stimuli',[43] sonically emitted from a laboratory metronome.[44]

It needs little recounting how the rhythmic epistemology endorsed by Wundt, James, Bolton and others was antithetical to that of skilled musicians living in the same age. For experienced musicians, performing 'in time' was not an act of cerebral chronography. Unlike 'artistic' musical traditions, psychologists' attention experiments usually required a laboratory subject to accept a completely passive physical and mental attitude prior to perceiving the mechanical sound source.[45] Under this passive state of suggestibility, subjects were often asked to perform additional mental tasks in order to engender those fictitious rhythmic patterns. In *A Beginner's Psychology*, Titchener detailed one typical variant on the metronomic attention experiment:

> If you listen to a metronome beating, say, 15 in the minute, you will be able *with practice* to hold six successive strokes in the focus of attention, but not more; if you try to group the seventh stroke with the preceding six you become confused; the series breaks, and cannot be welded together again. As the speed of the metronome is increased, the beats fall of themselves into groups of twos and threes; and you can still grasp and hold six of these rhythmical impressions. When the speed has reached some 200 in the minute, the rhythmical grouping becomes more complicated; as many as eight single beats may be bound together in a rhythmical unit; and the attention is adequate, *again after practice*, to five of these complex groups.[46]

Titchener divulges a new scientific training method for the purpose of verifying a new scientific epistemology of rhythm. Not merely intended to document a subject's ability to mentally group metronomic clicks into patterns, the process challenged the laboratory subject to comprehend various rhythmical groupings within the redundant series of mechanical-sound stimuli, regardless of whether the subject initially perceived the groupings or not. Neglecting the rhythm familiar to creative and interpretive 'artists' entirely, he instructed:

> Set a metronome beating, with an interval of about a quarter of a second between stroke and stroke. Try to throw the beats into all the different possible rhythms, trochaic, iambic, etc. You will find it quite easy to change from rhythm to rhythm, especially if you use movement to assist you, – moving foot or hand when the [metronomic] beats come which you wish to emphasize. Then see how complex a foot or measure you can construct in the various rhythms [as dictated under metronomic sound].[47]

These techno-rhythmic exercises amounted to more than disinterested empirical explorations into the mind. Experimental psychologists tested what mental rhythmical groupings the subject could comprehend and sustain over time when exposed to unwavering metronomic regulation. The individual's mental capacity to ascertain this laboratory rhythm often amounted to listening, memorizing and responding to the metronomic sound-source alone – a practice that more closely mirrored the work of observatory astronomers than skilled musicians in the nineteenth century (see Figure 4.2).

Titchener did acknowledge the inescapable fact that rhythms occurred in nature unencumbered by metronomic reference or chronographic regulation. In *An Outline of Psychology* (1902), he recollected many of the paradigms of rhythmic action known to pre-modern societies, which included the variable paces and

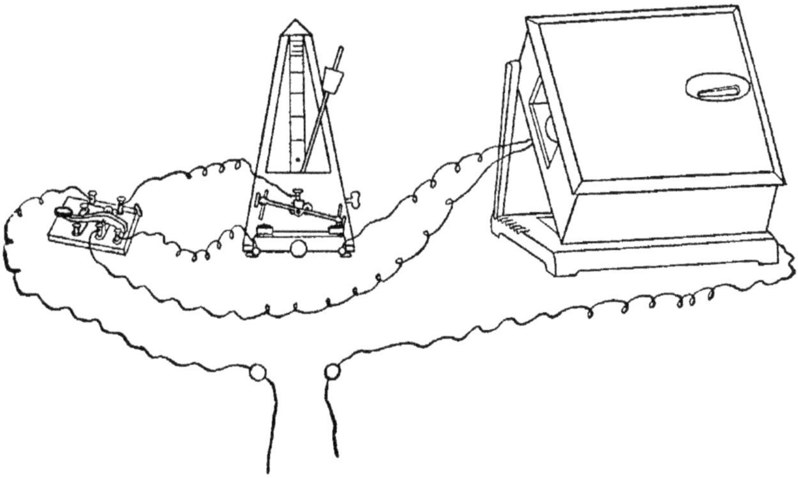

Figure 4.2 The 'Apparatus for the Serial Exposure of Nonsense Syllables'.[48]

pulses witnessed in vital human movement, breathing and speaking. Regardless of these anti-scientific, pre-metronomic truths, Titchener faithfully promoted experimental psychology's modern chronographic concept of rhythm, with the evaluation,

> Sounds are, indeed, better material for the idea of rhythm than are tactual complexes; for the limbs are fixed to the trunk, and can therefore do no more than oscillate to and fro, pendulum fashion, giving of necessity the most rudimentary form of rhythm, – *beat′ beat / beat′ beat*, – whereas a series of [metronomic] sounds can be divided into groups of any complexity. The [metronome-derived] rhythm: *beat″ beat beat / beat′ beat beat / beat″ beat beat / beat′ beat beat /* could not be formed [in the mind] from tactual impressions.[49]

Given their loyalty to chronography, these technoscientists could not abide the variability and nuance seen in the rhythms of nature. Even a swaying tree, as Titchener described it, seemed to be a silent and imperfect clockwork metronome. Since embodied rhythmical references and practices were especially incommensurate with these scientific theories, people as active and expressive agents of truly complex tactual rhythm remained far from Titchener's own epistemology. He intimated that the 'better' rhythmic stimulus was founded on metronomic sound, rather than the visual gestures, pulse-sensations or the many other movements of living beings. For the new psychologists, the natural rhythms conceived and experienced well outside the realms of scientific experimentation held negligible weight when compared to the limited laws justified through their metronomic methods and apparatuses.

Sensing the Sound of Scientific Rhythm

As the chronographic bias of these scientists stipulated, non-auditory rhythmic sources – seen for example in an orchestra conductor's beat, a dancer's stride, workers' collective motions or a tree's swaying limbs – were undesirable and inappropriate material for deriving the precise psychological laws of rhythm and pulse. Once it was determined that metronomic sound alone could help to uncover the mental laws of rhythm, many experimental psychologists reasoned that more precise metronomes would yield more accurate laws. Thus rhythm researchers logically needed to improve metronomic technologies in order to define the laws of rhythm with ever-increasing degrees of sonic precision. Beyond Wundt's metronome revisions, other experimental psychologists not only improved the rapidity and stability of the mechanical beat, but altered its timbre as well. For some, the more unnatural, unfamiliar and monotonous the sound, the better the rhythmic reference for their experiments. In 1901, Titchener recommended tampering with the common clockwork metronome to reduce the very prominent and distractive click, which was found to negatively influence a subject's reactions during simple tests. He explains:

> The customary tick of the metronome, as it stands upon table or piano, may be described as a noisy clang or a metallic clack. To deaden the sound, i.e., to eliminate the clang elements, E [the experimenter] must remove the floor of the clock-chamber (it comes away when a button is turned), and set the instrument upon a layer of thick felt. The tick then becomes a mere click or crack.
>
> The table on which the metronome stands must be placed in such a relation to O [the observer, i.e. the subject] that there is no echo or reverberation from the walls of the room. The ticks are to form a series of sharply separated dead sounds.[50]

According to Titchener – who perhaps remained unaware of Hofmann's opposition to 'deadlike' metronomic rhythm in the music room – 'sharply dead sounds' from a metronome *were* the most appropriate temporal references in the laboratory.

Nineteenth-century clockwork metronomes, as most musicians and experimental psychologists from Beethoven to Titchener well understood, were plagued with technical problems.[51] Piano pedagogue Adolph Friedrich Christiani, who described to his readers how the metronome scale was based on astronomical and horological constants,[52] concurred that the typical Maelzel machine was quite imprecise even when compared to household time-tellers.[53] 'In practice', Christiani reminded, 'a metronome is seldom as correct as a good clock'. Yet, in contradistinction to technoscientists such as Titchener, he attested that such sonic precision had no place in music education and performance, adding that the accuracy of a clock 'is not necessary'.[54] Christiani's anti-metronomic aesthetic was indicative of nineteenth-century musical culture: state-of-the-art metronomic precision was not a predominant factor when conceiving musical time. Even an accurate clockwork metronome was unnecessary for Christiani and others because musical temporality transcended the redundancy and precision of automatic sound-machines.

During Wundt's lifetime, Paderewski keenly ascertained the insufficiency of even an accurate clockwork metronome when he concluded, 'To be emotional in musical interpretation, yet obedient to the initial tempo and true to the metronome, means about as much as being sentimental in engineering. Mechanical execution and emotion are incompatible'.[55] Yet, Wundt's rhythm was grounded in 'mechanical execution'. Rhythm, as a scientific rule, was the result of an objective sound imposed upon the imperfect, subjective mind. His experimental 'rhythm' represented the passive laboratory subject's *ex post facto* reactions and responses to continuous metronomic 'impressions'.[56] These laws logically transferred to the understanding of artistic time itself. Indeed, Wundt defined rhythmic 'feeling' in dance and music as a wholly reactionary process born directly from hearing objective mechanical 'stimulations', divorced from the volatile acts of artistic creation, movement or expression. In his widely published *Lectures on Human and Animal Psychology* (2nd ed., 1892), he explains:

> The feeling of rhythm, which is the single psychological motive in dancing, and ranks with harmony and disharmony as a psychological motive in musical composition, contains the elements both of expectation and satisfaction. The regular repetition in rhythmical sense-excitation makes us expect every succeeding stimulation, and the expectation is immediately followed by satisfaction. Rhythm therefore never involves strain, or if it does, it is simply bad rhythm.[57]

Such theories seem entirely unrelated to the world outside of Wundt's laboratory, where it would be untenable to assume that mental-emotional strain resulted from hearing an irregular or 'simply bad rhythm', if the more challenging nineteenth-century works by Beethoven, Berlioz, Liszt or countless others from Wundt's own century are considered.[58] Under Wundt's definition, in which personal strain equated to 'bad' musical rhythm, and 'regular repetition' equated to 'good' musical rhythm, experimental psychology might prove Beethoven's latter-period compositions, among thousands of other challenging works by hundreds of other composers, to be rhythmical rubbish.[59]

Prominently highlighting this epistemological paradigm shift in modern musical time, Wundt described and documented his species of good, accurate and regular rhythm – metronome-derived rhythm – through standard music terminology and notation. The infallible metronomic regulator in his laboratory now became the temporal constant of musical time in performance.[60] In *Outlines of Psychology* (1896), Wundt exposes this foundational ideological transference, in which the objective, mechanical rhythm of psychological experimentation quietly replaced the traditional, subjective time of musical action. Using language that carefully removes both the mechanical sound-stimuli and the passive auditor-subjects from his scientific rules of rhythm, Wundt explains:

> When the [metronomic] rate passes the upper limit of about one [click] every 0.12 sec., the formation of distinctly defined temporal ideas is impossible because the [subject's] attention can not follow the [clicking] impressions any longer. The most favorable [beat-per-minute] rate is a succession of [metronomic] strokes, one every 0.2–0.3 sec. With this rate and with the simplest rhythm of 2/8 time which generally arises of itself when the [subject's] perception is uninfluenced by any special objective conditions, as a rule, 8 double or 16 single [metronomic] impressions can be just grasped together [in the mind]. The best rhythm for the perception in one group of the greatest possible number of single impressions is the 4/4-measure with the strong accent on the first [metronomic] stroke and the medium accent on the fifth.[61]

For Wundt and his school, the redundant metronomic rhythm of his attention experiments precisely equated to musical metre, in '2/8 time', '4/4 measure' and so on. In a radical epistemological shift that served his chronographic method above all else, Wundt objectified musical metre and measure as phenomena independent of human agency. Musical time amounted to a series of precisely calculable metronomic objects to be exactly replicated, untainted by the variable traits of individual rhythmical expression. In Wundt's own mind, the metronome was no meagre initial pulse reference, no artificial approximation of the living musical beat – it was a chronographic regulator for all musical perceptions and performances.

Wundt's musical chronography directly translated into a new, objectivist interpretation of music notation, in which each metronome click precisely defined the movement of the invariable eighth note – for each and every time signature commonly seen in print. He refashioned a once-sensory epistemology, which scores of musicians previously considered unquantifiable, variable and fundamentally human in origin into a new scientific law of musical pulse. Supported by his metronomic apparatus, Wundt literally mechanized musical time and notation:

> If we denote the metronome-strokes by quavers, and the bell-signal by an accent placed above them, an experiment consisting of two successive series may be represented in this way:[62]

Wundt strikingly infused conventional music notation with his modern chronographic values. Justified by the metronomic noise in his laboratory, music metre became part and parcel of the attention experiment. Out of the repetitive stream of clicking eighth notes, variations of volume-intensity alone created the phenomenon of rhythmical groupings. Wundt continues to explain how passive metronomic exposure served to test a subject's mental tolerance for perceiving 'good rhythm':

> If we give ourselves quietly up to the apprehension of the [metronomic] impressions, we observe that the separate strokes are not all alike [in the mind], even though they are really and objectively perfectly equal. We alternately accentuate and slur them, just as we do in marking time in speech, by a voluntary and regularly alternating intensity of accentuation. If we denote the accented impressions by points placed above them, we have the two series of our former figure as they usually occur in reality [in the mind]:[63]

> A certain degree of complexity may result simply from the effort to hold as many impressions as possible together in consciousness. You may quite easily obtain the following system, e.g., in which the different degrees of accentuation are again denoted by points, the strongest by three, the next by two, and the weakest by one:[64]

Through these attention-study examples, Wundt actually reduced rhythmical notation to 'really and objectively perfectly equal' impressions formed through mechanical clockwork sound. The musical times had indeed changed, straying far from 'artistic' sensibilities. Given this new rule in which musical temporality ran in a constant unbroken stream of metronomic impressions, Wundt then refashioned once-interpretive musical and poetic metres, which housed those previously 'sentimental' rhythms, into rigid mechanical-psychological rules:

> If we look at the metrical forms employed in music and poetry, we find again that the limit of three degrees of accent is never exceeded. The absolute amount of accentuation may, of course, be very different in different cases. But in immediate perception these different degrees are always arranged in three principal classes, which alone are of any real importance in metrical division as a basis of classification for rhythmical forms … a time like the 6/4 is one of the most complex of the rhythms employed in music. Its scheme is the following:

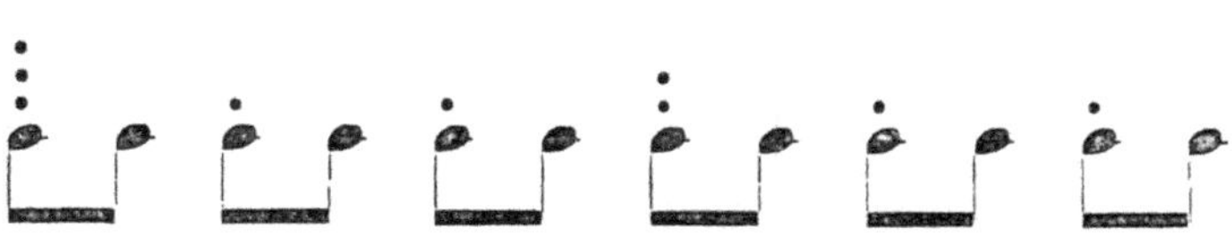

> It contains, you see, only twelve simple impressions.[65]

From Wundt's novel metronome experiments emerged a mechanized musical temporality. The perception of 'simple metronomic impressions' in the laboratory became the reified rhythms and 'metrical forms' of notated music and poetry. In both concept and text, Wundt and following experimental psychologists reduced artistic interpretation to a mere 'single psychological motive'. Thus musical and poetic rhythms became simple sonic objects to be precisely replicated by the passive subject of scientific rhythm.[66] Once Wundt reduced all time signatures to a succession of automatically clicking eighth notes while disregarding affect, melodic, harmonic or formal structure – without reference to physical or rhetorical gesture – he notated a new precision-oriented quality for the time of musical performance. It was in the nineteenth-century laboratory where musical temporality in the Western tradition first came to be heard and seen as a lifeless, mechanically grounded series of note values. Correspondingly, it was also where the embodied sensations of pulse and movement, traditionally gleaned from the musical metre, conceptually died.[67]

Musician as Automatic Metronome

The Wundt School of experimental psychology seemed not to consider – or perhaps did not want to consider – the nuanced and variable interpretations of 'great artists' from their century, or how the very meanings and practices of playing 'rhythmically' likely changed over the course of history for different cultures and repertoires. In their laboratories, rhythm was a precise, objective law of perception and action for all peoples, places and centuries. Their studies suggested that divergence from such scientific sound-boundaries constituted human error, regardless of the cultural or historical circumstances. Thus Wundt and his followers delved even further into the realm of aesthetics to challenge the creative authority of those traditional agents of musical rhythm, 'artistic' performers themselves, since musicians, no matter their levels of creativity or intuition, still reeked of personal-equation problems: those egregious 'errors' of variable action made evident through the invariable sound and motion of laboratory machines.

Charles H. Sears in 1902 quantified 'practiced' musicians' performances of standard (pre-metronomic) hymn melodies and tunes with precision 'down to hundredths of a second', showing how the length of all printed half, dotted quarter, quarter and eight notes were variable in practice, a fact that would come as no surprise to skilled interpreters. Describing the mechanical imprecision of his musician-subjects, Sears reported, 'the measures vary not only with the general rate but among themselves for the same subject and the same selection', and perhaps to his chagrin, 'there is no clearly evident order in the variation of the measures except a retardation when nearing the end of the selection'.[68] This chronographic bias, the need to assess living performance against an unstopping metronomic constant, led others to similarly document temporal imperfections through ever more precise mechanical measurements.[69]

Prominent Yale University psychologist E.W. Scripture also replaced musicians' performance intentions with his own desire for regulating and reducing the personal equation. Advanced instrumentalists 'must learn to strike the notes at regular intervals', Scripture said in 1895, since 'the quarter notes should all be about the same length; equal measures should be completed in equal times'.[70] He also presumed to understand the musical intentions of a skilled conductor, 'who must beat time with his baton. It does not make much difference just how fast he beats, provided he beats regularly'.[71] According to Scripture, even a trained military 'drummer himself may have something the matter with him' when dictating accurate time, since percussionists usually failed to play a monotonous 'series of clicks' as exactly as Scripture's precise metronomic apparatus.[72]

These highly desirable, precision-striving performance practices, Scripture noted, were only verifiable through his chronographic references and regulations. He thus lamented that music teachers in the 1890s, lacking access to his superior machines, were perpetually ill-equipped to inculcate students with truly accurate musical rhythm.[73] Assuming Scripture's mantle in the twentieth century, music-psychologist Carl Emil Seashore similarly believed that promising music students should gain access to laboratory equipment in order to test their 'precision in the time of rhythmic action'.[74] Considering such rhythmic action to be the a priori standard of musical behaviour, he dubbed professional performers' occasional alteration from the metronomic beat – a performance practice he could not realistically deny – the 'Artistic Deviation' from 'the pure, the true, the exact, the perfect, the rigid, the even and the precise' in mechanical-musical time.[75] What nineteenth-century musicians understood as artistry was now defined against the metronomic standard as psychologically deviant behaviour.

Even though typical musicians lacked the state-of-the-art time-dictating and recording equipment found in his Yale psychological laboratory, Scripture had indeed recognized the increasingly popular household tool currently transforming average students into scientists' brand of 'successful player': the performer with the lowest levels of personal variation in musical time, who behaved with the most sonically precise rhythm – who was, in other words, the least artistically deviant. It was the machine indispensable to experimental psychologists in their own laboratories, as Scripture explains: 'By practice with the metronome successful players are able to reduce their *MV* [mean variation, i.e., personal equation of rhythm] till it does not disturb the playing. It is not known just how far this may be carried, as no one has ever taken the trouble to make accurate measurements' (see Figure 4.3).[76] By continuously using the clockwork metronome at home, the modern musician could finally enjoy the same effects known to astronomers through the chronographic method.

The Wundt School, in order to define and then reduce these personal-equation problems of musicians – so they do 'not disturb the playing' of a perfectly metronomic interpretation – consistently devalued, and in most cases outright ignored, the fundamental humanity exhibited through musical thoughts and actions. In *Outlines of Psychology*, Wundt scientifically diminished the intuition and education of the instrumentalist,[77] and at times reduced a musician's agency over a performance to simple techno-psychological descriptions. This generally disparaging attitude towards musicians resonated throughout his scientific discipline, well into

Figure 4.3 In Scripture's Yale laboratory, chronographically 'Measuring the Simultaneity in Actions of a Piano-player' using an electrical-contact keyboard connected to a recording barrel-drum, which also marked the metronomic standard of musical time.[78]

the twentieth century through the work of later researchers such as Meumann, Scripture and Seashore. They further described novice and 'practiced' musicians functioning identically to typical subjects in psychological laboratories.[79] And without learning musical composition or performance to any advanced degree, Wundt diminished musicians' actions across history to ritualistic, naïve and simple '*ceremonial* games' that displayed 'regular rhythmical movements performed in unison'. Wundt reasoned: 'Music itself is nothing but a game of this order, so highly developed that it has grown to be a form of art'.[80] For Wundt, musicians – no matter how advanced, skilled and accomplished they were – required the essential methods and definitive apparatuses of the new experimental psychology to act appropriately, to play their childishly 'amusing' games, their music, within his rules of 'time'.

Noting the authority with which Wundt and other psychologists talked about the nature of musicality, it must be stated that their vast body of research on the science of rhythm produced no definitive answers, no chronographic or metronomic absolutes for the exceptional music or minds in their century or any other. Wundt's science neglected the compositions and performances of musical artists spanning the age, those creative individuals who did not conform to the reductive, mechanical paradigms found exclusively in experimental psychologists' laboratories. And while evidence suggests that Wundt did not compose, improvise or perform music with any semblance of experiential knowledge or skill, he scientifically presumed to define a trained pianist's abilities, in which,

> In most cases there are no acts of intelligence involved at all, but only associations; and in any case intelligent action must have been reduced to association

> before it could become mechanized. The piano-player has first of all to form a stable association between the printed note and the movement of touch. But this association gradually lapses from consciousness, and the interconnection of movements becomes purely mechanical.[81]

Few skilled nineteenth-century performers, teachers or composers would likely have recognized or condoned Wundt's radically objectivist theories on musical thought or performance. Living motion 'in rhythm' was logically reduced to simple chronographic explanation, as Wundt states in *Outlines of Psychology*, 'The *mechanical* properties of the limbs are important physiological bases for the rise of these ideas … we have movements in which the voluntary energy of the muscles is operative only so far as it is required to set the limbs oscillating in their points and to maintain this movement – *rhythmical* movements'.[82] Through his extensive and influential publications, Wundt and his school continued to refashion rhythmical action with mechanically objective attributes. 'With rhythmical movements', he believed, 'their significance for the psychological development of time ideas is due to the same principle as that which gives them their importance as physiological organs, namely, the principle of *the isochronism of oscillations of like amplitude*'.[83] By describing the paradigmatic human, and human musician, as an automaton who projected mechanically redundant beats with exactitude, precision and isochronism, Wundt seemed at times to equate the perfectly rhythmical musician to the perfectly regulated clockwork metronome.

Metronomic Rhythm as Musical Reality in Twentieth-Century Culture

The pronounced Wundtian refashioning of rhythm represented a distinct paradigm shift in which the sound of metronomic machines infallibly dictated time and pulse for musical, and indeed many cultural, performances in the modern twentieth century. There can be no doubt that the ideals of performing with ever-increasing degrees of mechanical precision became popularized in modern culture only since experimental psychologists' seminal laboratory research on rhythmic perception and action. It must also be recognized that although metronomic 'thinking, feeling, and doing' originated in the psychological laboratory, these practices did not end there. Some of the very scientists who refashioned rhythm in the laboratory directly prescribed their rhythmic epistemology to musical culture at large. Carl Emil Seashore, for example, brought the benefits of scientific rhythm to music educators and performers through the very first issue of *The Musical Quarterly* in 1915.[84] Over his career, which extended into the 1940s, Seashore prescribed ever greater mechanical-rhythmical values to the music periodical's readership. In his *Musical Quarterly* article 'The Sense of Rhythm as a Musical Talent' (1918), his science established a novel 'musical talent' founded on metronomic mimicry, one intended to replace traditional musical thought and action, when he stated: 'First, the capacity for rhythm rests upon certain fundamental powers which can be measured serviceably in various forms by methods now being introduced through experimental psychology'.[85] He then related verbatim the technoscientific laws first devised in Wundt's laboratory:

> There are two fundamental factors in the perception of rhythm: an instinctive tendency to group [objective, metronomic] impressions in hearing, and a capacity for doing this with precision in time and stress. The subjective tendency is so deeply ingrained, on account of its biological service, that we irresistibly group uniform successions of sound, such as the tick of the clock, into rhythmic measure.[86]

By the 1920s, experimental psychologists' rhythm had pervaded musical culture, beyond the immediate sphere of their laboratories. In discussion with Camille Saint-Saëns, composer and conservatory professor Vincent d'Indy directly paraphrased the meaning of psychological 'rhythm', in which metronomic sound was the absolute source and measure of a largely reactionary mental phenomenon.[87] Igor Stravinsky, perhaps the first eminent composer to intend his tempi to move in continuous clockwork clicks (in opposition to those sentimentalist rhythms and subjective tempo words of the past), likewise described musical time as an objective, equal series of measurable sound-impressions, ideally unaffected by the volatilities of human agency.[88] Even a treatise for mechanical pianola manipulators, *The Art of the Player-Piano* (1922), provided a training exercise for developing the performer's 'metrical sense' in 'accurate' rhythm, taken directly from psychologists' now-common attention experiments with the metronome.[89]

While the turn towards objectified, metronomic rhythm also appears in the compositions and writings of other notable modern musicians during the 1920s including Bartók, Ravel, Antheil and Hindemith, perhaps the most significant and influential cultural champion of chronographic twentieth-century 'Rhythm' was Émile Jaques-Dalcroze (1865–1950), who, as early as 1906,[90] prescribed and promoted a precision-oriented pulse to Western culture *en masse*, a musical time founded expressly upon experimental-psychology traditions from previous decades. His Eurhythmics inculcated students with a laboratory species of 'good' rhythmical behaviour – for the mind and body – through an externalized and constant beat standard. 'My pupils themselves began to realize', Jaques-Dalcroze recollected in 1915, 'that they were not merely being taught music, but were being generally educated. They were undergoing psycho-physical transformations'.[91]

In 1914 Jaques-Dalcroze's scientific pedagogy seemed so uncanny to behold that a *New York Times* reviewer remarked, 'it all seems absurd enough until you get the idea', partly because – disregarding metre, phase, harmony or affect – 'the quarter note is translated into a single movement and is the unit of the rhythm'.[92] With the objective, unwavering quarter note (and occasionally eight note) regulating students' performances – just as in Wundt's nineteenth-century laboratory – living rhythmical actions were reduced to a scientific process of exacting perception and response to repetitive sound-impressions (see Figure 4.4).[93]

Jaques-Dalcroze himself often echoed Wundt's precision-oriented performance ideals in which musicians were isochronously oscillating machines – and admitted as much in his article 'Initiation into Rhythm' (*c.*1907), stating, 'By means

Figure 4.4 An excerpt from Jaques-Dalcroze's many Eurhythmics exercises for the training of precise rhythmic performance, in which the mathematical-metronomic consistency of small note-unit 'divisions' was of paramount importance to maintain. Similar to Wundt's physiological experiments, theories and musical notation from the previous century, Jaques-Dalcroze's now-objectified 'rhythm' was de-contextualized from musical phrase, harmony, along with the traditional, physical sensations of metre, rhetorical gesture and affect.[94]

of movements of the whole body, we may equip ourselves to realize and perceive rhythms'. Jaques-Dalcroze explains:

> Consciousness of rhythm is the faculty of 'placing' every succession and combination of fractions of [mechanically referenced] time in all their gradations of rapidity and strength. This consciousness is acquired by means of muscular contractions and relaxations in every degree of strength and rapidity.[95]

In 1913, early Eurhythmics promoter Percy B. Ingham reiterated the technoscientific foundations of Jaques-Dalcroze's educational philosophy, in which 'rhythm in music consists of a regularly recurring series of accented sounds, unaccented sounds, and rests, expressed in rhythmic gymnastics by movements and inhibitions of movements'.[96] Through Jaques-Dalcroze's pedagogy, experimental psychologists' chronographic 'rhythm' – 'a regularly recurring series of sounds' – transitioned into 'rhythm in music'. Once confined to a scientific philosophy, Wundt's mechanical-rhythmical performers – who habitually played ritualistic, synchronized 'games' with the least amount of mental-physical strain or reflection – found creative realization through Jaques-Dalcroze's Eurhythmics (see Figure 4.5).[97]

Figure 4.5 Jaques-Dalcroze's young Eurhythmics students perform an exercise in his neo-Hellenic school at Hellerau, Germany – otherwise known as the Temple of Rhythm.[98]

In his role as cultural critic, Ezra Pound (1885–1972) asserted in the 1920s that trained orchestral performers' 'ability to count, their metronomic ability, has engulfed them, and they have become insensitive to shape'.[99] Pound implied that metronomic-sound regulation, made increasingly familiar in his lifetime through scientific-musical pedagogies such as those promoted by Seashore and Jaques-Dalcroze, reduced musicians' understanding and sensitivity to the variable, pre-metronomic rhythm of the past – the 'sentimental' rhythm of Bach, Beethoven and Brahms. (Yale psychologist Scripture, we recall, considered this very same effect, the metronomic reduction of personal-rhythmical variations, to be musically and scientifically beneficial.) In 1929, *New York Times* critic John Martin seemed to confirm this metronomic turn in performance practices when he strikingly claimed, 'Rhythm is becoming constricted into ever-narrower meanings; it is already used almost as a synonymous term for the regularity of the audible beat'.[100] Martin might have recognized that this regulative and constricting 'beat' popularized in the twentieth century was audibly defined through continuous metronomic noise. Given these realizations, that a type of mechanical hegemony was overtaking the qualities of rhythm in the twentieth century, it should come as no surprise that musicologist Curt Sachs, in his monumental musical-time study *Rhythm and Tempo* (1953), asserted with scientific certainty the recently established truth of modern musical temporality: 'Ours is a mathematically counting notation' in which 'the quarter note is our motor unit'.[101]

The popularity of Eurhythmics and later metronomic pedagogies such as the Suzuki method seems to confirm the existence of the metronomic turn: mechanical-mathematical precision defines musical rhythm today more than at any point during Wundt's lifetime. The one-time pejorative behaviour, acting to a constant metronomic sound-source, has become a pejorative no longer, but a natural expectation to 'good' and 'correct' musical performance for many students and professionals in our age. In 1920, Harriet Moore Brower, one of the many twentieth-century American pedagogues influenced by the ideals of experimental psychology, voiced this objectivist value system still operative today: 'I cannot urge too strongly the constant effort to play in time ... we must have a just sense of the mathematical values of notes. ... The only way to have a metronome in one's head, that is, to have a thoroughly grounded sense of rhythm, is to make good use of this valuable little monitor. Let us stand up for this tireless little "policeman"'.[102] To have the sound of 'a metronome in one's head' as many educators, performers and scholars maintain to this day, is the mental and physical norm of musicality.

As Pound, Martin and Sachs claimed generations ago, rhythm in Western culture has largely metamorphosed into a technological sound-series. This sound-series, ever increasing in precision and constancy, was once heard in clockwork clicks and now more commonly in computerized click-tracks. Past critics insisted that this metronomic sound-series, by its very usage, could only confine an individual's musical-rhythmical sense to constricting beats-per-minute impressions. Can it be any wonder why Man Ray titled his first artwork metronome in 1923, *Object to be Destroyed*?[103] The act of metronomic reference, once an arcane, scientific process of perception, has become a prevalent and largely accepted aspect of Western performance practices since the twentieth century. Unbeknownst to those musicians today who subscribe to daily practice routines dictating that every printed note must synchronize to a metronomic sound (driving an unstoppable 'motor unit' of musical time), their training methods mirror a nineteenth-century laboratory experiment first devised to measure attention span.

While ignoring skilled, creative musicians as they truly existed outside the laboratory, experimental psychologists' reinterpretation of rhythm represented a revolutionary, twofold paradigm shift, a *metronomic turn* for modern performance practices beginning in the 1890s: The musician became the observer, a passive subject of a standardized action, as musical notes now represented precise, mathematically gauged objects – outside of the mind and body, outside of the individual's will, outside of the personal sense of pulse – to be observed against absolute chronographic-metronomic regulation. First through Wundt's laboratory procedures, musical performance became a mental-physical experiment of mechanically efficient, exacting and reproducible action. In order to fix the 'problems' of the musician's personal equation, these scientists refashioned the ideal performance 'in time' to be one of passive perception and reaction to an unwavering mechanical rule, a new truth received through technoscientific sound-impressions, alternately heard as 'sharply dead' noise or that now-ubiquitous, inexorable 'tick-tack-tick' of metronomic rhythm.

Notes

1 This study was assisted by a New Faculty Fellows award from the American Council of Learned Societies, funded by The Andrew W. Mellon Foundation. Research support was also provided by the Department of Music at Duke University.

2 See Alexander Evan Bonus, 'The Metronomic Performance Practice: A History of Rhythm, Metronomes, and the Mechanization of Musicality', PhD Dissertation

(Case Western Reserve University, 2010) for an extended history of metronome technologies and scientific values of mechanical precision as they extended to pedagogies, performance practices and compositions through the so-called Machine Age of the twentieth century.

3 These conventional definitions are kept alive not only in print but on the Internet as well, through sources such as *Webster's Revised Unabridged Dictionary* (1913): http://machaut.uchicago.edu/websters and Wordnik: www.wordnik.com.

4 The 1815 British announcement for Maelzel's first metronome patent is purposefully ambiguous, suggesting that the invention was not yet fully conceived or perfected: 'John Malzl, of Poland-street, Middlesex, Machinist; for an instrument or instruments, machine or machines, for the improvement of musical performance, which he denominates a *Metronome*, or musical time-keeper. Dated December 5, 1815'. See *The Repertory of Arts, Manufacturers, and Agriculture* XXVIII, Second Series (London: J. Wyatt, 1816), pp. 127–8. Note that the machine, in whatever form Maelzel had in mind, was originally intended as an aid for musical education, and not as an objective indicator of any *a priori,* mechanically absolute musical time.

5 From Thomas Robert May, *The John Adams Reader: Essential Writings on an American Composer* (Pompton Plains, NJ: Amadeus Press, 2006), p. 273.

6 The scope of this chapter cannot address in depth the significant changes in twentieth-century compositional aesthetics, which moved musical-temporal values towards greater metronomic regulation. The metronomic aesthetic of composition, as I call it, chronologically proceeded from much of the historical evidence presented in the current study, and it can be witnessed in the creations and performance practices of Jaques-Dalcroze, Stravinsky, Ravel, Bartók, Antheil and other pivotal champions of modernist and Machine-Age culture.

7 See Richard Osborne (ed.), *Conversations with Karajan* (Oxford: Oxford University Press, 1998), p. 97. Late in life Karajan proclaimed his ingrained, mechanical-tempo skills: '[My brain] is not a computer [but] I trained it with metronomes. And I still test myself. I can walk in 120 [bpm] and sing in 108; and if you ask me to sing in 105 now, I will manage it. If I get it wrong, I feel it with my whole body. And in the orchestra, if a solo comes in slower or faster [in metronomic time], I sense it right away; it makes me feel uneasy'.

8 The distinction should be made between metronomes and timepieces such as watches or clocks, since they each serve different techno-cultural purposes. Clockwork metronomes are rhythm and tempo dictators beating at a given rate per clockwork minute. Clocks and watches are conventionally used as intermittent temporal references of the day, and do not usually serve as sonic beat-per-minute regulators. It should be noted, however, that in the early nineteenth century cheap pendulum clocks were sometimes modified to serve as replacements for Maelzel's metronome. For further discussion on the initial purposes of early metronome technologies, see Bonus, 'The Metronomic Performance Practice', pp. 81–101.

9 For an extended discussion with additional historical evidence on scientific rhythm studies, see Bonus, 'The Metronomic Performance Practice', pp. 223–333.

10 See M. Norton Wise (ed.), *The Values of Precision* (Princeton: Princeton University Press, 1995); Theodore Porter, *Trust in Numbers: The Pursuit of Objectivity in Science and Public Life* (Princeton: Princeton University Press, 1995); Lorraine Daston and Peter Galison, *Objectivity* (New York: Zone Books, 2007), for histories that document the precision-oriented ideals driving nineteenth-century technoscience. An apt parallel history to the present study, *Objectivity* concerns the visual depiction of scientific information in the nineteenth century, as imagery becomes increasingly abstracted and standardized through the use of machines, and further distanced from 'truth-to-nature' representations, which were often the product of subjective interpretation and artistry.

11 This chapter is concerned primarily with the psychologically invasive sound produced by automatic metronomes, and not the visual swing of simple tempo-pendulums,

devices that some twentieth-century tempo scholars erroneously considered to be interchangeable in purpose and value. Silent and simple tempo-pendulums, appearing from the seventeenth to the twentieth centuries, were finite references for what remained a largely variable and embodied sense of musical time.

12 See Ludwig van Beethoven, *Letters*, collected, trans. and ed. Emily Anderson (London: Macmillan; New York: St. Martin's Press, 1961), vol. III, p. 1295. As Anton Schindler among others maintained, the master's living expression of musical time was far more fluid and wilful than the clicking of Maelzel's metronome. See Anton Schindler, *The Life of Beethoven: Including His Correspondence with His Friends, Numerous Characteristic Traits, and Remarks on His Musical Works*, ed. Ignaz Moscheles (London: Henry Colburn, 1841), vol. II, pp. 104–6, 111, 114, 116–17.

13 See Adolf Bernhard Marx, *Introduction to the Interpretation of Beethoven Piano Works*, trans. Fannie Louise Gwinner (Chicago: Clayton F. Summy Co., 1895), p. 74. For additional comments on freely expressive musical time as occasionally depicted in music notation, see Bonus, 'The Metronomic Performance Practice', pp. 59–81, 163–4, 171–2.

14 See Franz Liszt, *Letters of Franz Liszt*, ed. La Mara, trans. Constance Bache (New York: Charles Scribner's Sons, 1894) vol. II, p. 75. The quote seems to be a recurring motive in the nineteenth century. See also Marx, *Introduction to the Interpretation of Beethoven Piano Works*, p. 68; and Franz Kullak, *Beethoven's Piano-Playing*, trans. Theodore Baker (New York: G. Schirmer, 1901), p. 22.

15 Schindler, *The Life of Beethoven*, vol. II, p. 111 n.

16 Hector Berlioz, *A Treatise on Modern Instrumentation and Orchestration*, trans. Mary Cowden Clarke (London: Novello, Ewer and Co., 1856), p. 246. It often goes unacknowledged that, prior to the twentieth century, composers rarely if ever necessitated or expected constant metronomic adherence, either in the practice or performance of their music.

17 See James Francis Cooke, *Great Pianists on Piano Playing: Study Talks with Foremost Virtuosos* (Reprint, Philadelphia: Theodore Presser Co., 1917), p. 213. During an age in which the machine was quickly becoming a commonplace fixture in music education, Rachmaninoff comments, 'The metronome itself must not be used "with closed eyes", as we should say it in Russia. The player must use discretion'. For more of Paderewski's comments, see Henry T. Finck, *Success in Music and How it is Won* (New York: Charles Scribner's Sons, 1909), pp. 321, 455–6.

18 Albert Ross Parsons, *The Science of Pianoforte Practice* (New York: G. Schirmer, 1886), p. 24.

19 Parsons, *Science of Pianoforte Practice*, pp. 21–2. Emphasis added.

20 Julia E. Crane, *Music Teachers Manual*, Seventh Edition (Potsdam, NY: Elliot Fay & Sons, 1915), p. 70.

21 Joseph Hofmann, 'Playing the Piano Successfully', *The Ladies' Home Journal*, (Mar. 1905): 2–14 p. 9.

22 'How any musician could ever play with a metronome, passes my humble understanding', balked Constantin von Sternberg in 1917, 'It is not only an inartistic, but a downright antiartistic instrument'. See Constantin von Sternberg, *Ethics and Esthetics of Piano-Playing* (New York: G. Schirmer, 1917), pp. 79–80. He concludes: 'From all this it is easily seen that the impulse leading to the invention did not come from an artistic temperament'.

23 See James McKeen Cattell, 'The Psychological Laboratory at Leipsic', *Mind*, 13 Jan., 1888: 38–9; also O. L. Zangwill, 'Wundt', *The Oxford Companion to the Mind Online*, http://www.oxfordreference.com (accessed 22 September 2007). For an extensive history of experimental psychology written by a noted practitioner of the science, see Edwin G. Boring, *A History of Experimental Psychology* (New York: Appleton-Century Company, 1929). For an unparalleled collection of studies about Wundt, see

Robert W. Rieber and David K Robinson (eds), *Wilhelm Wundt in History: The Making of a Scientific Psychology* (New York: Plenum, 2001).

24 Wilhelm Max Wundt, *Lectures on Human and Animal Psychology*, Second German Edition, trans. J.E. Creighton and E.B. Titchener (London: Swan Sonnenschein & Co., 1894), p. 1.

25 Decades prior to Wundt's experimental psychology, the field of physiology utilized electro-magnetic and mercury-contact metronomes to regulate experimental surgeries. See Bonus, 'The Metronomic Performance Practice', pp. 190–203. It has been argued that besides astronomy and physiology, the industrializing landscape of Germany, and Leipzig in particular, may have influenced Wundt to incorporate new, automatic laboratory machines. See Sven Dierig, 'Engines for Experiment: Laboratory Revolution and Industrial Labor in the Nineteenth-Century City', *Osiris*, 18 (2003): 116–34.

26 Cited in C.A.B., 'George Grove's Analyses of Beethoven', *The Musical World*, 3 November 1888: 844–74, p. 850.

27 The so-called 'eye and ear method' of astronomical recording, *Stoddart's Encyclopedia Americana* (1884) reported, 'required the observers' powers of attention to be distracted beyond what they are able to bear'. Printed during the rise of experimental psychology, *Stoddart's Encyclopedia Americana* presents an exceptional account of the eye-and-ear method, which shows the methodological links between Wundt's metronomic experiments and past astronomical observations: 'The usual plan for noting the time of any celestial phenomenon was for the observer to listen to the beats of his sidereal clock; to keep count of the minutes, and estimate the space passed over by the object during the interval between the beats; to mentally divide this space into tenths, and record in his note-book the minutes, seconds, and tenths, meanwhile counting the clock-beats that not one be lost. All this must be repeated at the bisection of the object with each of the several wires (spider-lines) in the field of view of his telescope, sometimes as many as ten or more being employed. Even by this wearisome method skilled observers, after years of practice, have been able to record the time of transits to within one-tenth of a second, though it was more frequently to one-fifth of a second'. See J.M. Stoddart, *Stoddart's Encyclopedia Americana* (New York: J.M Stoddart, 1884), vol. II, p. 88.

28 E.W. Scripture, *Thinking, Feeling, Doing* (Meadville: The Chautauqua-Century Press, 1895), p. 40.

29 Stoddart, *Stoddart's Encyclopedia Americana*, p. 88. Chronography was so successful in diminishing the astronomer's variable perceptions and actions in time that many experimental psychologists including E.C. Sanford and Edwin Boring suggested that the machine in large part 'did away with' the need for the observer himself. See Boring, *A History of Experimental Psychology*, p. 133 and E.C. Sanford, 'Personal Equation', *The American Journal of Psychology*, 2 November 1889: 3–38, at p. 24.

30 For a more comprehensive introduction to chronography and the personal equation, see Bonus, 'The Metronomic Performance Practice', pp. 180–90. For a history of chronography as it affected timekeeping in nineteenth-century society, see Ian Bartky, *Selling the True Time, Nineteenth Century Timekeeping in America* (Stanford: Stanford University Press, 2000). A scientific viewpoint concurrent with Wundt's appears in Sanford, 'Personal Equation'.

31 Reporting on these Leipzig laboratory procedures, Cattell reiterated, 'these times [found through chronographic testing] are of further and great use … in analyzing complex mental phenomena, and in studying the nature of attention, volition, &c'. Cattell, 'The Psychological Laboratory at Leipsic': 45.

32 Wundt explained, 'Experiment is accompanied by *measurement*, step by step. … Measurement reveals the *constants* of nature, the laws that regulate phenomena. … [Numbers] are the indispensable means of arriving at its true object, for only numbers can reveal law'. Quoted in Théodule Ribot, *German Psychology of To-day:*

The Empirical School, trans. James Mark Baldwin (New York: Charles Scribner's and Sons, 1886), p. 192.

33 Cattell, 'The Psychological Laboratory at Leipsic': 45.

34 Edward Bradford Titchener, *A Beginner's Psychology* (New York: The Macmillan Company, 1916), pp. 82–3.

35 These tests were significantly influenced by Karl Vierordt's early mental experiments with a typical Maelzel metronome – which concluded that it was nearly impossible for the nineteenth-century mind to accurately recollect metronome beats (a phenomenon otherwise known as Vierordt's Law). Vierordt's Law, in the nineteenth century at least, seemed an absolute, scientific fact. Yet some contemporary musicians could perhaps make claims to challenge its veracity through their extended and rigorous training with vastly more-precise modern metronomes. The nearly exact recollection of beats-per-minute tempi is no longer deemed impossible for many modern musicians. Indeed, Vierordt's experiments in the 1860s seem quite simple when compared to the rigours of modern musical study, especially in the training of conductors. Nineteenth-century psychologist Ribot recounted Vierordt's arguably antiquated Law of metronomic perception and recollection: 'To study the duration of these small intervals, Vierordt causes the subject to attend for some time to the beating of the metronome; then he [the subject] is to reproduce the beats as fast as he heard them. Now, the repeated intervals are too short when the real intervals are long, and too long when the intervals are short. The individual variations on both sides of the exact point are large'. See Ribot, *German Psychology of To-day*, p. 273; and Karl Vierordt, *Der Zeitsinn nach Versuchen* (Tübingen: H. Laupp'schen Buchhandlung, 1868).

36 Wundt's prominent former student, Edward Bradford Titchener recounted in 1901 that Wundt was indeed 'the first to propound a psychological theory of rhythm' through his laboratory research on the human 'range of consciousness'. See Edward Bradford Titchener, *Experimental Psychology: A Manual of Laboratory Practice*, vol. I (New York: The Macmillan Company, 1901), p. 355.

37 Wundt, *Lectures*, pp. 260–61.

38 Wundt's design was far more constant and precise in its clicking than typical store-bought metronomes intended for use by nineteenth-century amateur musicians, who were often instructed never to employ the machine as a continuous regulator of musical time, rhythm or expression. From his published *Lectures on Human and Animal Psychology* (Second edition, 1892), p. 261.

39 For further discussion see Wilhelm Max Wundt, *Outlines of Psychology*, Fourth Edition, trans. Charles Hubbard Judd (Leipzig: Wilhelm Engelmann, 1902), pp. 230–36.

40 From William James, 'The Perception of Time', in *The Principles of Psychology* (New York: Henry Holt and Co., 1890), vol. 1, Chap XV, reprinted in Charles M. Sherover (ed.), *The Human Experience of Time, the Development of its Philosophic Meaning* (Reprint, Evanston: Northwestern University Press, 2001), p. 373.

41 See Thaddeus L. Bolton, 'Rhythm', *The American Journal of Psychology*, 6 (January 1894): 145–238, at p. 205. Establishing the provenance of his own rhythm experiments, Bolton recounted the initial discovery at Leipzig University: 'In the work undertaken by Dietze in Wundt's laboratory upon the *Umfang* of consciousness, this rhythmical grouping of the sounds of the metronome was observed and employed to determine the length of the mental span. The grouping was accomplished [mentally from the subjects] by intensifying voluntarily certain sounds and subordinating others to it. By grouping the sounds first by eight and then the groups of eight by five, it was possible to grasp forty sounds … Wundt refers this grouping to the ripening of the concept on the wave of apperception'.

42 Bolton, 'Rhythm': 232–3.

43 Charles Samuel Myers, *A Text-book of Experimental Psychology* (New York: Longmans, Green & Co., 1909), pp. 314–15.

44 In 1901, Titchener, the great codifier of Wundt's procedures, defined the essential 'materials' of increasing precision employed over the past decade for experiments on the subject's 'rhythmic' perception: 'The instruments which give the rhythm stimuli, the variously modified sound series, must have a high degree of mechanical accuracy, and are correspondingly expensive. The most useful appliance for investigation is, probably, Meumann's "time-sense" apparatus, consisting of Baltzar kymograph, time-disc, set of contacts, and sound-hammers … The metronome recommends itself, for the initial experiment, both by its cheapness and by its wide range of rate … The metronome is, in general, a very reliable instrument. Nevertheless, these rates should be tested, on principle, by counting the beats with a stop-watch'. See Titchener, *Experimental Psychology*, p. 339. For more on the 'Scientific Metronome' as used in astronomy, physiology and experimental psychology, see Bonus, 'The Metronomic Performance Practice', pp. 203–21.

45 For a discussion of pre-metronomic, pre-psychological epistemologies of musical metre and pulse, see Bonus, 'The Metronomic Performance Practice', pp. 14–42. Neither Wundt nor Titchener, with their limited musical education or experience, devised attention tests to measure the subject's interpretive rhythm while actively performing works of music, poetry, dance or drama. For more on the non-experiential basis of Wundt and Titchener's music theories, see Bonus, 'The Metronomic Performance Practice', pp. 257–9.

46 Titchener, *A Beginner's Psychology*, pp. 103–4. Emphasis added.

47 Edward Bradford Titchener, *An Outline of Psychology* (New York: The Macmillan Company, 1902), p. 184.

48 As depicted, the battery-powered Maelzel metronome regulated the rate of this psychological performance-experiment intended to test memory, habitual recollection and response time. From Edward Bradford Titchener, *A Text-book of Psychology* (New York: The Macmillan Company, 1910), p. 381.

49 See Titchener, *An Outline of Psychology*, pp. 182–3. I have provided hash marks where Titchener originally printed commas, in order to better clarify his textual depiction of rhythmical sound-groupings.

50 Titchener, *Experimental Psychology*, vol. 1, p. 175.

51 The mechanical inconsistencies found in store-bought metronomes were so prevalent that Titchener instructed experimental psychologists on the proper way to shop for a scientifically acceptable machine, since 'an instrument whose clacks sound approximately equal on the music-counter (and such an one should be found among the first half dozen tried) will give still more nearly equal ticks in the laboratory, when resonance is ruled out'. Ibid., p. 339. Titchener even commented that he 'has not been able to find a bell-metronome which answers the purpose' of some of his basic rhythmic-sense experiments.

52 Adolph Friedrich Christiani, *The Principles of Expression in Pianoforte Playing* (New York: Harper & Brothers, 1885), p. 262.

53 See Bonus, 'The Metronomic Performance Practice', pp. 193–8.

54 Christiani, *Principles of Expression*, p. 262.

55 Finck, *Success in Music*, p. 455.

56 Wundt, *Outlines of Psychology*, 4th ed., p. 187.

57 Wundt, *Lectures*, pp. 376–7.

58 Issues of 'strain'-versus-'relaxation' in the 'psychological motive' constituting rhythm derive not from traditional or contemporaneous musical theories, compositions or performance practices, but from Wundt's philosophy about the 'tri-dimensional' qualities underpinning human feeling. His reductive 'tridimensional theory of feeling' stated as a psychological rule: 'We can distinguish *six types or classes of these sense-feelings: the agreeable and disagreeable, the exiting and subduing, and the straining and relaxing*'. See Titchener, *A Beginner's Psychology*, p. 81. Adolph Friedrich Christiani posed a strikingly antithetical, anti-metronomic understanding of artistic 'rhythm'

from 1885. Christiani explained the innate, sensory phenomena that help musicians and listeners distinguish musical metres and expressive accentuations, asserting: 'In fact, any kind of diversion – is generally a welcome change, a relief. Harmony, after discord, is a new pleasure; sunshine, after rain, gives fresh enjoyment. And so with rhythm. A break in the rhythmic form gives more real animation to a movement and stronger evidence of artistic spirit, than strict observance of uniformity, or of positive rules, could possibly do'. See Christiani, *Principles of Expression*, p. 69.

59 See Bonus, 'The Metronomic Performance Practice', pp. 239–41 for additional commentary on Beethoven's rhythm and its historical reception in relation to Wundt's speculative theories.

60 For a contemporaneous discussion of Wundt's motivations for quantifying the creative mind, see Ribot, *German Psychology of To-day*, p. 229.

61 Wundt, *Outlines of Psychology*, 1st ed., p. 215; 4th ed., p. 236.

62 Wundt, *Lectures*, p. 262.

63 Ibid., p. 263.

64 Wundt, *Lectures*, p. 264.

65 Ibid., p. 265.

66 A select group of twentieth-century experimental psychologists, including P.F. Swindle, began refuting Wundt's rhythm methodologies. Their arguments seemed not to influence contemporaneous music pedagogies to the same extent as those by Wundt and his followers. See Bonus, 'The Metronomic Performance Practice', pp. 320–29.

67 Titchener, inheritor of Wundt's rhythmical epistemology, also translated auditory metronomic impressions into music notation, whereby the eighth-note assumed metronomic consistency. Titchener added dynamic indications to show the duple and triple groupings as it related to the 'intensity', or volume, of each metronomic eighth-note click. See Titchener, *An Outline of Psychology*, pp. 182–3.

68 See Charles H. Sears, 'A Contribution to the Psychology of Rhythm', *The American Journal of Psychology*, 13 (January 1902): 28–61, at pp. 38 and 41–2.

69 See for instance Bolton, who employed a telephonic 'noise' apparatus, with which he correspondingly discovered a more-exacting scientific law of musical temporality: 'This general principle may be stated: *The conception of a rhythm demands a perfectly regular sequence of [metronomic] impressions within the limits of about 1.0 sec. and 0.1 sec. A member of the [mechanical] sequence may contain one or more simple impressions. If there are a number of impressions, they may stand in any order of arrangement, or even in a state of confusion, but each member of the sequence must be exactly the same in the arrangement of its elements*'. Bolton, 'Rhythm': 237. Influential rhythm psychologist Ernst Meumann offered similar definitions, which were cited well into the twentieth century.

70 E.W. Scripture, 'Some Principles of Mental Education', *The School Review*, 3 November 1895: 537.

71 Scripture, *Thinking, Feeling, Doing*, p. 263.

72 Ibid., pp. 255–6.

73 Ibid., pp. 33.

74 Carl Emil Seashore, 'The Measurement of Musical Talent', *The Musical Quarterly*, 1 January 1915: 146; and Carl Emil Seashore, *The Psychology of Musical Talent* (Boston: Silver, Burdett and Company, 1919), pp. 203–4. For further discussion, see Bonus, 'The Metronomic Performance Practice', pp. 456–7.

75 Carl Emil Seashore and Milton Metfessel, 'Deviation from the Regular as an Art Principle', *Proceedings of the National Academy of Sciences of the United States of America*, 11 (September 15, 1925): 538–42, at p. 538; and Carl Emil Seashore, 'Artistic Deviation as an Esthetic Principle in Music', *The Scientific Monthly*, 54 (February 1942): 99–109, at p. 104.

76 Scripture, 'Some Principles of Mental Education': 537. For more on Scripture's theories and procedures, see Bonus, 'The Metronomic Performance Practice', pp. 272–83.

77 Wundt, *Outlines of Psychology*, 4th ed., p. 314. 'This "enigma of the instincts" ceases to be an enigma when we come to look upon instincts, as we have done above, as special forms of impulsive action, and consider them as analogous to the simple impulsive acts of men and animals, for which we have a psychological explanation … Such reduction can be easily observed in the case of man, as, for example, in the habituation to complex movements in learning to play the piano'.

78 Scripture, *Thinking, Feeling, Doing*, p. 31.

79 For instance Wundt described players as unthinking creatures who translated 'the printed page mechanically into the proper movements … [through a process that] extends to a whole series of actions, and each particular one takes place without effort and without knowledge'. See Wundt, *Lectures*, pp. 395–6. Wundt's belief in the mechanized musician perhaps originates from his physiological training, in which human actions were considered the habituated outcropping of muscular reactions to nerve impulses. Physiologist William Carpenter wrote nearly identically of impulsive action in 1874: 'The Human organism may be likened to a keyed instrument, from which any music it is capable of producing can be called-forth at the will of the performer, who may compare a Bee or any other Insect to a barrel-organ, which plays with the greatest exactness a certain number of tunes that are set upon it, but can do nothing else'. See William Carpenter, *Principles of Mental Physiology*, Fourth edition (New York: D. Appleton and Company, 1883), p. 61. Also see Carpenter, *Principles of Mental Physiology*, pp. 217, 218 for more theories on the automatic musician. James reprints parts of Carpenter's discussion in *The Principles of Psychology*, p. 117.

80 Wilhelm Max Wundt, *Ethics: An Investigation of the Facts and Laws of the Moral Life*, 2nd German Edition (1892), trans. Edward Bradford Titchener, Gulliver and Margret Floy Washburn (London: Swan Sonnenschein & Co., Lim., 1902), vol. 1, pp. 209–10. If music was a trifling cultural phenomenon, then so too were human musical performers. Wundt's disdain for innate creativity and subjective musicality is remarkable: games were for children, and children, after all, needed rules, structure and a method by which to behave with 'regular rhythmical movements'.

81 Wundt, *Lectures*, p. 398.

82 He continues: 'Every single period of oscillation in such a movement is made up of a continuous succession of sensations which are repeated in the following period in exactly the same order'. Wundt, *Outlines of Psychology*, p. 159. Wundt's original emphasis makes the relationship between automatical machines and the epistemology of scientific 'rhythm' all the more striking.

83 Ibid., p. 160.

84 Seashore, 'The Measurement of Musical Talent': 129–48.

85 Carl Emil Seashore, 'The Sense of Rhythm as a Musical Talent', *The Musical Quarterly*, 4 October 1918: 507–15, at p. 513.

86 Seashore, 'The Sense of Rhythm as a Musical Talent': 507.

87 Saint-Saëns reported: 'When we hear successive sounds of equal duration like those of the metronome, one of the two has more intensity that the other; we can at will, M. d'Indy tells us, attribute to the more intense sound the odd numbers … or the even numbers. "The possibility we have of choosing by a mere effort of will one or the other of these inequalities, clearly proves that rhythm proceeds not from the [mechanical] sounds themselves but from a necessity of our own mind" [in choosing a pattern from those mechanical sounds]'. See Camille Saint-Saëns, trans. Fred Rothwell, 'The Ideas of M. Vincent d'Indy', *The Musical Times*, 61 (Apr. 1, 1920): 244.

88 See Igor Stravinsky, *Poetics of Music in the Form of Six Lessons*, trans. Arthur Knodel and Ingolf Dahl (New York: Vintage Books, 1956), p. 28. For the historical issues in performing Stravinsky's precise metronomic indications, see Robert Fink, '"Rigoroso (♪ = 126)": "The Rite of Spring" and the Forging of a Modernist Performing Style', *Journal of the American Musicological Society*, 52 (1999): 299–362.

89 See Sydney Grew, *The Art of the Player-Piano* (London: Kegan Paul, Trench, Trubner & Co., 1922), p. 32. Grew writes: 'The clear steady ticking of a clock helps counting. A metronome is better than a clock, because the speed of its beats can be varied'. He continues with the scientific definition of rhythm: 'A clock seems to tick naturally in duple-time *–tick-tack, tick-tack*. But by concentration we can make the clock present its ticks in triple-time, quadruple-time, six-time, and even in five-time. Our metrical sense develops quickly when we train it so to accentuate the ticking of the clock. The metronome can, by means of its adjustable bell, be made to mark off metrical groups of two, three, four, and six'.

90 Percy B. Ingham, 'The Jaques-Dalcroze Method', in Maynard Small (ed.), *The Eurhythmics of Jaques-Dalcroze* (Boston: Small Maynard and Company, 1913), pp. 31–45, at p. 35.

91 Comments by Dalcroze. Quoted in Grace Smith, 'The Eurhythmics of Jaques-Dalcroze', *Journal of Proceedings and Addresses, National Education Association*, (16–27 August 1915): 871.

92 'Dalcroze Eurhythmics Have Come to New York', *New York Times*, 25 January 1914: SM5.

93 In 1914, another New York reporter similarly described: 'The Dalcroze method makes bodily movements depend on the exact musical values, so that the music and the bodily expression are so closely related that together they formed a whole … the Dalcroze method depends on exact interpretation of note and measure duration'. See Margaret Naumburg, 'The Dalcroze Idea, What Eurhythmics is and What it Means', *Outlook*, 17 January 1914: 130.

94 From Jaques-Dalcroze, *Rhythm, Music, and Education* (New York: G. P. Putnam's Sons, 1921), 'Musical Supplement', p. 8.

95 Emile Jaques-Dalcroze, 'The Initiation into Rhythm', in *Rhythm, Music and Education*, trans. Harold F. Rubinstein (New York: G. P. Putnam's Sons, 1921), pp. 79–92, at pp. 79–80.

96 Ingham, 'The Jaques-Dalcroze Method', p. 42.

97 For further discussion of Jaques-Dalcroze and his Eurhythmics, see Bonus, 'The Metronomic Performance Practice', pp. 440–50; 470–74.

98 From Ethel Ingham, 'Lessons at Hellerau', in *The Eurhythmics of Jaques-Dalcroze* (Boston: Small, Maynard, and Company, 1913), unmarked plate, after p. 52.

99 Ezra Pound, *Antheil and the Treatise on Harmony* (Reprint, New York: Da Capo Press, 1968), p. 132.

100 John Martin, 'The Dance: Listening In', *New York Times*, 31 Mar. 1929: 108.

101 Curt Sachs, *Rhythm and Tempo* (New York: W.W. Norton & Co., 1953), pp. 168, 173. While he documented musical time throughout history based on mathematically 'additive' and 'divisive' formulations, Sachs eschews any substantial discussion of Jaques-Dalcroze or Eurhythmics, a pedagogy founded two generations before Sachs's study appeared in print.

102 Harriette Moore Brower, *Self-Help in Piano Study* (New York: Frederick A. Stokes Co., 1920), pp. 16–17, 120–21.

103 Perhaps even more prophetic are his many variations currently displayed around the world, each of which holds the same title: *Object Indestructible*.

Section II

Sound Politics

Introduction

Ian Biddle and Kirsten Gibson

The three chapters in this section all consider the politicization of sound; the manipulation or, conversely, suppression of sound as a political act and marker of power. Sound plays a significant role in the assertion of, or challenges to, hegemonic political power. Indeed, as Jacques Attali has argued, 'any organization of sounds', which might be understood as a 'tool for the creation or consolation of a community, of a totality', is 'an attribute of power in all of its forms'.[1] If we are to think about sound historically, then, such exploration inevitably demands that we consider uses and manipulations of sound, and the sonic by-products of human activity, that mark out power relationships, political forces and identity politics. The chapters are, again, scattered both historically and geographically – eighteenth-century Paris, *fin de siècle* British Ceylon and *fin de siècle* Habsburg Bosnia-Herzegovina – but are bound together though their collective illumination of sound as a fundamental force to be manipulated, controlled, unleashed or suppressed in the establishment, maintenance or resistance of particular power relations.

Chapter 5, by Hedy Law, takes as its focus the aural consequences of theatre censorship in eighteenth-century Paris. The theatrical privilege system operating in Paris during the Old Regime had the power to grant to official theatres the exclusive rights to the uses of certain media, and thus by extension to suppress certain kinds of media, and therefore certain kinds of sound (dialogue, monologue, song, for instance), in those not granted official rights. Focusing on pantomimes performed at the theatres of the Paris Forains (fairgrounds), Law considers not only the innovative ways by which such performances worked to attract, engage and communicate with audiences within these aurally restricted legal limits, but also how various re-workings of the popular Orpheus myth in this context were created in response to such 'silencing'. In particular, Law argues, these versions of the Orpheus story, reflecting 'the politics of silence' engendered by theatre censorship, highlighted the tension between speech and silence, a tension that is subsequently heightened in Christoph Willibald Gluck's 1774 Paris adaptation, *Orphée et Euridice*, of his 1762 opera *Orfeo ed Euridice*. These explorations of sound and silence, or at least 'silencing', on the stage reflected a broader cultural turn towards silence in eighteenth-century Paris, as the work of James H. Johnson has shown.[2] Law's exploration of acts of silencing – and representations of 'silence' – on the stage thus elucidates broader questions about the ways in which silence became politicized in eighteenth-century Paris. In particular, this chapter asks how representations of silence on the stage, and the concomitant falling silent of opera audiences in the second half of the eighteenth century, were underpinned by enactments of power and resistance.

While the first chapter in this section deals with the politics of silence, the following two chapters are concerned with the politics of sound. Both focus on the politics of sound in colonies of large, powerful empires. Chapter 6, by Jim Sykes, considers sound, particularly drumming, in the religious processions and festivities of colonized communities in British Ceylon, and the colonial authorities' attempts to regulate them. While Sykes is concerned with British attempts to regulate the noisiest aspects of such processions and festivities, and what these colonial laws tell us about the underlying 'base assumptions' regarding musical aesthetics and the ontologies of sound of those making the laws, this chapter draws a much more complex picture of the ways in which power relationships between colonizer and colonized and between the three main religious communities in British Ceylon were mediated and articulated through sound. Thus the two case studies at the heart of this chapter focus on the violent clashes that occurred between Ceylon's Sinhala Buddhists and the minority Ceylonese Christian and Muslim communities, the result of the minority groups' attempts to enforce silence on Buddhist processions via legal avenues, growing Buddhist resentment of the minority groups' social and economic successes and the fundamental misrecognition in British colonial law of the musical gift. Sykes's analysis of the role of the British authorities – and the colonial laws relating to 'music' and 'noise' in religious processions – in these case studies has far-reaching implications. Through his exploration of the misrecognition of the musical gift in colonial Ceylon at the turn of the twentieth century, Sykes examines the impact of European musical aesthetics on non-Western societies during the colonial era, while also critiquing the underlying concepts that have become so deeply entrenched in the musicological language of modernity.

The final chapter in this section, Chapter 7 by Risto Pekka Pennanen, shifts the focus to Habsburg Bosnia-Herzegovina. Drawing primarily on the Bosnian press and the German-language press of the Habsburg Empire, Pennanen draws a detailed picture of the hitherto unexplored uses of sound – in particular cannon, rifle fusillades and church bells – to articulate and reinforce Habsburg colonial power in Bosnia-Herzegovina. Like the previous chapter in this section, the soundworld of Habsburg Bosnia-Herzegovina was marked by the presence of three main religions – Catholicism, Islam and Serbian Orthodoxy – and was thus aurally more diversified than elsewhere within the Empire or the Balkans. Pennanen's exploration elucidates uses of sound in state-related celebrations, particularly in the neo-traditions created around 'the Emperor cult', and the superimposition of state policy and ceremonies onto all three main religions. As Pennanen shows, Habsburg policy aimed to promote equality between the three religions, to promote its own legitimacy and power and to pave the way for future Austro-Hungarian expansion in the Balkans, and sound played a significant, and unavoidably palpable, role in the promotion of such policies. Through military bands, cannon shots, rifle fusillades and the ringing of church bells the Habsburg authorities both marked and negotiated their presence and power in Bosnia-Herzegovina.

Cutting across all these of these chapters are narratives not only of broad political hegemony, and its resistance, but also identity politics on a more subtle, complex and individual level. Sound, as we see in each of the ensuing chapters, acts as a marker for articulations of class, gender, religion, ethnicity and nationhood, themes that resonate with the chapters scattered across the collection as a whole.

Notes

1 Jacques Attali, *Noise: The Political Economy of Music*, trans. Brian Massumi (Minneapolis and London: University of Minnesota Press, 1985; reprinted 1999), p. 6.
2 James H. Johnson, *Listening in Paris: A Cultural History* (Berkeley: University of California Press, 1995).

5 Orphée at the Forains

Silencing and Silences in Old Regime France

Hedy Law

> Hark! hark! The dogs do bark.
> The beggars are coming to town.
>
> Old chairs to mend. Old chairs to mend.
> I never would cry old chairs to mend.
> If I'd as much money as I could spend.
>
> Christmas is coming. The geese are getting fat.
> Please to put a penny in an old man's hat.[1]

For Robert Darnton this nursery rhyme bears witness to the fact that the early modern city was noisy.[2] Dogs barked; peddlers shouted; beggars sang. But what interests him most about this nursery rhyme is that it is a written trace of an oral culture. Like other vernacular tales, nursery rhymes stand the test of time and echo the early modern soundworld. In an age before the invention of sound recording technology, this soundworld is often taken to be a lost world. But sometimes, bits and pieces of this lost world can be retrieved. Talking and writing criss-crossed at times and formed nodes of a complex circuit of communication.[3]

It is unhelpful, of course, to assume that this nursery rhyme portrayed the early modern soundworld accurately or in its entirety. In eighteenth-century Paris, for example, the city had its quiet moments. I am not referring simply to the off-peak hours of the day, nor the downtimes of the year, say, Lent. Rather, I'm referring to the longer period of time towards the end of the century when Parisians grew introspective, a period in which refraining from speaking became a symptom of what historian James H. Johnson has called the 'epidemic of sensibility', or put simply, 'the sentimental'. Struck by the affective power of Gluck's expressive music, Parisian audiences in the 1770s fell silent.[4]

This chapter shows that the period of silence identified by Johnson had important precedents. Throughout the first half of the eighteenth century, the theatres of the fairgrounds of Saint Laurent and Saint Germain, known generally as the Forains, were suppressed. By censoring speech, the enemies of the Forains thought they would have made it difficult enough for them to survive. But on the contrary, suppression encouraged the Forains to avoid speech. Attempting to attract audiences while trying to work within legal limits, the Forains exploited a range of non-verbal media including large cardboard placards that were read by audiences; acrobatics and tightrope dancing; freaks and animal shows; popular vaudevilles sung by the audiences; and even pantomime. These media placed special demands on audiences, for they did not just listen passively, but were invited to participate actively by reading the placards, singing the vaudevilles, interpreting gestures in the dumb shows, as well as commenting on, interacting with, and sometimes even booing, performers. Far from eradicating speech, theatre censorship expanded the idea of speech, detached it from the stage and moved it to the parterre. As censorship unintentionally promoted types of non-verbal communication, it also introduced types of silence.

Against this backdrop of censorship, adaptations of the Orpheus myth featured the politics of silence. Of the 1,133 shows performed at the Forains, the Orpheus myth appears in four titles: *Orphée aux enfers* (1689, 1707, 1711); a pièce par écriteaux (roughly translated as 'a play with placards') *Arlequin Orphée le cadet* (1718); a pièce à la muette *Arlequin Orphée* (1736); and a marionette *Orphée et Euridice* (1742).[5] In addition, the opéra-comique *Le rappel de la foire à la vie* (1721) was also based on the Orpheus myth. Inspired by the success of the Forains, the Comédie-Italienne produced the ballet-pantomime *Orphée* in 1738. Remarkably, the 1736 Forain version and the 1738 Comédie-Italienne version modified the Orpheus myth in Ovid's *Metamorphoses* by emphasizing Orpheus' speechless state. Created in response to censorship, these versions highlight tensions between speech and silence and illuminate crucial differences between Gluck's *Orfeo ed Euridice* (1762) and his *Orphée et Euridice* (1774).[6] By the time Gluck's *Orphée et Euridice* premiered in Paris in 1774, audiences grew quiet not only because they were struck by Gluck's expressive music qua music, but also because they identified with the politics of silence.[7]

The Politics of Silence

The history of silence in eighteenth-century Paris had much to do with theatre censorship. It was a history of rivalry, which led to decades-long bribery, threat, negotiation and, at times, collaboration. The history stems from the privilege system that offered the official theatres exclusive rights to use certain media during performances. In 1672, the Académie Royale de Musique obtained exclusive rights to produce works with sung texts. In 1697 Louis XIV expelled the Italian comedians and the Comédie-Française obtained exclusive rights to use spoken dialogue in their productions. The rights to use speech onstage posed a particular problem for the Forains. In 1706, upon the request of the Comédie-Française, dialogue was banned at the Forains, and they started using monologue as a defensive tactic a year later. To make their performances more entertaining, the directors of the Forains, Charles Alard and Maurice the widow (née Jeanne Godefroy), negotiated in 1708 with the Académie Royale de Musique and purchased the rights to use scene changes, singers and dancers. Yet on 17 April 1709, the Comédie-Française made it illegal for the Forains to use ballets, dances, machines and decorations. The Comédie-Française kept pushing. In 1710, it banned dialogue and monologue at the Forains, and as a result, the Comédie-Française-veteran-turned-Forains-playwright Alain

René Lesage invented a type of mute performance called *pièces à la muette*. The performances at the Forains, with the use of a mixture of media, gave rise to the new genre of opéra-comique, which was authorized in 1714 and was to become one of the most influential popular theatrical genres in eighteenth-century France. The Comédie-Italienne felt threatened by the Forains when it returned in 1716 after the death of Louis XIV. The Forains were banned subsequently in 1719–21, 1722–23 and 1745–51. Yet, despite rounds of censorship, the Forains managed to produce works almost continually until 1762, the year they merged with the Comédie-Italienne. Against this backdrop of theatre censorship, the politics of silence could be said to operate on two levels. On the one hand, it refers generally to the suppression of the Forains. On the other, it refers more specifically to the censorship of different forms of speech at the Forains, such as dialogue and monologue.[8]

Monologue, as it was performed at the Forains, exemplified the types of silence that were portrayed there, and would go on to influence writers including Jean-Jacques Rousseau and Denis Diderot in the 1750s.[9] One form of monologue involved one actor speaking while other actors used bodily gestures to express the meaning of the monologue. In another form, actors would take turns to say their lines onstage then run to hide in the wings. Informed audiences would have understood the fragments of monologue as a dialogue and would have tried to piece together a story on their own. Sometimes, the actors spoke in the wings loud enough so that audiences could overhear their speech. Sometimes, an actor repeated loudly what another actor had spoken softly.[10] None of these tactics involved silence as the absence of sound, but the monologues demanded audiences listen to speech fragments in order to piece together a coherent story. They had to eavesdrop; they had to pay close attention to sounds that sometimes came from the stage, sometimes from offstage, sometimes from one actor, sometimes from multiple actors. They had to differentiate the meanings of loud voices from those of soft voices; they had to seek out meanings of words gently repeated; they had to decide for themselves the dramatic significance of what we now call diegetic noises. Exposed to a carefully orchestrated range of diegetic and non-diegetic sounds, audiences with different levels of literacy would have enjoyed the sonic sophistication of these performances. The monologues at the Forains cultivated advanced listening and hermeneutic skills.[11]

Some works at the Forains made thematic and narrative use of their rivalry with the theatre; Lesage's *La querelle des théâtres* of 1718, for instance, placed theatre rivalry centre stage. These ingenious, biting plays with self-explanatory titles gained such broad appeal that even the Royal family wanted to see them.[12] The Royale Madame invited the troupe to perform Lesage's opéra-comique, *Les funerailles de la foire*, at the Théâtre du Palais-Royal on 6 October 1718. Le duc d'Orléans admired its resilience: 'The opéra-comique resembles a swan, one that never sings more melodiously except when it is about to die'.[13]

In 1719 and 1720 the Forains were banned except for marionettes and tight-rope dancing, but in 1721 the royal ordinance approved a new contract between the Opéra and the Opéra-comique. As a result, the Forains were now allowed to use music and sung text. To chronicle this success, *Les funerailles de la foire* was performed again, followed by Lesage's new opéra-comique *Le Rappel de la foire à la vie* on 1 September 1721, the opening day of the Saint-Laurent fair.[14]

To prick audiences' interests, the Forains had, since the mid-seventeenth century, employed elements we find in modern circuses and zoos. Audiences saw freak shows that featured giants, dwarfs and wild animals, followed by acrobatics, gymnastics and tightrope dancing mixed with farces or marionettes. Some performances also incorporated animals. In 1718, for instance, the Saint-Germain Forain featured a 'flying donkey'. To be sure, the donkey in question did not fly like a bird. It was merely suspended in the air for two weeks and pulled up and down during the performance from one end of the theatre to the other.[15] The noises and looks made by these animals, combined with audiences' fascination and disgust, further enriched the already complex soundworld of the Forains.

This type of crude entertainment led to more sophisticated plays. Lesage's *Les animaux raisonnables* was performed on 25 February 1718 at the Saint-Germain fair. As indicated by the title, the plot is about Ulysse's search for an animal with the power of reason. He starts with bears, pigs, donkeys, then a wolf, a chicken, a bull, a linnet and even a fish. At the end, Ulysse finds a dolphin that can reason. The dolphin stands for the Dauphin, who had long been a strong supporter of the theatre.[16] He wields the magic wand and turns the dolphin into Arlequin, who tells Ulysse that numerous great musicians live in the sea. Ulysse turns these musicians into human beings. *Les animaux raisonnables* brought forth music that complicated the soundworld and there are seven vaudevilles in the short Scene 1 alone. In Scene 2, the animals make all kinds of noises. The multiple types of sound involving animal gibberish and human speech, sung and spoken text, prepared the audiences for the politicized meanings of this type of seemingly unsophisticated entertainment: 'Do not judge if absolutely necessary … Be reasonable animals'.[17] Asking the audiences to identify with the animals, these verses politicize animal gibberish by turning it into a type of politicized silence. Much like Jimi Hendrix's iconoclastic performance of the 'Star-Spangled banner', cacophony at the Forains betrayed the Forains's deep discontent with the censors.

The Comédie-Française probably felt threatened by this politicized form of entertainment. But they also understood that this type of sophisticated plot depended on performance, not just printed text, and that these plays required experienced actors with excellent improvisatory skill. In another round of censorship in 1731, the Comédie-Française successfully asked the Parlement to restrict actors at the Forains to child actors and to restrict their performance to the genre of pantomime. But again, the child pantomimes were so popular that the troupe was for the first time invited to perform for the king at Versailles.[18] Once again the Forains outwitted the Comédie-Française and showed exceptional resilience. Sound, music, noise, booing, singing, mocking, speaking, roaring, animals hanging in the air and even children gesturing with limited acting skills presented types of politicized silence that further intensified theatre rivalry.

The politics of silence at the Forains represents a category of repertories that exemplified creative reactions to censorship and the mechanism of silencing.

Arlequin muet par crainte was, for instance, performed numerous times between 16 December 1717 and 17 December 1779. In Alexis Piron's monologue *Arlequin-Deucalion* (1722), Deucalion-Arlequin is the only character who speaks onstage. He speaks in Scene 2 to an invisible voice, 'La Voix', which identifies itself as a tiny parrot, but one that speaks in a way that Michel Chion would characterize as an eighteenth-century *acousmêtre.*[19] These works show that silence was employed as a reaction against censorship during a period when the Forains fought hard for survival. Within the context of theatre censorship and politicized entertainment in early eighteenth-century Paris, as I shall demonstrate, even the time-honoured Orpheus myth was adapted to present the politics of silence at the Forains.

To Look or Not To Look

In order to understand how the Orpheus myth dramatized politicized silence in eighteenth-century Paris, we might start with Ovid's version in book 10 of the *Metamorphoses*. Orpheus is not allowed to look back until he leaves Hades. The challenge for the couple is that Orpheus must believe in Euridice, who is supposed to follow him. The myth hinges on blind faith. Fittingly, there is hardly any description of sound, except for two instances: one, when Orpheus starts walking on 'the soundless path'; and secondly, after Orpheus's fateful look at Euridice, when she says 'farewell' to Orpheus one last time before she perishes.[20] Plato's *Symposium* follows in general Ovid's version except the ending, where Orpheus is killed by Dionysus' followers.[21] Virgil's *Georgics* portrays Euridice as more vocal than the Euridice in the *Metamorphoses*. She protests before she disappears: 'Orpheus', she cries, 'what terrible folly has brought perdition / to you and hapless me? Look! Again the merciless / Fates call me back, and sleep seals my eyes, swimming with tears. / Farewell. I am taken back now, cloaked in the night, and, oh, / reaching out to you with powerless hands that are no longer yours'.[22] As shown from these three versions, whether or not Euridice says anything before she perishes the second time is a minor issue. Her fate depends exclusively on Orpheus' look.

Adaptations of the Orpheus myth had a long tradition in seventeenth-century French theatre. François de Chapoton composed the tragedy *La descente d'Orphée aux Enfers* in 1640, which was revived in 1648 with the new title *La grande journée des machines ou le mariage d'Orphée et Euridice, ou la grande journée des machines.*[23] *Orfeo*, written by Francesco Buti, and set to music by Luigi Rossi, was performed in Paris in 1647. Jean-Baptiste Lully wrote a *concert et rédit d'Orphée* in his ballet *Les muses* (1666) and compared himself to Orphée.[24] Marc-Antoine Charpentier wrote the cantata *Orphée descendant aux enfers* in 1683 and Louis de Lully wrote a three-act tragédie lyrique *Orphée* in 1690.[25] In 1699, André Campra wrote a one-act opera *Orfeo nell'inferni* and used it as Act 3 of *Le Carnaval de Venise.*[26]

Early eighteenth-century French composers employed the Orpheus theme in the new genre of cantatas. The first book of French cantatas was published in 1706.[27] Louis-Nicolas Cléramrault included a setting of *Orphée* in his first book of cantatas (1710).[28] Jean-Philippe Rameau's *cantate Orphée* (1727–28) for solo voice follows Ovid's version by emphasizing the sense of sight.[29] In the opening recitative, the soloist describes in the third person the power of Orphée's music. Equipped with the irresistible charm of sweet harmony, Orphée tries to rescue Euridice. In keeping with Ovid's version, Rameau's version features the fateful look while Pluto specifically asks Orphée to restrain his tender looks. In the air gracieux, Orphée sings, 'J'ai pour témoin de ma victoire / Les beaux yeux qui m'ont enflammé. / C'est le seul prix, la seule gloire / dont mon cœur puisse être charmé' ('I have as witness of my victory / The beautiful eyes that excited me. / This is the only prize, the only glory / of which my heart can be charmed'). In the following recitative, the soloist returns to the third person and recounts the fateful scene. When Orphée cannot resist turning around and seeing Euridice, the soloist speaks to Orphée in the second person: 'Attends, fais sur ton coeur encore quelques efforts' ('Wait! make more effort in your heart!'). In the next verse, the soloist switches back to the third-person narrative 'c'en est fai ... et ses yeux ont vu ceux d'Euridice!' ('It is the end, and his eyes have met those of Euridice!'). Their eyes lock. Orphée fulfills his wish, but Euridice is pulled by a 'cruel hand' back to Hades.[30]

Similar to Rameau's *cantate*, Lesage's three-act *Arlequin Orphée le cadet* with placards (1718) performed at the St. Germain Forain features blind faith. Fed up with his demanding boss, Columbine, Arlequin wants to hang himself. He meets a philosopher, who wants Arlequin to reconsider his suicidal thoughts and refers him to the magician Urgande, who tells Arlequin that he is in fact Orphée, son of Apollo. Urgande tells him that from now on he must expect to be as fortunate as he has been miserable.

As with *Les animaux raisonnables*, *Arlequin Orphée le cadet* celebrated the new deal the Forains had negotiated with the Opéra. As indicated by the title, Arlequin enjoys his upgraded status as Orphée, the son of Apollo, a dramatic kinship that represents the Forains as the Opéra's new political subsidiary. Orphée asks to see Apollo, who descends from the sky and gives Orphée his lyre. Orphée's performance charms a monkey, a monster and even the difficult Columbine, who enjoys Orphée's music so much that she abandons her lover Pierrot for Orphée. Here the Orpheus myth takes a new turn. Pierrot dies with a broken heart and his ghost comes back to haunt Columbine, who faints and dies. Columbine's death prompts Orphée to go to Hades, in the same way as Orphée attempts to bring back Euridice in the 'original' myth. He comes across ghosts in the underworld that 'make a thousand kinds of surprising gestures' and they try to intimidate Orphée. Pluto is impressed by the song sung by this new Orphée and allows him to take Columbine. But Orphée must not look at her until she steps out of the underworld. Following Ovid's version, Orphée does, however, look back and loses Columbine once again.[31]

In order to get across the complex plot, Lesage's *Arlequin Orphée le cadet* used placards on which vaudevilles and dialogues were written. Audiences who attended performances with large placards would have undergone an experience not unlike watching a silent film or an opera with subtitles.[32] Some literate audience members would read them silently but some would say the words written on the placard out loud. Those with limited literacy would repeat the words they heard other members of the audience say. From the perspective of sound, the absence of speech onstage, the use of placards and a complex plot based on a well-known myth would have encouraged active audience participation. Even though speech was banned onstage, placards encouraged audiences to read, to sing and in general produce sound in the parterre. The performance space and the audience space became integrated. The lack of speech onstage would be compensated by speech made by the audience.

Lesage likewise produced *Le rappel de la foire à la vie*, also based on the Orpheus myth. Here, Opéra goes to Hell, the infernal court, in order to bring its cousin, La Foire, back to life. Pluto and the dog Cerberus are charmed by the music sung by Opéra.[33] Making use of the audiences' knowledge of the Orpheus myth, Lesage and his collaborators told the story as if Opéra came to the rescue and revived La Forain. With its return, nothing is 'old, serious, and boring'.[34] In scene 13, La Comédie-Française appears pedantic and denigrates La Foire as an 'odious monster'. The funniest part of the play presents the confrontation among the two comedies, La Foire and Le Public onstage. La Comédie-Italienne attempts to give La Foire a hug but La Foire avoids it. La Comédie-Française sings a vaudeville called 'censorship' that threatens to take La Foire back to 'the eternal night'. La Comédie-Italienne follows up with another vaudeville that accuses La Foire of 'disgusting' Paris with its opéra-comique. Then Le Public, a gentleman La Foire deeply respects, comes to congratulate the return of La Foire. Informed audiences must have taken great delight in these not-too-subtle plays that chronicled recent theatre history. In the period that preceded the publication of the daily newspaper *Le journal de Paris* in 1777, the public would have gone to the Forains not just for entertainment, but also in order to get updates on issues relating to the politics of theatre rivalry.

Compared with *Arlequin Orphée le cadet*, *Le rappel de la foire à vie* presents a different soundworld. In 15 scenes, this play had 104 vaudevilles that made good use of music. By the end of Scene 16, La Comédie-Italienne and La Foire sing the lines 'laire-la, larre lan-laire, / laire-la, lan-la'.[35] These catchy lines encouraged audiences to sing along, but the meaning of these vaudevilles did not reside in the somewhat meaningless lyrics. With these popular tunes, cacophony at the Forains was organized. In a manner not unlike modern rock concerts, the vaudevilles helped build a community – a fan club – by means of communal singing.

Over their lifespan, the Forains did not just react to external circumstances; the playwright Claude-Florimond Boizard Pontau also brought political stability to the Forains. Between 1731 and 1743, the year Louis-Armand-Eugène de Thuret took over, the Forains enjoyed more than a decade of peace. But it was also the period when the audiences became so powerful that they seemed to own the parterre. They got to decide which plays would be performed, and they would boo the actors they deemed to be falling short of their expectations.[36] Emboldened by the audiences' support, Lesage produced *Arlequin Orphée*, a pièce à la muette on Wednesday 27 June 1736, the first day of the Saint-Laurent Forain.

This performance in turn became part of an elaborate four-act play with prologue called *L'Histoire de l'Opéra-Comique ou les Métamorphoses de la foire*. Lesage makes it clear that this play is about the innovations the Forains had thus far introduced to the theatre. Act 1 consists of a parade, *Arlequin chirurgien de Barbarie*, and the farce, *Le mensonge véritable*. Act 2, *Pierrot valet de magicien*, is a pièce en monologues, followed by *Arlequin Orphée*, a pièce à la muette. Act 3, *Ariadne et Thésée*, a one-act vaudeville, was performed with écriteaux. The fourth and last act is *Les ennemis reconciliés*, a new Opéra comique with vaudevilles.[37]

This cycle was written to tell the history of theatre rivalry. An anonymous writer, probably Lesage himself, wrote the following preface to the manuscript of the play, which explained the otherwise episodic play as a coherent victim narrative:

> This play was devised by Lesage, who wrote the beginning acts and the prologue. It gives an idea of the various changes endured by the opéra-comique since its establishment, and reveals the type of fury its rival theatres had persecuted it with. In an attempt to attract the audiences that had always strayed away from them and had gone to the opéra-comique, these theatres wanted a total destruction of opéra-comique.[38]

As with numerous plays that Lesage had produced, the plot dramatizes suppression and resistance. The prologue, *Arlequin chirurgien*, is a parade. The French officer Scaramouche is shot in the battle of Parma, and asks the surgeon Arlequin for medical help. He says the bullet is in the right arm and Arlequin amputates it immediately. But then Scaramouche says that his left arm hurts. Arlequin then amputates his left arm as well. This plot sets the comic tone for the show. Actors come out and announce that the show is about to begin.[39] Despite its comic appearance, the prologue subtly introduces a subtext of complaint.

The theme of recent theatre rivalry continues throughout *Le mensonge véritable*. Doctor Balouard betroths his daughter Isabelle to Polichinelle, a rich merchant in Marseille. But he goes back on his word, after learning that his future son-in-law, Polichinelle, has lost all of his goods in a storm. The distressed Polichinelle goes to see the schemer Mézétin and promises to give him half of the dowry if he can help change Doctor Balouard's mind. Mézétin then sends Pierrot, disguised as a messenger, to Balouard and tells the old man that Polichinelle's vessels have arrived at the port and they are loaded with diamonds and gold powder. It is a lie, but Balouard believes him anyway. Luckily Pierrot's lie, like the one in *Oedipus Rex*, turns out to be a self-fulfilling prophecy. The captain of the vessel arrives and confirms Pierrot's words. Hence, the paradoxical title, the truthful lie.

Yet during the celebration of Polichinelle's wedding, a bailiff arrives uninvited and shows the Forains actors an Arrête that bans the use of dialogue at the Forains. The unexpected turn of events turns the audience's attention from the fictional drama onstage back to the all-too-familiar reality of theatre rivalry. In order to conform to the restriction, the actors in the play proceed to perform the monologue, *Pierrot valet de magicien*.[40] Pierrot is the valet of a magician. One Sabbath his master has gone out and he opens his master's book of spells and summons the devils. Similar to Dukas's *The Sorcerer's Apprentice*, Pierrot orders the devils to do house chores for him. Yet another bailiff shows up out of the blue and shows the Forains actors yet another Arrêt that bans their use of monologue. In the reactionary spirit of the Forains in the 1720s, Pierrot and Arlequin decide to follow the increasingly stringent terms. They then proceed to perform a pantomime *Arlequin Orphée*.[41]

Despite their compliance with various Arrêts, the Forains made their complaints heard. *Arlequin Orphée*, piece à la muette, follows in general Ovid's version. But because it is a piece à la muette, this version marks a significant stylistic breakthrough. It extends Orphée's blind faith in Euridice to his mute state. Arlequin goes to Hades and asks Pluto for his wife. Pluto agrees to give her to him, but Arlequin must not turn around and look at her until they both leave Hades. But he does so anyway. The play ends with the women from Thrace trying to console him; Orphée dismisses and thus offends them, and they dismember him.[42] In this pièce à la muette, Orphée is silenced multiple times, on multiple levels. Orphée is not just forbidden to see, but he is also forbidden to speak. Within the context of theatre rivalry, Orphée here clearly stands for the suppressed Forains.

The Orphée in *Arlequin Orphée*, as a mute lover, must have fascinated the theatre world, especially the financially struggling Comédie-Italienne, which attracted as few as 18 audience members for a performance on 17 August 1735.[43] Two years later, the Comédie-Italienne borrowed the idea of the dumb show from the Forains. Antoine-François Riccobini, who had choreographed the ballet *Pygmalion* in 1734, produced an elaborate ballet-pantomime, *Orphée*, performed by the Comédie-Italienne on 12 December 1738, as the fourth part of Pierre de Morand's play *Les muses*.[44]

The play starts with Orphée grieving over the death of Euridice. The furies take turns to frighten him and they, in turn, are disarmed by his music. Pluton, Proserpine and all the other gods of the underworld are surprised to find a mortal who wants to bring Euridice back to life. Melodramatic expressions were key to this play. Riccobini's synopsis describes Orphée's gestures in vivid detail: 'he throws himself to his knees; he prostrates himself in front of Pluto; he declares very pathetically to him the cause of his distress and his wishes'. He then turns to Proserpine. Both are moved and explain to him the condition under which he can be reunited with Euridice. As soon as Orphée hears Euridice, he signals to her to follow him.

At this point diegetic sounds compensated for the lack of speech onstage. On their way out, Orphée and Euridice hear noises. Alarmed, the couple comfort each other and march on. But the noise gets louder and louder and Euridice becomes frightened by the noise. Suddenly, monsters appear and start attacking the couple. Concerned for Euridice's safety, Orphée turns around, only to see that the monsters quickly usher Euridice away from him. Orphée then realizes that the monsters, probably sent by Pluto, were to test his determination and that his care for Euridice has paradoxically taken away her only chance to live again. The monsters celebrate their victory by singing and dancing in front of the distressed Orphée. Without using any coherent speech, this ballet-pantomime staged a captivating work.

The second half of Riccobini's *Orphée* intensifies the violence of the first half. In the midst of Orphée's disbelief in the mistake he has made, Bacchantes appear and indulge in their earthly delights. Oblivious to Orphée's saddened state, they invite Orphée to join in with their merrymaking. Orphée is in no mood for frivolous socializing. Like the monsters that had tormented Orphée, the Bacchantes start bullying him. At this point melodramatic expressions turn into a full-blown physical assault. They grab him; he tries to escape; they follow him; he persists; they hit him; they become angry at him. Orphée rushes to the peak of a rock and they want to knock him unconscious. A vulture comes to the rescue. It takes Orphée's lyre, flies into the sky, and places it among the heavenly signs. The ballet-pantomime ends here. The action-filled synopsis offers plenty of opportunities for a sensational performance. No wonder Parfaict found this ballet-pantomime 'unique'.[45]

As with the classic Orpheus myth that had been set repeatedly and most recently at the Forains, Riccoboni's *Orphée* makes speech unnecessary by directing the audience's attention to sounds, actions and gestures that were immediately understandable. Like watching silent film, audiences were invited to make sense of the story based on the plot that was presented to them. Since audiences were familiar with the story, the play delighted the audiences not because of its story per se, but because of its story-telling techniques and novel sound effects.

Not all performances at the Forains followed Lesage's formula. Adrien-Joseph le Valois d'Orville produced the marionette play *Orphée et Euridice* at the Saint Germain Forain in 1741 and 1742. His version follows the conventional model by featuring blindness. After the merger of the Comédie-Italienne troupe and the Opéra-comique company on 3 February 1762, the new opéra-comique did not stage a play with a title referring to the Orfeo myth from 1762 to 1800.[46] The lack of plays based on the Orpheus myth is confirmed by another source: no work was based on the Orpheus myth as shown from Brenner's repertory list of the Théâtre italien from 1716 to 1793. The lack of titles in these two lists does not mean that the Orpheus myth was absent from the Comédie-Italienne. *Orphée et Euridice*, for example, was performed as part of *Les muses* in 1738. But these two lists nonetheless show that Orphée was a specific topic found more often at the Forains than at the Comédie-Italienne. From 1718 to 1721, and from 1736 to 1742, the Forains survived various attempts at censorship. Time and again, the Orpheus myth was used to chronicle the history of the Forains.

To Speak or Not to Speak

From around 1750 to the end of the Old Regime, the politics of silence at the Forains gave way to sentimental culture. Silence emerged as a sign of sentimentality. At the Comédie-Italienne the playwright Louis de Boissy wrote the influential comedy *Le prix du silence* in 1751, which showcased sentimental silence. The new sentimental plays in the theatres invited spectators to identify with characters they saw onstage. Instances of silence marked the sentimental moments when spectators were called upon to sympathize with performer-characters.[47] In the combat scene in Act 2 of Michel-Jean Sédaine's *Le roi et le fermier* (1762), for instance, Pierre-Alexandre Monsigny set music to 'sad and profound silence'.[48] André-Ernest-Modeste Grétry highlights sentimental silence in his comédie *Lucile* (1769), his comédie-parade *Le tableau parlant* (1769), his comédie-ballet *Zémire et Azor* (1771) and his comédie *Les méprises par resemblance* (1786).[49] By the time the opéra-comique *Nina, ou la folle par amour* was staged at the Comédie-Italienne in 1786, the Parisians fully recognized silence as a symptom of the sentimental.[50]

Gluck made sentimental silence a theme in the 1762 version of *Orfeo ed Euridice* and introduced the politics of silence into the 1774 version of *Orphée et Euridice*. Indeed, these two types of silence co-exist in *Orphée et Euridice*. In the 1762 version, Love appears and offers Orfeo a chance to bring Euridice back to life. In the aria of Act 1 Scene 2, Love tells Orfeo to refrain from looking at Euridice as well as staying quiet. 'Gli sguardi trattieni / affrena gli accenti' ('Restrain your glances, / refrain from words'). In the second stanza, Love explains the extension from sight to speech: 'Non sai che talora / Smarriti, tremanti / Con chi gl'innamora / Son ciechi gli amanti, /Non sanno parlar' ('Do you not know / that sometimes lovers, / bewildered and trembling, / are blind to those they love, / and do not know how to speak?').[51] As Bruce Alan Brown has noted, Gluck's librettist, Ranieri de' Calzabigi, probably wrote these verses in light of Act 2, Scenes 8–9, of Metastasio's *L'Olimpiade*, in which Amore says the same lines, 'Son ciechi gli amanti / non sanno parlar'.[52] Gluck emphasized this verse 'non sanno parlar' ('cannot speak') by writing a pause that stops the music in the middle of the bar. To stress the importance of reticence, Gluck used this verse four times, of which three include pauses that illustrate silence as a sentimental symptom. Gluck must have believed that these pauses and repeats signified simple and natural expression.[53]

In the 1774 Paris version, Gluck kept the same music, sung by Amour, but instead of making silence an extension of not looking, Gluck's librettist moves away from Ovid's version and emphasizes fashionable sentimental culture: 'Soumis au silence / contrains ton désir' ('Subject to silence, / restrain your desire'). In the second section, Amour describes the sentimental condition: 'Tu sais qu'un amant / discret et fidèle, muet et tremblant / Auprès de sa belle, / en est plus touchant' ('You know that a lover / who is discreet and faithful, / mute and trembling, / is to his beloved / only more touching').[54] In this version Gluck uses the same music but the pause falls on the word 'touchant' rather than on the words 'non sanno parlar'. Hence, whereas the 1762 version features silence as a sentimental symptom, the 1774 version shifts the focus from silence to an explanation of silence.

But here the politics of silence enters the picture. Orfeo's real test comes in Act 3, Scene 1, after he successfully charms the furies. In the 1762 version, Orfeo promises to tell Euridice how it is that he is able to rescue her: 'Saprai tutto da me. Per ora non chieder più!' ('You will learn everything from me, but do not ask more now'). Euridice accuses Orfeo of insensitivity. Confused, she demands an explanation. Orfeo can only reply 'giammai dirò perchè' ('I will never speak why').[55] The duet ends with a series of repetitions, both complaining that the situation is unbearable to them. In the next recitative, Euridice misinterprets Orfeo's silence. Her complaints then disarm Orfeo, in a similar way that Orfeo's song charms the beasts in Act 2. In the next recitative, Euridice, consumed with sorrow, asks Orfeo to leave her, and requests an embrace and a final farewell. Feeling that his reason is slowly leaving him, Orfeo turns around and looks at her. This final look is, of course, the fateful look that annihilates Euridice.

The 1774 version reinforces the tension between speaking up and remaining silent. In the same scene, Orphée tells Euridice to follow him. Euridice asks for a look from Orphée, who ignores her. Euridice says, 'parle, contente mon envie!' ('speak, satisfy my desire!'). Orphée replies, 'Dût-il m'en coûter la vie, / non, je ne parlerai pas!' ('even at the cost of my life / No, I will not speak!').[56] Euridice asks him again to speak up. Orphée again rejects her. His non-emotional, robotic repetition intensifies the tension. Unlike in the Italian version, Euridice in the 1774 version notices his silence and asks for an explanation. She interprets his silence not as a sign of his speechless love for her, but as a sign of his insensitivity. Similar to the 1762 version, Euridice, feeling rejected, describes her feelings in terms of stock sentimental signs: 'Je frémis … je languis … je frissonne … / je tremble … je pâlis … mon coeur palpite, / un trouble secret m'agite' ('I shiver … I languish … I shudder … / I tremble … I turn pale … my heart throbs, a secret confusion troubles me').[57] In the next duet, Orphée, speaking to himself, expresses his insufferable state of silence: 'Que dire? Que faire? Elle me désespère' ('What do I say? What do I do? She is driving me to despair'). Just as Orphée remains silent, Euridice complains that she is running out of things to say. Her silence then weakens Orphée, who wants to tell her the truth – 'Rassure-toi, je vais tout dire:/ Apprends …' ('Console yourself, I am going to tell you all: Tell …') – but he catches himself and does not finish his sentence.[58]

Feeling deceived once again, Euridice gives up and says farewell to Orphée. Contrary to Ovid's version, Euridice is not supposed to say farewell at this point. Her precocious farewell terrifies Orphée. Unable to control himself, he turns around and looks at her. As soon as he turns around, Orphée hears Euridice's last words: 'Orphée … o ciel! je meurs …' ('Orfeo, heavens! I'm dying …').[59] Her precocious farewell prophesies her own death. The deadly silence Orphée describes in the last aria, 'mortel silence' ('deadly silence'), represents suffocating emptiness.

In the 1762 version, Euridice is displeased with Orfeo's insensitivity. In the 1774 version, Orphée has much more agency. He finds his own discretion unbearable. He is torn not between looking at her and not looking at her, as in Ovid's version, but rather between whether or not to tell her the truth. While the Italian version features Orfeo's action – to look or not to look – the Paris version features Orphée's speech – to speak or not to speak. Forms of silence and silencing as well as their misinterpretations are intensified in the Paris version.

Even though Gluck incorporated the politics of silence in the 1774 version, one might want to ask how it is related to theatre rivalry at the Forains? To be sure, it is unclear to what extent the politics of silence in *Orphée et Euridice* was a direct result of theatre censorship at the Forains. After all, the Comédie-Italienne merged with the Forains in 1762 and the Orphée myth had since then ceased to be staged there. But Charles-Simon Favart introduced to the Viennese public his opéra-comique *Le coq de village*, which had opened the Saint-Laurent Foire in 1743, and Gluck set Lesage's 1718 *Le monde renversé* as his first opéra-comique in Vienna.[60]

Yet the more important issues of censorship and resistance in the Orphée adaptations at the Forains were still very much alive through to the end of the Old Regime. This culture was sustained partly by Alexis Piron, who knew many of the theatre anecdotes compiled in the three-volume *Anecdotes dramatiques* (1775) by Jean-Marie Bernard Clément and abbé Joseph de Laporte. Piron died on 21 January 1773 at the age of 83, and his seven-volume *Oeuvres complettes* was published in 1776. In the same year, the three-volume *Dictionnaire dramatique* was published. These publications reminded Parisians of the type of suppression the Forains had endured. While it would be precarious to draw a direct connection between Gluck and the politics of silence at the Forains, it would also be precarious to underestimate the endurance of the politics of silence in Paris in the 1770s. After all, *Arlequin muet par crainte*, first performed in 1717, was still being performed as late as 1779. If anything, its more than six-decade-long performance history was testimony to the history of censorship and its resistance in Old Regime France.

That said, perhaps Gluck did seem to overwrite the confrontation scene between Orphée and Euridice. Whereas the review published in the official Viennese newspaper *Wienerisches Diarium* noted in 1762 that Orfeo in the Vienna version was portrayed as a tender husband, the review of Gluck's 1774 version published in *Mercure de France* called the confrontation between Euridice and Orphée 'tedious' ('languissante').[61] Likewise, the critic who wrote a review for the *Journal des beaux-arts* found this scene unconvincing: 'Is it probable that Orphée, who has had the courage to go to Hell and, when it is time to go, procrastinates for a such a long time in order to listen to Euridice and to hesitate to look at Euridice?'[62] Whether or not the French found this scene palatable, Gluck clearly incorporated the politics of silence into the 1774 version.

Conclusions

The Orphée adaptations in eighteenth-century French culture show that silence was framed as a type of sound that had a history of politicized meanings. The Orpheus myth dramatizes a refusal to look at Euridice and, by extension, a refusal to acknowledge her. The journey from the underworld dramatizes not just blind faith, but also deafness, for Orpheus was deaf to Euridice's complaints. Gluck's versions excel at showing that the Orpheus myth helped sensitize the Enlightenment by extending seeing to listening.[63] If the Ovid version features the limits of blind faith, then the 1736 pantomime, the 1738 ballet-pantomime and Gluck's 1774 *Orphée et Euridice* illustrate that blind faith alone was not enough in Old Regime France. Reticence, too, was a desirable virtue.

In *Listening in Paris*, James H. Johnson attempted to answer a simple question: 'Why did French audiences become silent?'[64] This chapter has shown that the politics of silence taught the Parisians throughout the eighteenth century to be tactful. In Old Regime France, speaking sometimes had a hefty price to pay. The politics of silence compelled generations of suppressed playwrights to complicate the time-honoured Orpheus myth in ways that Ovid himself had never imagined. By the time the Parisians saw Gluck's *Orphée et Euridice*, they were reduced to silence not only because of the strong emotion provoked by the music, but also because they could sympathize with Orphée's silence.

Notes

1 Iona and Peter Opie (eds), *The Oxford Dictionary of Nursery Rhymes*, 2nd ed. (New York: Oxford University Press, 1997), pp. 178–9.
2 Robert Darnton, 'Mademoiselle Bonafon and the Private Life of Louis XV: Communication Circuits in Eighteenth-Century France', *Representations*, 87/2 (2004): 102–24, at p. 102.
3 On tales, see Robert Darnton, *The Great Cat Massacre and Other Episodes in French Cultural History* (New York: Basic Books, 1984), Chapter 1, 'Peasants Tell Tales: The Meaning of Mother Goose', pp. 9–74.
4 James H. Johnson, *Listening in Paris: A Cultural History* (Berkeley: University of California Press, 1995). Johnson used the term 'epidemic of sensibility' in an abridged essay of his book. See James H. Johnson, 'Listening and Silence in Eighteenth- and Nineteenth-Century France', in Mark M. Smith (ed.), *Hearing History: A Reader* (Athens: University of Georgia Press, 2004), pp. 169–83, at p. 170; James Chandler, 'The Languages of Sentiment', *Textual Practice*, 22/1 (2008): 21–39.
5 Isabelle Martin, *Le théâtre de la foire: des tréteaux aux boulevards* (Oxford: Voltaire Foundation, 2002), pp. 320–78.
6 In this chapter I use the name 'Orpheus' to indicate Ovid's version. 'Orfeo' indicates Gluck's *Orfeo ed Euridice* and 'Orphée' refers to various French adaptations of the Orpheus myth.
7 Ellen T. Harris, *Handel as Orpheus: Voice and Desire in the Chamber Cantatas* (Cambridge, MA: Harvard University Press, 2001), pp. 31–2; Vanessa Agnew, *Enlightenment Orpheus: The Power of Music in Other Worlds* (New York: Oxford University Press, 2008), Chapter 2, 'Music's Empire', pp. 73–120; Carolyn Abbate, *In Search of Opera* (Princeton: Princeton University Press, 2001), Chapter 1, 'Orpheus, One Last Performance', pp. 1–54.
8 See, for example, Martin, *Le théâtre de la Foire*, pp. 28–43; Raphaëlle Legrand and Nicole Wild, *Regards sur l'opéra-comique: trois siècles de vie théâtrale* (Paris: CNRS, 2002), pp. 11–44.
9 Jacqueline Waeber, *En musique dans le texte: Le mélodrame de Rousseau à Schoenberg* (Paris: Van Dieren, 2006), pp. 37–9.
10 Alain René Lesage, *Oeuvres choisies de le Sage*, vols 13–16 (Paris: Leblanc, 1810), vol. 13, p. v.
11 On eavesdropping see Mary Hunter, *The Culture of Opera Buffa in Mozart's Vienna* (Princeton: Princeton University Press, 1999), p. 88.
12 Franco Moretti, 'Style, Inc.: Reflections on Seven Thousand Titles (British Novels, 1740–1850)', *Critical Inquiry*, 36/3 (2009): 134–58, at p. 135.
13 'L'opéra comique ressemble au cygne qui ne chante jamais plus mélodieusement que quand il va mourir'. Lesage, *Oeuvres choisies*, vol. 13, p. xiv.
14 Claude Parfaict, *Dictionnaire de théâtres de Paris*, 7 vols (Paris: Rozet, 1767), vol. 2, p. 659.
15 Claude Parfaict and François Parfaict, *Mémoires pour servir à l'histoire des spectacles de la foire*, 2 vols (Paris: Briasson, 1743), vol. 1, p. 204.
16 Don Fader, '"The Cabale du Dauphin", Campra, and Italian Comedy: The Courtly Politics of French Musical Patronage Around 1700', *Music & Letters*, 86/3 (2005): 380–413.

17 'Ne jugez pas à la rigueur … Soyez animaux raisonnables'. Alain René Le Sage [Lesage], *Le théâtre de la foire*, 10 vols (Paris: Étienne Ganeau, 1721), vol. 3, pp. 3–35, at p. 33.
18 Robert S. Isherwood, *Farce and Fantasy: Popular Entertainment in Eighteenth-Century Paris* (New York: Oxford University Press, 1986), p. 94.
19 Alexis Piron, *Oeuvres complettes d'Alexis Piron*, Jean Antoine Rigoley de Juvigny (ed.), 7 vols (Liège: C. Plomteux, 1776), vol. 3, pp. 11–13; Michel Chion, *The Voice in Cinema*, trans. and ed. Claudia Gorbman (New York: Columbia University Press, 1999), Chapter 1, 'The Acousmêtre', pp. 17–30.
20 Ovid, *Metamorphoses*, 10.1–105 and 11.1–66, at 10.72. Translation taken from Ovid, *Metamorphoses*, trans. Charles Martin (New York: Norton, 2004), pp. 341–4, 369–71, at p. 343.
21 Plato, *Symposium*, 179d., trans. Christopher Gill (New York: Penguin, 1999), p. 11.
22 Virgil, *Georgics*, 4: 494–7. Translation taken from Virgil, *Virgil's Georgics*, trans. Janet Lembke (New Haven: Yale University Press, 2005), p. 76.
23 James R. Anthony, *French Baroque Music from Beaujoyeulx to Rameau*, rev. ed. (Portland, OR: Amadeus, 2003), p. 82.
24 Georgia J. Cowart, *The Triumph of Pleasure: Louis XIV and the Politics of Spectacle* (Chicago: The University of Chicago Press, 2008), pp. 77–8.
25 Cowart, *Triumph of Pleasure*., pp. 146–50.
26 For an introduction to the Orfeo productions in seventeenth-century theatre see Lois Rosow, 'Power and Display: Music in Court Theatre', in Tim Carter and John Butt (eds), *The Cambridge History of Seventeenth-Century Music* (Cambridge, New York: Cambridge University Press, 2005), pp. 197–239, at pp. 205–18.
27 Jean-Paul Montagnier, 'Charles-Hubert Gervais's "Psiché burlesque" and the Birth of the Comic "Cantate française"', *Journal of Musicology*, 17/4 (1999): 520–45, at p. 521.
28 David Tunley, *The Eighteenth-Century French Cantata* (Oxford: Clarendon Press, 1997), pp. 125–32.
29 Mary Cyr, 'Towards a New Chronology of Rameau's Cantatas', *The Musical Times*, 124 (1983): 539–41, at p. 539.
30 'Cantates', in Jean-Philippe Rameau, *Cantates. Canons. Airs*., Opera Omnia, series 3, vol. 1, ed. Jean-Paul C. Montagnier (Paris: Bärenreiter, 2008), p. L, ll. 31–4 and 37–41.
31 Parfaict, *Dictionnaire*, vol. 1, pp. 264–8.
32 Daniel Heartz, *From Garrick to Gluck: Essays on Opera in the Age of Enlightenment*, John A. Rice (ed.) (New York: Pendragon, 2004), Chapter 8, 'Terpsichore at the Fair: Old and New Dance Airs in Two Vaudeville Comedies by Lesage', pp. 135–58, at p. 138.
33 Isherwood, *Farce and Fantasy*, p. 92.
34 Lesage, *Oeuvres choisies*, vol. 14, p. 176.
35 Lesage, *Oeuvres choisies*, vol. 14, p. 187.
36 Jean Monnet, *Supplément au Roman comique, ou Mémoires pour servir à la vie de Jean Monnet,* 2 vols (Paris, 1772), vol. 2, pp. 45–8.
37 Parfaict, *Dictionnaire*, vol. 3, pp. 94–5.
38 'Cette pièce fut imaginé par M. Lesage; qui en composa les premiers actes et le prologue, pour donner une idée des différents changements que l'opéra comique avait souffert depuis son établissement jusqu'alors et pour faire voir avec quel acharnement ce spectacle avait été persécuté par les autres théâtres ses rivaux, qui voulaient son entière destruction pour rappeler chez lui le public qui s'en écartait tous les jours, pour affluer à l'opéra-comique'. Quoted in Martin, *Le théâtre de la foire,* p. 42.
39 Parfaict, *Dictionnaire*, vol. 1, pp. 208–9.
40 Parfaict, *Dictionnaire*, vol. 3, pp. 393–4.
41 Ibid., vol. 4, p. 144.
42 Ibid., vol. 1, pp. 263–4.
43 Clarence D. Brenner, *The Theatre Italien: Its Repertory, 1716–1793* (Berkeley: University of California Press, 1961), p. 109.

44 The synopsis of the ballet-pantomime *Orphée* was published in Pierre de Morand, *Théâtres et oeuvres diverses de M. de Morand*, 3 vols (Paris: Sébastien Jorry, 1751), vol. 2, p. 120–26; for a succinct account of Riccoboni and his family, see Rebecca Harris-Warrick and Carol G. Marsh, 'The French Connection', in Rebecca Harris-Warrick and Bruce Alan Brown (eds), *The Grotesque Dancer on the Eighteenth-Century Stage: Gennaro Magri and His World* (Madison: University of Wisconsin Press, 2005), pp. 173–98, at pp. 176–80.
45 Parfaict, *Dictionnaire*, vol. 4, p. 45.
46 Nicole Wild and David Charlton, *Théâtre de l'opéra-comique Paris: Répertoire 1762–1927* (Sprimont: Margada, 2005), pp. 33–60.
47 Downing A. Thomas, *Aesthetics of Opera in the Ancien Régime, 1647–1785* (New York: Cambridge University Press, 2002), Chapter 7, 'Music, Sympathy, and Identification at the Opéra-Comique', pp. 201–64.
48 Phillipe Vendrix, *L'opéra-comique en France au XVIIIe siècle* (Liège: Margada, 1992), p. 97.
49 David Charlton, *Grétry and the Growth of Opéra-comique* (Cambridge: Cambridge University Press, 1986), pp. 46, 50 and 115.
50 Stefano Castelvecchi, 'From *Nina* to *Nina*: Psychodrama, Absorption and Sentiment in the 1780s', *Cambridge Opera Journal*, 8/2 (1996): 91–112, at p. 95.
51 Christoph Willibald Gluck, *Orfeo ed Euridice*, Sämtliche Werke, eds Anna Amalie Abert and Ludwig Finscher, series I, vol. 1 (Kassel: Bärenreiter, 1963), p. 45.
52 Bruce Alan Brown, *Gluck and the French Theatre in Vienna* (Oxford: Clarendon Press, 1991), pp. 363–4.
53 See Calzabigi's letter to Prince Kaunitz dated 6 March 1767, quoted in Patricia Howard, *C.W. von Gluck: Orfeo* (Cambridge: Cambridge University Press, 1981), p. 22.
54 Gluck, *Orphée et Euridice*, Sämtliche Werke, ed. Ludwig Finscher, series I, vol. 6 (Kassel: Bärenreiter, 1967), pp. 47–8.
55 Gluck, *Orfeo ed Euridice*, pp. 117 and 127.
56 Gluck, *Orphée et Euridice*, pp. 188–9.
57 Ibid., p. 198.
58 Ibid., pp. 202 and 209.
59 Ibid., p. 210.
60 Brown, *Gluck and the French Theatre in Vienna*, pp. 196 and 219–24. For an excellent overview of theatre censorship see M. Elizabeth C. Bartlet, 'Beaumarchais and Voltaire's "Samson"', *Studies in Eighteenth-Century Culture*, 11 (1982): 33–49.
61 Patricia Howard (ed.), *Gluck: An Eighteenth-Century Portrait in Letters and Documents* (Oxford: Clarendon Press, 1995), p. 62; *Mercure de France*, September (1774), pp. 190–98, cited in Christoph Willibald Gluck, *Orphée et Euridice = Orpheus und Eurydike: (Pariser Fassung von 1774): tragédie opéra (drame héroïque) in drei Akten*, in Ludwig Finscher (ed.), *Christoph Willibald Gluck: Sämtliche Werke*, ser. 1: Musikdramen, 11 vols (Kassel, New York: Bärenreiter, 1967), vol. 6, p. xi.
62 'Est-il vraisemblable qu'Orphée ait eu le courage de descendre aux Enfers, & que, lorsqu'il est au moment d'en sortir, il s'amuse si long-temps à écouter Euridice, à hésiter s'il la regardera ou s'il ne la regardera pas?' *Journal des beaux-arts* III Septembre (1774), pp. 520–47, quoted in Gluck, *Orphée et Euridice*, p. xiii.
63 On a recent discussion of the meanings of what Orpheus hears and sees, see Daniel K.L. Chua, 'Untimely Reflections on Operatic Echoes: How Sound Travels in Monteverdi's *L'Orfeo* and Beethoven's *Fidelio* with a Short Instrumental Interlude', *The Opera Quarterly*, 21/4 (2006): 573–96, at p. 574.
64 Johnson, *Listening in Paris*, p. 1.

6 Sound as Promise and Threat

Drumming, Collective Violence and Colonial Law in British Ceylon

Jim Sykes

Citizens do not make the laws; laws, in part by codifying the legal categories of belonging, make citizens.[1]

Even when the tom-tom was viewed more benignly as native exotica, its penetrating and disorderly capacities intruded upon the British mentality as a symptom of that which was uncultured and ill-disciplined. Through the tom-tom the Wild Other rattled through the Briton's bones.[2]

Figure 6.1 The Kandy Perahera, Ceylon, *c.* 1900. Two dancers are visible, lower right; drummers trail the elephants.

Colonial Governmentality and Expressive Culture

The colonial theatre was an important forum for the emergence of European musical modernity. At the very least, a musicological analysis of the colonial encounter makes available certain base assumptions embedded in European musical modernity that are less noticeable when the musicological lens is focused solely on Europe. Far from a clash of aesthetic civilizations, the colonial theatre mandated strategies of navigation between competing ontologies of sound. The maintenance of native musical traditions became a problem, however, when they appeared to threaten public order. Religious processions and festivals, for example, were key sites for the emergence of colonized religious communities in the public sphere; when the sounds of such festivities appeared unruly, overly noisy and disruptive to public order, the colonial authorities sought to regulate them. Thus an important avenue for the navigation between ontologies of sound in the colonial theatre – perhaps the one that comes down to us as the most visible record of aesthetic debates between colonizer and colonized – is the legal debates that occurred on the regulation of native religious practices in public space.

In such debates, European aesthetic ideologies were not only written down, they were written into law. In the British Empire, the legal regulation of colonized sacred sounds amounted to the use of an aesthetic idea – the notion that sacred musics express feelings of personal devotion – to write into law modern notions of the interrelations between the sacred and the secular, public and private and self and society. It could do this because embedded within the notion that sacred musics are a type of personal expression is the idea that their noisy public manifestations are an unnecessary amplification of what could be achieved more safely and properly in homes and private places of worship. This definition of music as a form of expression fundamentally misreads the traditional function of sacred musics in many of the colonized communities of the British Empire, particularly the Hindu and Buddhist communities of South and Southeast Asia. The failure of the colonial authorities lay in their inability to acknowledge that for some religious communities, the public performance of sacred sounds functioned not merely as a form of 'expression' but also as a type of collective action, undertaken so as not to anger the gods, to ensure a good harvest, to ward off drought and chicken pox and so forth. For such communities even today, divine authority is not benign but judicial, and the failure to give the musical gift can lead to harsh divine retribution.

For some communities living under the British Crown, then, public space and political authority were not wholly secular, and sacred musics were not only expressive but also a kind of sanctified gift exchange. Thus one can imagine a type of aesthetic misunderstanding commonly occurring in the British Empire, with possibly harsh effects on the relationship between colonizer and colonized. From the perspective of the colonial authorities, the sacred musics of the natives became 'noise' when they appeared to threaten public order; in such cases, they had to be restrained for the public good. Regulation would cause no hardship for the natives, it was presumed, because their music was not being silenced for good, but was simply being told to operate in a respectful manner. To the natives, however, regulation was not a matter of keeping musicians quiet, but was a matter of life and death. In 1913, for example, within the context of an ongoing legal debate on musical processions I will describe below, Sinhala-speaking Buddhists in Ceylon (the country now called Sri Lanka) refrained from holding their grand annual musical procession, the *Esala Perahera*, in the town of Gampola: that year, a prominent ritual officiant (*kapurala*) died, and the town was racked by floods. These events were blamed on the failure to hold the procession.[3] On a more general level, however, for Sinhala Buddhists (who are my protagonists in this chapter), musical offerings (*sabda pujava*)[4] are productive and not merely defensive in nature: they are held not just to ward off disaster, but also to acquire merit (*pin*). By offering certain sounds to those who are above your station (like the gods), you acquire merit, the store of which determines your *karma*; upon your death, *karma* determines your station in the next life. The failure to give the musical gift, then, could not only ruin this life, but also your next one.

In this chapter, I explore the Buddhist musical gift in the British Empire at the turn of the twentieth century, by situating it at the confluence of political, legal, violent, sacred and communal actions, and by placing it within competing discourses on community, crowds, governance, the arts and public space. My focus, however, is not solely on British attempts to regulate the Buddhist musical gift through colonial law; rather, my discussion centres on the violent clashes that occurred between Ceylon's Sinhala Buddhists (73 per cent of the population) and two of the island's minority communities (the Christians and the Muslims, 10 per cent and 7 per cent of the population respectively) over a span of 32 years (1883–1915). These clashes were the result of the minority communities' attempts to use *legal channels* to enforce silence on Buddhist musical processions (*peraheras*), so that they would be forced to halt their drumming when marching past churches and mosques. The clashes, however, were also the result of a growing Buddhist resentment of Christians and Muslims, due to the social and economic successes these minorities achieved before and during this period. It seems that the Buddhists achieved a newfound awareness of the ability of their musics to *irritate* during this time, for they staged extra loud and prolonged processions, in order to publicize a vigorous (and, we will see, burgeoning nationalist) Buddhist revival in public space. However, these clashes would also not have been possible without the colonial authorities' *wavering* on the issue of whether the Buddhists had the *right* to play music while marching past non-Buddhist places of worship. This indecision, reflected in legal statements and in comments on native music made by colonial officers, was based on the failure of officers and the courts to acknowledge that 'tom-toming' in religious processions was anything more than a 'harmless amusement'.[5]

The belief that drumming was simply a form of amusement is, of course, part and parcel of what musicologists call the ideology of 'autonomous music', a set of beliefs that categorize music as merely reflective and expressive (rather than constitutive) of sociopolitical processes.[6] In this chapter, I argue that the ideology of autonomous music played a prominent role in instigating communal violence in colonial Ceylon, through its ability to influence a consistently clumsy application by the police of a law regulating music in religious processions. What was missing from the British colonial attempts to regulate sacred music in Ceylon was the idea that the Buddhists, or any other community, could *intentionally* use or manipulate 'music' to foment violence, even though the British did develop a legal channel through which 'noise' was acknowledged as a disturbance. Therefore, 'music' was assumed harmless or only 'accidentally' violent, in which case it became legally redefined as 'noise'. These ideas, mixed with the faulty definition of the Buddhist musical gift as a form of expression, clouded the authorities' abilities to act on their otherwise stringent (and universalizing) liberal need to regulate crowds for public safety. This produced a state of ambiguity with regard to the meaning and power of music in processions that, I argue, allowed Ceylonese communities the opportunity to *manipulate* the legal code for their own desired ends. Thus, the liberal desire to regulate sound for the sake of public safety created legal avenues for the creation of difference between communities, thereby facilitating avenues for unrest, rather than the peace and quiet the British so desired. My focus in this chapter, then, is not on the regulation of sound through processes of colonial governmentality per se, but on the emergence and institutionalization of British-derived *liberal* views on cultural identities, public space, music, noise and order, on the one hand, and the capacity of the musical gift to *threaten*, via musicians' abilities to manipulate the basic tenets of liberalism and its accompanying musical aesthetic (the ideology of autonomous music), on the other.[7]

While my narrative here is about events in the colonial theatre, my goal is to draw out aspects of the interrelations between aesthetics and liberal modes of governance, in order to query their implications for (ethno)musicology.[8] Acknowledging the rapprochement between musicological language and political liberalism, I will show, has significant ramifications for understanding the relationship between music and the self, for the ethics of our 'music and identity' paradigm (particularly its linking of music and ethnicity) and for our tendency to refuse the possibility that music serves as a 'causal player' in events of historical importance, such as in the riots described here.[9] Thus, my labour in this chapter is to use a historical study of the misrecognition of the musical gift to perform a critique of our musicological languages of modernity, in the process stressing the *sociopolitical* impact of European musical aesthetics on non-Western societies. *The very idea* that music is autonomous from society, I will show, had a great impact on colonized societies, through colonial law.

Sound and *Resentiment*, Part One: Kotahena, 1883

On Easter Sunday, 25 March 1883, 2,000 Catholics with white crosses painted on their foreheads and backs attacked a Sinhala Buddhist procession (*perahera*) in Kotahena, a Christian-majority ward of Colombo (the capital city of Sri Lanka: then Ceylon).[10] The attack happened shortly before 1pm; all of the bells in Kotahena's cathedrals were rung at once, a sonic cue that assembled the angry mob, who wielded clubs. The Buddhists were prepared to meet them, however, for a half-mile up the road they had heard rumours that a monk had been assaulted, and they armed themselves with makeshift weapons taken from a nearby timber yard.[11]

This incident must have been quite an aural spectacle, as well as a visual one. The Buddhist *perahera*, with dozens of devotees marching to the sound of vigorous drumming (*bera*), was interrupted by the cascading sound of church bells coming from all quarters. These two sacred sounds, bells and drums, each with a powerful symbolic value, overflowed into public space in this event, instigating actions that veered from celebration, devotion and sacred music offerings (*sabda pujava*) to hatred, resentment, fear and violence. In the end, two people were killed (one Buddhist and one Catholic) and 30 wounded, including 12 police officers.[12] Despite the small number of fatalities, the event was the most significant communal disturbance in British Ceylon up to that point, with ramifications felt throughout the country in the weeks that followed.[13]

The *perahera* that was attacked was part of a *netra pinkama*, a merit-making ceremony held to celebrate the completion of a reclining Buddha statue located in Kotahena's Buddhist temple, the Dipaduttama Vihara.[14] The auspicious day for the event was originally determined to be 12 February but the government asked for the ceremony to be delayed because of a smallpox scare then affecting the city. Rather than hold off the ceremony until the conclusion of the Catholic holy season (Lent through to Easter) as the colonial authorities had requested, the head monk of the Kotahena temple, Migettuwatte Gunananda, scheduled a number of unusually large and loud celebrations to coincide with the Christian calendar. The start of the Buddhist celebrations was rescheduled for 8 February (Ash Wednesday), the eye-setting ceremony for 23 March (Good Friday) and the concluding festivities for 31 March, after Easter Sunday. The concluding festivities were set to include 500 monks chanting 500 sections of the *Tipitaka* (the Pali-language Theravada Buddhist scripture) in a single day, as the monks were presented with 500 *atapirikara*, the eight articles that constitute the personal property of a Buddhist monk.[15] In order to obtain these articles, Buddhist devotees had to bring them in from Colombo and the surrounding villages, a movement which necessitated daily *peraheras* with loud drumming moving past St. Lucia's Cathedral, which lay only 400 yards from Gunananda's temple. As if this was not enough, Gunananda scheduled nightly sessions of *pirit* (Buddhist chant) and *bana* (Buddhist sermons), to begin on 8 February.[16]

Although the Catholics suffered these festivities throughout Lent, they soon realized the Buddhists had no intention of stopping during Holy Week. Thus what began as an annoyance came to appear as an intentional provocation. As the Catholics annually stage their own public celebrations during Holy Week, it became clear there would be 'competitive and irreconcilable festivities taking place simultaneously'.[17] The Catholics wrote to the authorities on 6 March, requesting that the Buddhists cancel their processions planned for Palm Sunday, Good Friday and Easter Sunday. The British response is interesting: they replied that no one had requested a procession license for those particular days, and the first to do so would be granted such a license.[18] Although the Catholics then applied for and received such a license, the British bungled the job, for the Buddhists somehow also received a license to march on Palm Sunday. This Palm Sunday *perahera* was met with stones, although no one was arrested.[19] The authorities, after granting the Buddhists the right to march on Good Friday, realized the possibility of greater violence and called off the *perahera* the night before, rescheduling it for after Mass on Easter Sunday, between 1:30pm and 4pm.[20] It was right before this time that the famous confrontation took place.[21]

From the Buddhists' perspective, their scheduling of festivities was due to the determining of auspicious days by astrologers and to the full moon, which happened that year to fall on Good Friday. Furthermore, the request from the authorities to hold off celebrations until after 31 March seemed more to do with British sympathies for Ceylonese Christians than the fear of smallpox. The Buddhists, of course, had good reasons to resent the Christians. After the growth of Catholicism on the island during the first two colonial regimes (the Portuguese and the Dutch: 1505–1796), Protestantism gained ground in the nineteenth century from the influence of Anglican missionaries. These missionaries were seen to be particularly unwavering in their denigration of Buddhism. Tensions had grown especially in recent years between the religions because of the circulation of rumours that each performed actions demeaning to the other's religion.[22]

The need to counter Christian influence was one reason why a Buddhist revival blossomed during this period. The Buddhist revival drew on the institutional forms and propaganda techniques of Protestant missionaries, such as the printing press.[23] Gunananda, the monk at the Kotahena temple, is widely credited as the founder of this movement. Stanley J. Tambiah writes that Gunananda 'exemplified the new kind of monk spearheading the Buddhist revival that had overtaken the island since the 1860s':[24] a polemical orator and publisher of pamphlets denouncing Christianity, he was a successful fundraiser and organizer who founded the Society for the Propagation of Buddhism in 1862.[25] Gunananda's fame also rests on his staging of a series of debates between himself and Christian missionaries, wherein he demonstrated his rhetorical skills and allowed for the public to compare the benefits of Christianity and Buddhism.[26] Meanwhile, Gunananda's Kotahena temple had become a magnet for the Sinhala working class from the south, who increasingly moved to Colombo for work. As the number of Buddhists in Kotahena grew, so did the importance of Gunananda's temple. In the 1870s, after Catholics undertook numerous renovations to St. Lucia's Cathedral, Gunananda began renovations to his temple, probably in a spirit of retribution. The renovations to sacred sites in Kotahena became a 'symbolic struggle over sacred space' that culminated in the riot.[27] Kotahena,[28] a densely populated neighbourhood of Colombo that was over a half Catholic and a quarter Buddhist, was a prime site for Buddhist–Christian confrontation.[29]

The above description is a mainstream historiographical rendering of the 'causes' of the Kotahena riot, centring as it does on the Buddhists' motives for the prolonged festivities, and the Catholics' irritation at the Buddhist monopoly of *space*. Yet this perspective demotes the power of *sound* in these events to the status of an epiphenomenon, whose causal power is reduced to social and religious factors. Thankfully, Michael Roberts has challenged this perspective by highlighting how the Buddhists in Kotahena used 'noise as a weapon':[30] Sinhala drummers, he shows, used their music as a disruptive force in these and other *peraheras*, occasionally slowing down in front of churches, increasing the volume of their drumming, and shouting '*sadhu, sadhu*' ('a Buddhist hallelujah').[31]

If these incidents demonstrate that Buddhist drummers acquired a newfound consciousness of their music to irritate, this is an aesthetic shift that needs to be explained:[32] for there is no previous record of Buddhists using their drumming to annoy others, nor is there any previous record of others complaining about Buddhist processions.[33] This shift is certainly attributable in part to the re-signification of the Buddhist musical gift in modernity: Buddhist drums and voices appear to have been re-conceptualized by the time of Kotahena as akin to the sacred objects, sites and other practices that maintain a prominent link to the Sinhala Buddhist past and, by extension, an authentic Sinhala Buddhist identity. This means that *peraheras*, Buddhist chant (*pirit*), sermons (*bana*) and songs (*kavi*) came to seem like the Buddhist ruins, pilgrimage sites, temples and relics that function as 'sites where the ontic reality of a glorious Sinhala Buddhist past are made available for theorization and participation'.[34] The use of music in this manner in Kotahena is a shift away from the pre-modern ontology of the musical gift, which was *not* conceptualized as representative of a generic Sinhala Buddhist identity. By contrast, in its pre-modern form, the Buddhist musical gift was considered merely a set of sonic objects *owned* by specific castes of musicians and dancers, who alone had the right to offer them to the gods.[35]

This shift from a self*less* tradition to a self*ish* one is *the* major transformation of the musical gift in modernity, a shift that was enacted in part through the institutionalization of European musical aesthetics in colonial law. This can be demonstrated by examining the issue of permits, which, as we saw in Kotahena, was an ambiguous matter that led to the mixed signals between Buddhists or Christians and thus helped set the stage for violence. A discussion of the matter will allow us to see that the ambiguity of noise regulation hinged on two factors: the aesthetic disposition that music was 'actually' a harmless amusement, and the inability to legally determine how, when and where drumming in an event shifts from 'music' to 'noise'.

Roberts notes that after a 'rudimentary' police force was established in the mid-nineteenth century, a Police Ordinance written in 1865 stipulated *the legal possibility* for drumming to become 'noise' when it disturbed people and horses, particularly at night.[36] He draws our attention to clauses 69 and 90:

> 69. ...All officers of Police shall and may keep order in the public roads, streets, thoroughfares and landing places ... and prevent obstructions on the occasions of such assemblies and processions, and in the neighbourhood of places of worship during the time of public worship. ... *They may also regulate the use of music in the streets, when the same shall be allowed...*
>
> 90. All persons who shall, at any time, within any town or limits, either within any house or building, beat drums or tom-toms, or have or use any other

> music calculated to frighten horses, or who shall make any noise in the night so as to disturb the repose of the inhabitants, or shall, at any time, discharge fire arms, crackers or fireworks except under Military regulation, or unless they shall have obtained a license from the Police Magistrate of the district, or from the Superintendent of Police … shall be guilty of an offence, and be liable to a fine not exceeding Five pounds, or to imprisonment … not exceeding Three months.[37]

Roberts notes that these clauses 'embody within them the basic Western distinction between the private and the public', as well as the idea that the law is applicable equally to all populations on the island: thus it presupposes that Buddhists, Christians, Muslims and anyone else on the island must obey the same rules of behaviour in public space.[38] These clauses were part of a colonial attempt to enforce a universal moral code of behaviour in the colonies, centred on the legal conception of an abstract individual whose behaviour could be determined by the idea of tolerance and its limits. This notion of the de-cultured individual who must choose to act rationally in public space through the prism of toleration (to the extent that others tolerate him) is of course a cornerstone of liberalism, the dominant political philosophy in Britain at the time (and also in the present day). Political liberalism mandates a conceptual separation between politics, economics and the public sphere, on the one hand, and culture, community and the domestic sphere, on the other.[39] For liberals, one's identity is a private matter: even one's communal identity is assumed to develop away from the secular public sphere of the marketplace, in domestic and social spaces (such as places of worship). In this schema, public sacred music performances are an outward expression of an inward, pre-formed cultural identity. The over-publicization of one's communal identity is taken as an unwarranted intrusion that could disrupt the (supposedly) equal and harmonious relations between diverse individuals who congregate in the public square. This is what made native public religiosity dangerous in the colonial theatre, particularly when it was amplified by native sounds. It also shows that political liberalism, while *in theory* banishing the arts from its purview, had a significant impact on the development of the arts through its stipulation that law trumps culture in matters of public safety.

While clause 69 stipulates that no 'public worship' shall be allowed to obstruct other 'places of worship' in 'the neighbourhood', clause 90 explicitly mentions drumming as having the ability to cause a disturbance, through its ability to irritate. This irritation was due to the ability of drumming, via its rampant audibility, to cross the gaping chasm of liberalism (between the private and the public), but it was also due to the development of modern forms of labour and perceptions of time, which came to coexist with older traditions that had developed in another system.[40] Drumming in Sinhala rituals, for example, persists throughout the night, and even in the late nineteenth century drumming in such rituals commonly went on for days, with non-stop music and dance for at least 12 hours.[41] While the sound of drums wafting through the tropical air is described exotically, even romantically, by British diarists (and perhaps more so by Burghers, peoples of mixed Ceylonese-European descent),[42] the ability of drumming to ruin one's sleep before the workday threatened to ruin any peaceful night. The ability for drumming to become noise, then, was defined by the Police Ordinance according to its tendency to *transgress* the modern standards of behaviour related to time (the workday) and space (dense urban living quarters). However, the clauses *do* say that despite the ability of drumming to transgress these boundaries of modernity, drumming as noise *was* allowable, so long as 'they … obtained a license'. This *appears* to state that publicly performed drumming in processions that *does* disturb is perfectly legal, if it is performed under the guise of a valid license.[43]

A look at the actual form that licensed processions, however, shows ambiguities surrounding the legal right to disturb. For the actual license for processions was a 'Procession Form' that authorized processional music via the terms of the Police Ordinance: 'I, the undersigned, prescribe the route to be taken as above, in terms of Clauses 69 and 90 of Ordinance No. 16 of 1865'.[44] The legality of the procession is thus related to where the procession goes *and* to the stipulation that the sound of the procession does not become 'noise' in the terms set by the Ordinance. But this definition of an aural disturbance is expanded (and confused) by the next sentence: 'music must be stopped on the near approach of horses, or within one hundred yards of any place of Public Worship'. There are problems with this stipulation. In colonial Ceylon (as in modern Sri Lanka) there were *so many* places of worship – from Buddhist *viharas* and Hindu *kovils* to churches, mosques and multi-religious sites – that if the Buddhists were forced to silence their music as their processions passed by any one of them, they would be forced to cease their processional music altogether. This was, of course, especially true in dense urban spaces like Colombo.[45]

The British authorities, of course, recognized that enforcing silence in front of all places of worship and when approaching all horses was impossible. Furthermore, they were aware that the terms on the procession license form overstepped the terms laid out by the Kandyan Convention (the document signed in 1815 that granted sovereignty to Britain over the entire island, and which acted as the country's constitution): the Kandyan Convention required that 'the Religion of Buddhoo, its rites, ministers and places of worship [must be] maintained and protected'.[46] The enforcement of *so much* silence in processions, besides being impossible to monitor let alone enforce, would undoubtedly cause the resentment of the great majority of the Sinhalese population, for it would appear the British were trampling on the rights of the people.

Thus it was thought that in practice 'tom-toming' needed to be silenced or quietened only when its sound threatened to create, or had just created, a disturbance. But this was a leniency that stemmed also from the belief amongst many British that tom-toming, although a necessary part of Buddhist festivities and a right of the people, was 'really' just a form of entertainment that *should* be able to be taken away at any time. Drumming was to be tolerated, but only within limits. This difference between threat, on one side, and harmlessness, on the other, is mediated, however, by numerous statements made by Government Agents (G.A.s) that show that many British were simply annoyed by drumming: 'It is a scandalous thing, that with the liberty allowed them, they should be permitted to interfere with us [Christians] at even one minute when service is going on', says one officer; tom-toming is a 'great annoyance [to] the neighbourhood' and 'no religious body … is entitled to commit a public nuisance', says another; *yet*, for others (such as a judge named Layard), the Police Ordinance is 'inapplicable to the circumstances and surroundings of the villagers themselves', because drumming is a 'harmless amusement'. 'Why should there be a hard and fast license?' Governor West Ridgeway asks, 'Why should not the terms of the license alter according to circes [circumstances]? It is rather a tall order that processions are never to pass – with music – empty places of worship …'. Governor Arthur Gordon represents both extremes in his statement of 1884 when he claims that colonial 'interference' through the Police Ordinance with the 'harmless usages' of drumming happened in Colombo but not Kandy; still, 'the beating of native drums [is] an intolerable, and indeed, dangerous nuisance in crowded thoroughfares'.[47]

It was, then, the conflict between the universalizing tendencies of British liberal ideas as written into law, on the one hand, and the notion that the arts were a harmless amusement, on the other, that led to the ad hoc way in which music was judged capable as being a threat to order. But these extremes were mediated by an attitudinal hostility towards drumming that itself could shade into tolerance, on account of the belief that drumming was in and of itself harmless.[48] As long as the volume of the processions was not too loud, and as long as the procession moved in an orderly fashion, it seemed reasonable that all communities should allow the procession to proceed unmolested – and unlicensed. For the British were not about to push their luck with contravening the Kandyan Convention, and besides, another's music was (that is, unless it was a threat to order) something that must be tolerated.

In sum, then, there was a conceptual threshold where music became noise, which was simultaneously a transition from legal to illegal behaviour. This threshold was determined not only by decibel level, but also by the location of *some* sounds in *specific* spaces and times: noise in such times and places was a threat to order, while silence or orderly sounds in such times and places were a 'promise' to order, in that they showed the 'civilizing mission' of colonialism was working. Yet the whereabouts, volume and time that determined this threshold could only *really* be determined *after* noise had already become audible and visible as a 'disturbance': in such a case, the police could simply point to the natives' ignorance of the stipulation for silence on the procession form. The ambiguity of the noise threshold, as reflected in the ambiguity of enforcing the Police Ordinance and the stipulations of the Procession Form, thus created a malleable (physical and imaginative) space, wherein populations could push the limits and application of the law. It seems it was this noise threshold that the Buddhists were trying to locate and exploit in Kotahena, as it was this newfound illegality of 'noise' that the Christians appealed to in order to make the Buddhists halt their drumming.

The historiography of the Kotahena riot, then, needs to make room for musical *aesthetics* (rather than 'noise' per se) as an instigating force for violence: for it appears that the idea that music is autonomous from society (as expressed here through the dispositions that music was a 'harmless amusement', that it was a form of entertainment that expressed feelings of devotion and that it was simply annoying and unpleasant) operated to obscure British enforcement of a law that *did* acknowledge the ability of noise to instigate violence.[49] While the ability of noise to irritate was indeed a factor in the violence, the liberal view that aesthetics is a private matter unrelated to governance and law was also to blame. As Nira Wickramasinghe puts it, rather than enforce order, the Police Ordinance 'created a new space for conflict between religious communities that until then had not systematically considered drumming a "disturbance"'.[50]

Sound and *Resentiment*, Part Two: The Buddhist-Muslim Riots of 1915

If the events at Kotahena show the Buddhists pushing the legality of processions and the Christians reacting legitimately through confused legal channels, the disputes between Buddhists and Muslims over processions that occurred in the early twentieth century show that the legal ambiguity of the noise threshold gave minority groups a chance to manipulate legal channels, in order to assert themselves more aggressively in the public sphere.

By the late nineteenth century, a new group of Muslim immigrants from India had become mainstays on the island, and they complained loudly, through legal channels, about the music in *peraheras*. This group was the 'Coast Moors', or *Hambankariya* ('boat people', a name referring to their moving between Ceylon and India on boats). They were Wahhabis, followers of the conservative branch of Islam that originated in Saudi Arabia in the eighteenth century (Wahhabism is what is commonly referred to in the West today as 'Islamic fundamentalism').[51] To the Wahhabis, music is forbidden by the Quran and Hadith, as it is considered overly sensual and can lead one into bad behaviour, such as adultery; it is therefore considered *haraam* ('forbidden').

Sinhala Buddhists had many reasons to resent the Coast Moors. Unlike the Moors who had been living with them on the island for centuries, the Coast Moors were recent immigrants.[52] They were viewed as a particularly unfriendly bunch who kept to themselves and were more interested in their businesses and religion than fraternizing with their new neighbours.[53] The Coast Moors began to dominate the selling of essential foodstuffs, and 'Moorish boutiques' became a common sight in Sinhala-majority towns.[54] Thus, the Coast Moors were a small minority who achieved an economic dominance in a short time, which caused them to be resented by the majority population.[55] With the onset of World War I, a global recession necessitated a rise in prices, and this furthered resentment towards the Coast Moors, who were accused of inflating prices.

However, the resentment against the Coast Moors also had a religious component. There is documentation of 14 confrontations between Sinhala Buddhists and Moors over sacred music processions between the years 1899 and 1915. The three most significant, according to Roberts, are Galle (1899–1903), Gampola (1907–15) and Kurunegala (1912).[56] The Galle case is noteworthy for how the

Moors appealed to the Britons' liberal sensibilities – particularly their notion of tolerance – to win the case. In 1899, the southern Government Agent (G.A.) H.W. Brodhurst, with help from local officer Tranchell and American Buddhist activist Col. Henry Steel Olcott, forged an agreement that had the Buddhists halting their drumming while passing by mosques. But, 'an appellate decision of the Supreme Court constrained prosecutions against offending parties'.[57] Some Buddhists used this as an excuse to ignore the ruling, and drummers were found to repeat their tendency to slow down, drum loud and chant *sadhu, sadhu* – this time in front of mosques. In response, Brodhurst insisted on 'a strict application of the 1899 agreement backed by the legislative weight of the Police Ordinance'.[58] After the Buddhists petitioned against this ruling, the British assembled a meeting with Buddhist and Muslim representatives to forge an agreement. They decided all drumming would cease while passing mosques between noon and 1:30pm, and again between 6pm and 7:30pm. The Moorish community, however, found this intolerable. In a letter, their representative stressed that 'even the Koran could not be read so loudly as to make the other devotees hear it at their devotions':[59] thus the issue for them (or so they claimed) was volume and placement, not the time of the procession, and they appeared to respect the secular and shared aspect of public space. Unsurprisingly, this resonated with the British liberal belief that one's culture should be kept private and not intrude upon others. Accepting the argument, Brodhurst negotiated a space for silence that stretched 100 yards on either side of the mosque, but he said it could grow to 200 if the Buddhists misbehaved. However, the boundary would not apply to offerings of sacred music (*sabda pujava*) in houses within 100 yards of a mosque (except, it seems, after 11pm).[60]

The Galle case pales in significance, however, next to the procession dispute that occurred a few years later in the hill country town of Gampola. On 27 May 1907, an unlicensed *pinkama* (merit-making procession) was attacked while passing a mosque on Ambagamuwa Street, in the town of Gampola. The reason for the *perahera* was the annual celebration of Wesak, the holiday that honours the Buddha's birth, death and Enlightenment: a retinue of drummers, dancers, elephants and devotees march through the streets to the banks of the Mahaweli River and back. The Basnayake Nilame (Lay Officer) of the Walahagoda Devalaya (the temple that organizes this annual procession), 'pointed out that the route was a "via sacra" laid out by ancient custom and, he believed, guaranteed by the Kandyan Convention of 1815';[61] thus he felt the Buddhists had a *constitutional* right to play music in their procession. The trustees of the Coast Moors' mosque continued to lodge protests for two more years, however, resulting in the rerouting of the procession by the police. Despite this rerouting, the Coast Moors protested to the police yet again in 1911; it is unclear whether the protest was due to the Moors' ability to hear drumming from their mosque, or whether they were merely stretching the law to see if they could irritate the Buddhists and/or place further strictures on the *perahera*. In response, the British G.A. came up with an imaginative solution that created a physical avatar of the spectre of sound as a threat to order: he affixed actual posts on 50 yards of either side of the Coast Moors' mosque, so the Buddhists would know exactly where to be quiet.[62] In 1912, the Basnayake

Nilame applied for the annual procession license and was granted it, on condition that the official place for silence would be observed. The temple, however, decided to forgo this rule and reroute the procession – an interesting decision, which may show that continuous sound is more important for the Buddhist musical gift than the particular route a procession follows. But perhaps not, however, for the Basnayake Nilame's idea to reroute the *perahera* was met with protests from the temple's own constituency. Having run out of options, the Basnayake Nilame did what an American might do: he cancelled the *perahera*, and sued.[63]

Silencing the musical gift in its entirety was a drastic step. As mentioned in the introduction, the year that followed the silencing of this *perahera* saw a temple officiant (*kapurala*) die suddenly and the town become beset by floods. These events were attributed to Kataragama, the god of the Walahagoda Devalaya.[64] For the next several years, the Gampola *perahera* was not held, while the case was heard by the district court in Kandy. In June 1914, the Sinhalese had the good fortune of receiving a favourable decision from a Ceylonese judge, Paul E. Pieris. I present Ramanathan's summary in full:

> [The judge ruled that] the Esala Perahera was a rite of the religion of the Buddha, which was undertaken to be maintained and protected under the Convention of 1815; that the accustomed route of the perahera and the continuance of the music was an essential part of the rite; that the Kandyan Convention constituted a binding compact unalterable in all following times; that the plaintiff, as Basnayake Nilame … was entitled to conduct the procession with elephants to the accompaniment of drums and other musical instruments through all the streets of Gampola, including that portion of Ambagomuwa Street where the Hambayas had built in recent years a mosque; and that the acts of the Government Agent were a violation of the right conferred upon the plaintiff by the Kandyan Convention.[65]

Lest one think this legal decision was binding (coming as it did from a Ceylonese judge), the G.A. urged the Attorney General to file an appeal to the Supreme Court against the decision.[66] On 2 February, the Supreme Court ruled against the Buddhists. Their reasoning was that, while the government is required to honour the Kandyan Convention, its words on the safeguarding of Buddhist tradition 'were not intended to give this particular *perahera* any right to be conducted in a manner different to other religious processions in the Colony'; nor did the Convention allow any procession 'to be for ever conducted apart from the ordinary police supervision for the protection of the public peace and safety'; furthermore – and this is the important point – 'even supposing that the particular right claimed was reserved by the Convention to this particular *perahera*, such right was now controlled and varied by the provisions of the Police and Local Boards Ordinance'.[67] If the Police Ordinance was seen to contravene the Kandyan Convention, the Supreme Court argued, the Court 'had jurisdiction to enforce the Ordinance against the Convention'.[68] No law could trump native traditions, *unless* native traditions were a threat to order. This logic failed, however, to acknowledge that the denial of native traditions was itself a threat to order: for the ruling against the Buddhists was handed down in 1915 (the hundredth anniversary of British rule over the island), and frustrations with British rule were at an all-time high.

These Sinhala frustrations finally let loose in the city of Kandy, on 29 May 1915, although they were directed at the Moors, not the British. Rioting broke out during the Wesak *perahera* when Coast Moors stoned a Buddhist bullock cart carrying singers (not drummers). Over the next ten days, violence spread throughout seven of the country's nine provinces, with hundreds of Muslim shops burned down and dozens of mosques ransacked. In the end, 142 people were killed (mostly Muslim), and the British incarcerated upwards of 5,000 people in an attempt to quell the violence.[69] It bears repeating that the instigating moment for these riots – *the most widespread and historically important communal violence in colonial Sri Lanka* (1505–1948) – is traced to a group of Muslims pelting stones at Buddhist singers, as their musical procession proceeded down Castle Hill Street in the city of Kandy.

A modern component of the Wesak holiday is the singing of carols that honour the Buddha. In 1915, this was music performed standing up in bullock carts, by singers whose carts formed part of a larger *perahera*. The carts would stop outside the homes of prominent individuals who had donated money for the staging of the festivities, and singers would honour them with songs about the Buddha. In 1915, Wesak carolling was only about 30 years old: it was a genre derived from the Christian tradition of Christmas carols, but reworked according to a Buddhist ethos of generosity (*dana*). Armand de Souza put it nicely in his writings of 1916:

> The only common form of worship they [the Buddhists] understand is a procession, and the Wesak carols thus acquire a multitudinous character foreign to the associations of the English word. The Buddhist, who repudiates forgiveness of sins or propitiation of the Deity, pursues only one form of devotion – the acquisition of merit. To relieve want and suffering of any kind counts for high merit … Pious Buddhists, either individually or through organized societies collect funds and establish eating booths along the streets, where peasant pilgrims are given a simple meal free of charge. The booths so erected are called 'Dansalas' – dan signifying 'a giving' (Lat.: dare, donum) and sala a 'hall'.[70]

De Souza's text, which explains to an English-speaking readership the causes of the riots and the heavy-handed British reprisal, is marked by the author's Westernized idea of what Buddhism is.[71] De Souza's use of the word 'multitudinous' points to his perception that the carols are not just an expression of devotion, but are also undertaken for a purpose, to acquire merit. He then links carolling with related forms of acquiring merit, such as the institution of the *dansala* ('giving halls'), where pilgrims are offered free food on the Wesak holiday.[72] This points to the identity of the holiday as one large, merit-gaining festival, with music being merely one way of acquiring merit. The practice of carolling was meant to honour wealthy donors who had paid for the *perahera*, donations that were made because, besides providing financial assistance, they also allowed the giver to acquire merit. By honouring the righteous donors, the singers, too, acquired merit.

At about 1am, two bullock carts with carollers turned onto Castle Hill Street, as part of the larger *perahera*. The carollers had intended to honour a prominent Buddhist donor whose house lay across the street from the mosque. However, seeing that the Coast Moors' mosque was still lit up (for clearly it was well past the time for daily prayers) and seeing that a crowd of Moors stood outside on the street, the British Inspector Mr Cooray ordered the carollers to turn onto another street before reaching the mosque. The Moors at this point hurled stones and empty bottles at the singers. That was it. The Buddhists hurled the stones back, chased the Moors into the mosque, and proceeded to ransack the mosque and nearby Moorish boutiques.

I do not have adequate space to cover the historiography of the riots here, but a brief summary is relevant for the critique of music historiography I wish to develop. The literature has been preoccupied with whether the reasons for the riots were 'economic' or 'religious'. The economic reasons are, besides the Moors' dominance in foodstuffs, the rise in prices due to World War I.[73] The religious causes of the riots, by contrast, are described through the narrative of the 'procession disputes' I provide here.[74] However, like the Kotahena historiography, all of this literature (save, again, for the pioneering effort of Roberts)[75] *erases* the power of noise in inciting violence. More to our point, however, all of this historiography eliminates the central role of musical *aesthetics* in determining the facilitation (or misapplication) of colonial law.

My larger point is that the role of music in these events *as well as* their understanding in historiography can adequately be described as the misrecognition of the musical gift. The differences between the misrecognition of the gift, on the one hand, and the location of legal debates on music and the capacity of music to irritate, on the other, are not semantic. Rather, the connection between them has entirely to do with the incorporation of the ideology of autonomous music within mainstream historiography itself. Music, as a form of expression and a harmless amusement, is not thought to 'actually' be the 'cause' of the riots. Although music might *seem* to emerge in the events as a causal player in History, this 'leading role' is turned into a 'cameo appearance' once the 'actual' causes of violence are named: 'the immediate precipitating or triggering event leading up to the riots was the familiar one of religious processions playing music while passing Muslim mosques', says Tambiah, but ultimately,

> The riots were the culmination of developing antagonism between the Sinhalese population and the 'Coast Moors', recent Muslim immigrants from Southern India. The causes for this ill-feeling were mainly economic and religious. The Sinhalese resented the virtual monopoly enjoyed by Muslims over petty trade in foodstuffs. The War with its shortages and high prices caused considerable hardship to the people and it was easy to believe that these Muslim traders were exploiting the situation.[76]

This view was put forth philosophically by the legendary historian of the Haitian Revolution, C.L.R. James:

> In a revolution, when the ceaseless slow accumulation of centuries bursts into volcanic eruption, the meteoric flares and flights above are a meaningless chaos and lend themselves to infinite caprice and romanticism unless the observer sees them always as projections of the sub-soil from which they came.[77]

Needless to say, the determination of causality in the history of events is a fraught issue that will not be solved here. My point is more circumscribed and musicological, namely that the elimination of the possibility of artistic causal determination writes widespread and globally and historically important musical genres outside the purview of musicology. To put it in Karl Dahlhaus's memorable words, it banishes such musics from the 'history of Art', and relegates them to a 'History of art' which will never be interested in them.[78] Not only do we debunk such traditions through our historiographical silence, we allow for the mainstreaming of perspectives on the role of music in society that *also* misunderstand the form and function of such musics – as my narrative above shows, this is a misunderstanding with *legal* ramifications, and real effects on peoples' lives.[79] This is equivalent to the complaint Dipesh Chakrabarty[80] provides about mainstream historical writing in his book *Provincializing Europe*: gods and goddesses, he says, are erased as causal players from the pages of history, yet they 'cause' lots and lots of people to do things of historic importance. How do we write these musical players into history, particularly in a narrative history of the event? My labour here is not to perform such a historiography, but to demonstrate that European musical aesthetics, through its core belief in the arts as epiphenomenon, directly impacted sociopolitical processes in a colonized society.

Conclusion

At least one thing is missing from both historiographical engagements cited above and my own analysis: a sustained engagement with the aesthetics, musical content and traditional form and function of the musical gift itself. I have provided an in-depth analysis of it elsewhere,[81] but a cursory comparison of the Buddhist musical gift and British musical aesthetics will be useful for understanding the misrecognition of the musical gift in history and allow me to conclude by extending the above discussion to a critique of musicological language.[82]

For the British authorities, native sounds were necessarily linked with native selves. The idea that music is something that *inherently accrues* to a person – the notion that music and the self are inalienable – demonstrates, I think, that at the heart of the concept of music as a form of expression lies the Christian notion of the soul. In the West, music is viewed as having the power to corrupt the soul, by leading it astray (as in the Platonic tradition), or by contrast, music can 'enrich' it. Both of these views hinge on the notion of a pre-existing self that is affected by music: once music becomes attached to this self, it appears to be not so easily separated from it. It is this latter assumption, often unanalysed, which enables the existence of our 'music and identity' paradigm and its subset, the linking of music and ethnicity. By locating an essential connection between sound and bodies (and communities), communities are homogenized and ethnic differences are emphasized. In this conception, music is not something that comes to define meaningful relationship *between* peoples (or between ethnic communities, or between peoples and gods), but it comes to be something that defines people(s), full stop. This self*ish* rather than self*less* notion of music produces and legitimates the divisions inherent in ethnic conflict, a particularly troubling idea that finds its fruition in postcolonial Sri Lanka, a country where Sinhala Buddhist 'traditions' – including the musical gift described here – become part and parcel of a disenfranchising discourse on ethnonationalism.

In contrast to the Western self, for Sri Lanka's Sinhala Buddhists the self is inherently transitory, and sacred musics are understood to contain a type of power that, upon being offered to (divine and kingly) sovereign powers, influences the necessary transitions that a self must make. For Sinhala Buddhists, too, however, 'music' has the power to corrupt: non-sacred musics, that is, musics meant for dancing and entertainment, are considered so sensual that they inhibit one's path to Enlightenment (there is in fact a famous Buddhist precept that says, 'one must refrain from music and dance'). Yet Sinhala Buddhism is *filled* with practices that contain what Western ears would call 'music', particularly drumming and singing. While I have retained the word 'music' here, it is more proper to refer to these types of Buddhist sounds as forms of sacred speech: drummers, for instance, are told to 'speak' their drums (*kiyenne*), not to 'play' them (*gahanne*) (the latter verb is used to tell someone to play a guitar or piano). The musical gift is thus more properly understood as offerings of sacred speech that happen to be played on a drum. The Buddhist self is not enriched but *transformed* through the offering of sacred speech, through the musical gift's ability to acquire merit, which betters one's *karma*. In sum, then, it appears that the self*ish* foundation of our modern musicological framework obscures the very possibility of hearing the musical gift *as* a gift, and, I argue, it was this self-centred conception of music that the British applied through colonial law in my narrative above.

The missing discourse for both the British colonists and our historians is an *economic* one that would have waded through the defining chasm of liberalism to apply a theory of public, politicized, exchange to the understanding of a sacred artistic tradition, in order to develop a legal theory of artistic contract. It is impossible to tell whether the recognition of the musical gift in law would have averted violence, but a historiographical model could have been found. In her recent study of 'Indian vernacular capitalism' in the late colonial period, the historian Ritu Birla[83] describes the economic contracts written for Hindu temples, which legally defined a God or gods as having a stake in the offerings of gifts to a temple: 'The deity, to which Hindu endowments were consecrated, was rendered a legal subject and thus the beneficiary in the trusts' tripartite contract'.[84] From the perspective of colonial capitalism, the important legal component of merchant families' gifts to Hindu temples was the question of 'the intention of the gift':

> By asking the question of public or private intent, colonial authorities sought to identify charitable projects from those tied to family benefit, and thus to the family firm's profitable ventures. This juridical process privileged the a priori intentionality of contract and so recognized the purpose of customary gifts by reading their ritual dedications as contracts and interpreting them literally.[85]

Such a recognition of intent would mean recognizing the ability of the natives to *mean* to use their musics for desired ends in a literal exchange, which would be validated by contract. What such a juridical process did in the Indian Hindu case was legalize the tripartite relationship between giver-deity-community, a movement of monies from a donor to a deity or deities, to the worshippers who come to the temple, via the temple trustees (who use the money to refurbish the temple).[86] Such a contract applied to Sinhala Buddhist drumming might not have stopped the Christian and Muslim resentment of the Buddhist musical gift, but it would have at least provided a legal framework that acknowledged the musical gift's validity; thus it may have avoided the violence that emerged from the colonial authorities' ad hoc application of the Police Ordinance via the ideology of autonomous music. Perhaps, at least, such a contract would have provided grounds for a historiographical framework that would recognize the musical gift on its own terms and counter the liberal musicological bias that, even today, defines the musical gift as epiphenomenal to the history of the event.

Notes

1 Lisa Wedeen, *Peripheral Visions: Publics, Power, and Performance in Yemen* (Chicago: University of Chicago Press, 2008), p. 24.
2 Michael Roberts, *Exploring Confrontation: Sri Lanka: Politics, Culture and History* (Chur, Switzerland: Harwood Academic Publishers, 1995), p. 151.
3 Roberts, *Exploring Confrontation*, p. 202.
4 *Sabda Pujava* ('sound offering') is the term used to describe Sinhala Buddhist musical offerings in a range of contexts, from all-night rituals (*yak tovil, deva tovil*) and processions to smaller rituals, such as the offering of auspicious drumming (*magul bera*) to honour the Lord Buddha, performed on the occasion of greeting prominent guests.
5 The courts took the issue of legal precedence much more seriously than the ontological status of drumming: the Kandyan Convention of 1815, which granted sovereignty over the island to the British, stipulated that Buddhist traditions could not be molested; the legal debates on processions I discuss below often revolved around whether the rights of the natives to perform cultural practices could be trumped by the colonial requirement to maintain law and order, even if the aesthetic idea that music was harmless informed the courts' decisions.
6 In what follows, I draw out the often unacknowledged relations between the idea that music is a harmless amusement and the perceived ability of music to express personal and communal identity. One goal of this essay, explicitly confronted in the conclusion, is to link our current 'music and identity' paradigm to the ideology of autonomous music it nominally strives to debunk through its consideration of 'social context'.

The terms of this relation, which I view as due to the inability of our musicological language to separate 'music' from 'the self', is perhaps the most important musicological contribution made by an attention to the Buddhist musical gift.

7 While I focus here on *two* types of confrontation – that between Buddhists and Christians, and between Buddhists and Muslims – the events I describe must be situated in a broader field of communal disturbances and debates on musical processions in Ceylon during this period (the question of 'tom-toming', for example, was not limited to Buddhists, but extended to the issue of whether Hindu drumming at weddings could continue past 11pm); see Michael Roberts, 'Noise as Cultural Struggle: Tom-Tom Beating, the British and Communal Disturbances in Sri Lanka, 1880s–1930s', in Veena Das (ed.), *Mirrors of Violence: Communities, Riots and Survivors in South Asia* (Delhi: Oxford University Press, 1990), pp. 240–85, at p. 246. These debates are well known to Sri Lankanists and are adequately covered in that historical literature (see also John D. Rogers, *Crime, Justice and Society in Colonial Sri Lanka* (London: Curzon Press, 1987), pp. 176–202). My study here intentionally overlaps with Roberts's pioneering work: my goal is to summarize his and others' analyses of these events for a musicological readership, and to draw the relevant implications for musicology. A comparison of this study with Roberts's will show we have a difference in emphasis, for while he focuses on the capacity of noise to incite violence, I bring out the interrelations between musical aesthetics and colonial law, through my situating of the Buddhist musical gift (drawn from ethnographic fieldwork: 2004, 2006–08) in the discursive fields of musicology and liberalism.

8 The actual content, form and function of the Buddhist musical gift, its ability to transform attention to the gift in anthropology and literary theory and a consideration of how a study of the musical gift can contribute to theories of musical economy (Marxist and otherwise) are topics I discuss elsewhere. See Jim Sykes, 'The Musical Gift: Sound, Sovereignty and Multicultural History in Sri Lanka', PhD Dissertation (University of Chicago, 2011).

9 The events I describe here are the most significant communal disturbances Sri Lanka's entire colonial period (1505–1948). The fact they were 'caused' by music and by legal debates about music is a curious fact of history that I believe has a legitimate place in the global history of music.

10 The Catholics were of Ceylonese descent, the result of centuries of missionary work on the island.

11 Stanley J. Tambiah, *Leveling Crowds: Ethnonationalist Conflicts and Collective Violence in South Asia* (Berkeley: University of California Press, 1996), p. 51.

12 Nira Wickramasinghe, *Sri Lanka in the Modern Age: A History of Contested Identities* (Honolulu: University of Hawaii Press, 2006), p. 103. It seems that a sudden thunderstorm halted the violence.

13 Stirratt goes so far as to say that, 'By present-day Sri Lankan standards the level of violence [in the Kotahena episode] was low and injuries few. Yet at the time the riot was seen as being of major importance and it marked the beginning of a series of violent clashes between groups defined in terms of religious identity which were to last for the next 40 years'. R.L. Stirrat, 'Constructing Identities in Nineteenth-Century Colombo', in John R. Campbell and Alan Rew (eds), *Identity and Affect: Experiences of Identity in a Globalising World* (London: Pluto Press, 1999), pp. 39–63, at p. 40.

14 Properly speaking, such an occasion is not a celebration but a ritual: an eye-setting ceremony for a Buddha statue, it is believed that over the course of the ritual the statue learns its own life story (that is, the life story of the Buddha) and the Buddha's teachings (the *dhamma*) by listening to the recitation of monks. See Richard Gombrich, 'The Consecration of a Buddhist Image', *The Journal of Asian Studies*, 26/1 (November 1966): 23–36.

15 G.P.V. Somaratna, *Kotahena Riot 1883: Religious Riot in Sri Lanka* (Nugegoda: Deepanee Publications, 1991). Tambiah, *Leveling Crowds*, p. 49.

16 It is common in Sri Lanka to hold sessions of *pirit* that go on for a long time and repeat an auspicious number of times, for such sessions are particularly powerful, but especially long ceremonies would not typically be held in public. As one informant told me, the chanting of the Buddha's teachings (*dhamma*) is like an electric charge: when people and objects are connected to monks (*bhikkhus*) with a sacred thread (*pirit nul*), the protective power of the *dhamma* is channelled through the thread and to the people and objects, giving them protective power (the word for Buddhist chant, *pirit*, means 'protection'). See Sykes, 'The Musical Gift', Part One.

17 Tambiah, *Leveling Crowds*, p. 50.

18 This demonstrates that a legal structure for regulating processions was in place, but not enforced.

19 Tambiah, *Leveling Crowds*, p. 51; Somaratna, *Kotahena Riot.*

20 Rogers, *Crime, Justice and Society*, p. 177.

21 In the days after the incident a few other incidents occurred, such as the burning of a chapel in Dehiwala, another neighbourhood of Colombo (Tambiah, *Leveling Crowds*, p. 120). The day after the riot, the military turned back several previously planned *peraheras* as a precaution (Ibid.). Meanwhile, it was discovered that monks associated with Gunananda had staged processions on Easter Sunday in the towns of Ratnapura and Balangoda, with the intention of causing trouble for Christians in those towns. Thus the Kotahena celebrations were a planned confrontation after all. While the Sinhala Buddhist community blamed Catholics for the violence, the Commission of Enquiry, appointed by Governor Longden to look into the incident, decided the courts did not have sufficient evidence to try the Catholics accused as perpetrators. Kingsley de Silva writes that, 'although [the Commission] held that the Roman Catholics had indeed attacked the Buddhists, Gunananda's fiery speeches and "the fervour with which the Buddhists were conducting their activities" were described as factors that had contributed to provoking the riots'. See Kingsley De Silva, *A History of Sri Lanka* (Berkeley: University of California Press, 1981), p. 345.

22 Rumours played a prominent role in the Kotahena riot: Gunananda was rumoured to be seen stepping on the bible in his fiery sermons, and Buddhists were said to have marched in a *perahera* with an image of the Virgin Mary. 'At Kotahena, both sides believed that their respective priests had been attacked, and the Catholics believed that their sacred symbols were being insulted'. Tambiah, *Leveling Crowds*, p. 53.

23 Thus the revival has come to be called 'Protestant Buddhism', a term made famous by Gananath Obeyesekere in his 'Religious Symbolism and Political Change in Ceylon', *Modern Ceylon Studies*, 1/1 (1970): 43–63.

24 Tambiah, *Leveling Crowds*, p. 48.

25 In the years after Kotahena, two foreigners sympathetic to Buddhism (the American Buddhist convert Colonel Henry Steel Olcott and the Russian Theosophist Madame Blavatsky) played a key role in furthering the movement that Gunananda had started, particularly through their mentoring of the era's most influential Buddhist reformer, Anagarika Dharmapala. See especially George Bond, *The Buddhist Revival in Sri Lanka* (Columbia: University of South Carolina Press, 1988). Although I do not have space to consider him here, Olcott became personally involved in defending the Buddhists' case in the aftermath of the Kotahena riot. He went personally to London to ask them to rescind clauses 69 and 90 of the Police Ordinance, which stipulated the possibility of drumming to become legally defined as noise (pp. 15–16). For more on Olcott's role in the aftermath of Kotahena, see Stephen Prothero, *The White Buddhist: The Asian Odyssey of Henry Steel Olcott* (Bloomington: Indiana University Press, 1996), pp. 13–19 and Tambiah, *Leveling Crowds*, p. 54.

26 By all accounts Gunananda won these debates, although his audience was surely composed mostly of Buddhists. See Kitsiri Malalgoda, *Buddhism in Sinhalese Society, 1750–1900* (Berkeley: University of California Press, 1976), pp. 218–31.

27 Wickramasinghe, *Sri Lanka in the Modern Age*, p. 116. 'The Buddhist-Christian controversies' were vain in that, 'rather than furthering mutual understanding, they degenerated in opposition to each other, whether in favour of Buddhism or Christianity'. See also R.F. Young and G.P.V. Somaratne, *Vain Debates: The Buddhist-Christian Controversies of Nineteenth-Century Ceylon* (Vienna: Publications of the De Nobli Research Library, 1996), p. 217. The riots of 1883, Young and Somaratne argue, were the 'recoil of the *dhamma*', a phrase that refers to the Sri Lankan idea that those who use the Buddha's teachings (*dhamma*) for self-promotion 'will find that, like a snake grasped by the tail, it will turn and bite'. Rupert Gethin, Review of *Vain Debates*, in *Bulletin of the School of Oriental and African Studies, University of London*, 62/3 (1999): 574–5, at p. 575.

28 Kotahena, a neighbourhood famous even today for its Catholic population, traces its religious identity to the era of Portuguese colonial rule (1505–1650). During this period, the town emerged as part of a 'Christian littoral' on the island's west coast, whose hub was Negombo, a Christian-majority town forty kilometres north of Kotahena. The population of Kotahena (according to the 1881 census) was 26,696 (Wickramasinghe, *Sri Lanka in the Modern Age*, p. 117).

29 While Protestant missionaries may have been the significant instigating force for Buddhist resentment of Christians, Catholics also waged polemical attacks on Buddhists, for example through pamphlets denigrating Buddhism. Rogers remarks, 'that the first religiously-motivated violent clashes were between Buddhists and Catholics, not Buddhists and Protestants, may be partially explained by the relative size of the Catholic and Protestant communities. There was no numerically significant Protestant community with which to do battle'. Rogers, *Crime, Justice*, p. 176.

30 Roberts, 'Noise as Cultural Struggle', pp. 240–85.

31 Rogers, *Crime, Justice*, p. 180. Roberts documents this happening on the following occasions: Kotahena, 22 May 1883; Alutgama Church, 22 May 1889; Maggona Church, 25 May 1889; Holy Cross Church at Kalutara, 29 May 1889. Roberts, 'Noise as Cultural Struggle', pp. 257–8. All of these places are located on Sri Lanka's southwest coast, near Colombo.

32 For it appears as a historical rupture summed up in an event, akin to Alain Badiou's definition (described here by Richard Elliott): 'For Badiou, an event is something that can – but will not necessarily – occur at an evental site and that occurs as a complete break with the continuum of being. For Badiou the fields of human interaction where such events can emerge are those of science, art, politics and love. An event is a creative, assertive act that breaks with what has gone before and sets in motion a new truth that in turn creates subjects who show fidelity to it'. Richard Elliott, 'Popular Music and/as Event: Subjectivity, Love and Fidelity in the Aftermath of Rock 'n' Roll', *Radical Musicology*, 3 (2008), http://www.radical-musicology.org.uk/

33 Quite the contrary, it is known that Christians and Muslims (not to mention the island's Tamil Hindus) have their own processions containing music that are performed to this day. Muslims in Kandy used to haul a pagoda (*kudu*) through city streets in an annual procession (I have met older Muslim musicians in Kandy who remember this from their childhood), but this tradition has been stopped by the Wahhabis. Sufi Muslims in Sri Lanka continue to hold public processions today, however, such as the annual flag-raising ceremony at the Beach Mosque in Kalmunai, where Sufis proceed from a mosque in town to the Beach Mosque by the sea, singing with tambourines (Sykes, 'The Musical Gift', Part Two). Christians today in Sri Lanka march through city streets to honour Saints, processions that involve the singing of Christian hymns (Ibid.). All these processions, including the Buddhist ones, are probably descended from the Hindu temple festival tradition, which involves circumambulation of the boundaries of the village (*ur*) with the idol carried on a chariot accompanied by drummers. See Daniel E. Valentine, *Fluid Signs: Being a Person the Tamil Way* (Berkeley: University of California Press, 1984).

34 Daniel E. Valentine, *Charred Lullabies: Chapters in an Anthropography of Violence* (Princeton: Princeton University Press, 1996), pp. 43–71. This is *not* the same, however, as saying that these objects and places became important for the first time (far from it; rather, the point is their importance has been redefined). Sinhala Buddhist music in previous generations would have been less metonymic of tradition, and more associated with its function as a sacred music offering (*sabda pujava*), a point Daniel makes as well: 'The [*Esala Perahera*] processes through designated streets. That much is not new. It happened this way even in the days of the kings. But of course the British made it more efficient. The ways in which "crowds" (not believers or citizens) are "controlled" are new. The language ("crowd control") is that of a new disciplinary order, a new disciplinary space. The "spectators" are cordoned off onto the sidewalks; the streets are kept free for the "performers"'. Daniel, *Charred Lullabies*, p. 61. This statement points nicely to the rapprochement between governmentality (here the regulation of crowds) and the 'invention of tradition', which in this case necessitated the demotion of the gift from sacred offering to the music of 'performers' marching for 'spectators'.

35 The castes that traditionally own Sinhala music and dance (*Berava, Oli, Naeketi*) still exist in Sri Lanka today, but nowadays they do not have a monopoly on drumming and dance; these traditions have been institutionalized in universities and are open to all. See Susan Reed, *Dance and the Nation: Performance, Ritual, and Politics in Sri Lanka* (Madison: University of Wisconsin Press, 2010).

36 Roberts, *Exploring Confrontation*, p. 152.

37 Italics my own. This law was initially applied only to 'towns', and was first enforced only in Colombo (Roberts, *Exploring Confrontation*, p. 152). However, the law was extended island-wide after 1898, a growth in colonial governmentality that, in Roberts's view, was due to 'a hardening of British attitudes during the 1890s, 1900s, and 1910s': 'One can sense the growth of a Curzonian rigidity, an attempt to maintain imperial authority at all costs'. Roberts, 'Noise as Cultural Struggle', p. 255. Of course, the growth of the population and the increasing development of towns outside Colombo, particularly due to the development of the railway system, must also have played a role.

38 Roberts, *Exploring Confrontation,* p. 152.

39 In South Asia in particular, this division was gendered, so that the feminine domestic sphere came to be seen as the site of an authentic Indian identity untouched by colonialism. For an argument relating this to the Indian female voice, see Amanda Weidman, *Singing the Classical, Voicing the Modern: The Postcolonial Politics of Music in South India* (Durham: Duke University Press, 2006).

40 Roberts, *Exploring Confrontation*, p. 169.

41 For the classic studies, see Gananath Obeyesekere, *The Cult of the Goddess Pattini* (Chicago: University of Chicago Press, 1984) and Bruce Kapferer, *A Celebration of Demons: Exorcism and the Aesthetics of Healing in Sri Lanka* (Washington, D.C.: Smithsonian Books, 1983).

42 Michael Ondaatje, *Running in the Family* (New York: Random House, 1996), pp. 51–2: 'Ambalangoda was the centre for devil dances and exorcism rites, but this charmed group (upper class Ceylonese) were part of another lost world. The men leaned their chins against the serene necks of the women, danced a waltz or two, slid oysters into their partner's mouths. The waves on the beach collected champagne corks. … Deeper inland at midnight, the devil dances began, drums portioned the night. Trucks carrying horses to the next meet glared their headlights as they passed the group by the side of the road. The horses, drummers, everyone else, seemed to have a purpose. The devil dances cured sickness, catarrh, deafness, aloneness. Here the gramophone accompanied a seduction or an arousal, it spoke of meadows and "little Spanish towns" or "a small hotel", a "blue room"'. ('Devil dances' are colonial era English for the Sinhala rituals called *yak tovil*.)

43 Note that the discussion of regulation has mandated a switch in my own language, from 'gift' and 'sonic offering' to 'performance'.

44 Roberts, *Exploring Confrontation*, p. 169.
45 Not to mention that in the nineteenth century horses were a common sight in urban areas, and police officers on horses were hired to watch especially large *peraheras*.
46 De Silva, *A History of Sri Lanka*, pp. 230–31.
47 All of these statements are found in Roberts, *Exploring Confrontation*, pp. 155–6, 166 and 178.
48 This is shown by the statement on liability in the Procession Form: 'The applicant is warned that this Permit will not free him from liability to an action for damage, or to a charge of manslaughter, if any damage is done or life lost owing to anything the applicant may do or cause to be done during the Procession'. Cited in Roberts, *Exploring Confrontation*, p. 178. Legal blame falls on the organizer or representative of the organizer of the procession (one who signs the form), rather than, say, Buddhist drummers who incite hostility through their intentional raising of volume outside a church or mosque.
49 Thus there is a significant problem with the general tenor of the historiography, which describes the Kotahena riot as a 'symbolic struggle over *space*'. Wickramasinghe, *Sri Lanka in the Modern Age*, p. 120.
50 Wickramasinghe, *Sri Lanka in the Modern Age*, p. 120.
51 Roberts cautions us against assuming that all Coast Moors, and none of the indigenous population of Ceylon Moors, were Wahhabi: clearly there was overlap (Roberts, *Exploring Confrontation*, pp. 183–212). He does note, however, that many Coast Moors came from Kalyapatinam in South India, a town known at the time as home to many Wahhabis.
52 I stick to the term 'Moor' here, although in the present day they are referred to (and prefer) 'Muslim'. The term 'Moor', while having colonial connotations in the present day, was the term used to refer to Muslims until just the last generation.
53 Early commentators on the Buddhist–Muslim disturbances of 1915 compare the Muslims to the Jewish populations in Europe. See Armand de Souza, *One Hundred Days in Ceylon: Under Martial Law 1915* (Colombo: Young Socialist Publications, 1916, reprinted 2006) and Ponnambalam Ramanathan, *Riots and Martial Law in Ceylon, 1915* (London: St. Martin's Press, 1916).
54 This built on a much older tradition of Muslim businesses in the port cities, as traders from the Arabian peninsula are known to have settled in Sri Lanka well before the birth of Islam.
55 There was also a fear that the Coast Moors were seducing Sinhala women and converting them to Islam, which probably stemmed from the fact that the Coast Moors usually left their families in India, to do business on a temporary basis in Ceylon. 'In his village store', writes De Souza, 'the Coast Moor achieves acquaintance with Sinhalese young girls and either seduces them or, worse still degrades them into the condition of his mistress, eventually forcing Buddhist girls the status and the religion of Mohamedan women'. De Souza, *One Hundred Days in Ceylon*, p. 7.
56 Roberts, *Exploring Confrontation*, p. 160. These clashes were not just between Buddhists and the Coast Moors, but also occasionally between Buddhists and members of the indigenous Ceylonese Moorish community (although the former instances are my focus here).
57 Roberts, *Exploring Confrontation*, p. 160.
58 Roberts, *Exploring Confrontation*, p. 160.
59 Ibid.
60 Ibid., p. 167.
61 Charles Blackton in Tambiah, *Leveling Crowds*, p. 36.
62 *Leveling Crowds*, p. 36. This action demonstrates that noise has at least as much to do with *place* as it does decibel level.
63 Roberts, *Exploring Confrontation*, p. 202.
64 Kataragama is a god worshipped throughout Sri Lanka, not just by Sinhala Buddhists but also by Tamil Hindus, who call him Murugan or Skanda.

65 Ramanathan, *Riots and Martial Law*, p. 4.
66 It was during the period of waiting for this decision that the third major procession dispute of these years occurred, in Kurunegala. On 27 January 1915, a *perahera* held to honour a Buddha image was attacked, and some damage occurred to the Buddha image. Ibid., p. 5.
67 Ibid., pp. 5–6.
68 Ramanathan, *Riots and Martial Law*, pp. 5–6.
69 The significance of this event for Sri Lankan history is largely due to the perception that the British response to the riots was heavy-handed. Their incarceration of thousands, including many prominent Ceylonese civic leaders who had no role in the riots (out of a fear, incorrect at that, that the riots were anti-colonial in nature), fomented significant anti-British sentiment that legitimized the Ceylonese drive for self-governance. As the riots occurred during World War I, there was also fear the Germans were responsible for igniting the disturbances.
70 Ramanathan, *Riots and Martial Law*, p. 15.
71 For example, Lankan Buddhists do worship deities, and they do strive to 'repudiate sin', in as much as beneficent actions are considered highly esteemed and necessary for moving up the *karmic* ladder. De Souza's role is as a Western-educated, Ceylonese Christian who had grown up in the Buddhist-majority country, and he was acting as a mediator between East and West.
72 In fact, part of the resentment between Moors and Buddhists during this Wesak holiday was due to the Moors' feelings the Buddhists' free offerings of food in the *dansalas* was undercutting their business.
73 These were found to be a particularly strong factor for the spread of the riots to Colombo, where violence is documented as being instigated by a Moor's raising of the price of tea. See Kumari Jayawardana, 'Economic and Political Factors in the 1915 Riots', *Journal of Asian Studies*, 29/2 (February 1970): 223–33.
74 See particularly Robert Kearney, 'The 1915 Riots in Ceylon: A Symposium', *Journal of Asian Studies*, 29/2 (February 1970): 219–22; Charles Blackton, 'The Action Phase of the 1915 Riots', *Journal of Asian Studies*, 29/2 (February 1970): 235–54; P.T.M. Fernando, 'The British Raj and the 1915 Communal Riots in Ceylon', *Journal of Asian Studies*, 29/2 (February 1970): 245–55.
75 Roberts, 'Noise as Cultural Struggle' and *Exploring Confrontation*.
76 Tambiah citing Fernando, 'British Raj', *Leveling Crowds*, p. 56.
77 Cited in David Scott, *Conscripts of Modernity: The Tragedy of Colonial Enlightenment* (Durham: Duke University Press, 2004), p. 38.
78 This is Dahlhaus's famous terminology: 'history of Art' refers to musicology, and 'History of art' refers to the disciplines of history, sociology and so forth. Karl Dahlhaus, *Foundations of Music History* (Cambridge: Cambridge University Press, 1983).
79 One lesson from the story I present here is that the inability of the British to recognize the musical gift legitimized Muslim legal arguments in court.
80 Dipesh Chakrabarty, *Provincializing Europe: Postcolonial Thought and Historical Difference* (Princeton: Princeton University Press, 2007).
81 Sykes, 'The Musical Gift'.
82 Of course, the other glaring perspective missing from this study is the long history of theorizing the gift in anthropology (see again Sykes, 'The Musical Gift', for a consideration).
83 Ritu Birla, *Stages of Capital: Law, Culture, and Market Governance in Late Colonial India* (Durham: Duke University Press, 2009).
84 Birla, *Stages of Capital*, p. 72. Birla notes that the scholarship on Hindu institutions has generally accepted the colonial courts' claims for non-interference in Hindu institutions, therefore emphasizing 'the dissimulations of colonial sovereignty, which by resisting direct regulation over indigenous institutions performed its control benignly and circuitously' (Ibid., p. 73). Religion, then, like music, was perceived as lying in

the private domain of traditional culture, and thus the way colonialism *did* regulate religion – particularly through legal channels – has also been less recognized than it should be by scholars.

85 Birla, *Stages of Capital*, pp. 77–8.

86 This is a particularly strong notion in a Buddhist cosmology, where the idea of the Buddha or deities (who are future Buddhas) *consuming* a gift is ludicrous; as opposed to the sensuous enjoyment of music by Hindu deities, in the Sinhala Buddhist system the gods 'turn the gift down', so that the protective powers of sound may be offered to the whole community.

7 Cannons, Church Bells and Colonial Policies

The Soundscape in Habsburg Bosnia-Herzegovina

Risto Pekka Pennanen

Owing to their Ottoman past and to contemporary Austro-Hungarian colonial policies, Bosnia and Herzegovina[1] were exceptional areas in late nineteenth- and early twentieth-century Europe. In particular, their urban soundscapes were a mixture of Ottoman and Central European practices. The colonial administration, from the beginnings of the occupation, did not abandon the traditional Ottoman cannon signals for midday and fire alarms, for instance, while the presence of three main religions in Bosnia – Islam, Serbian Orthodoxy and Catholicism – meant that the spectrum of everyday religious soundmarks was more diversified than elsewhere.[2] Through standardized sets of signals, the colonial authorities provided these religions with a much more pronounced position in the soundscape than elsewhere in the Empire or in the Balkans. Another important component of the Bosnian urban soundscape was, however, common to all Habsburg lands: the sound signals of the imperial festivals and salutes for high-ranking guests.

This chapter addresses the hitherto unexplored themes of the use of church bells and the extra-musical sounds of rifle fusillades and cannon signals during religious and imperial holidays and other special occasions in Bosnia as well as their connection with Habsburg colonial policies and international politics. The contemporary Bosnian press forms the main body of sources for this study, but the German-language press of the Empire is also important for comparison and additional information. The nature of the available sources raises certain questions, however, and plotting such historical soundscapes is a difficult task even in a society with as rational and relatively predictable an administration as had Habsburg Bosnia. These sources do not, for instance, describe the attitudes of the native population towards drastic and sudden changes in their soundscape. Nor do they take account of the instability of the acoustic milieu, which was unstable since the climate and atmosphere would have contributed significantly to the acoustic profile of military salvos, bell ringing and the like; particularly in Sarajevo, thick mist and snow dampened sounds in wintertime.[3] Similarly, they do not comment on any systematic peaks in soundscape activity. As a crucial source-critical observation, it is thus important to note that over time newspaper reporters began taking salutes and cannon signals for granted and therefore seemingly ignored them.[4] Insomuch as the sound of church bells was taken for granted by writers from Catholic Croatia-Slavonia and Central Europe, the Bosnian

German-language press in particular failed to mention their ringing. Departures from previous practices in the descriptions of annual celebrations and their soundscapes, moreover, may be accounted for as either errors in reporting or a reflection of actual transitions in colonial policies. Yet, despite these issues, meticulous close readings of the sources enable many of the ambiguities to be resolved, and the sources nevertheless provide a rich, and previously untapped, account of this unique soundworld that was shaped by religious diversity, colonial policy and sonic articulations of colonial power. Habsburg Bosnia, then, provides an exemplary case study in the manipulation of sound in the imposition of colonial policy, and in the articulation of dominant political power in late nineteenth- and early twentieth-century Europe.

Historical Background

Bosnia and Herzegovina had belonged to the Ottoman Empire since 1463, but in 1878 at the Congress of Berlin, the Great Powers gave the Habsburg monarchy the right to occupy and administer the provinces. Consequently, Bosnia remained nominally under Ottoman suzerainty until 5 October 1908 when Austria-Hungary annexed the area and held it as crown land until the end of the Great War. Habsburg rule in Bosnia was imperialist in nature, and the officially pronounced goal of the colonial administration was to bring Bosnians over to the values of 'true' civilization, in other words Central European high culture. Simultaneously, however, the administration declared its desire to preserve the essence of Bosnian culture instead of Germanizing the provinces.[5] This policy notwithstanding, the pretension of respecting local traditions was not devoid of problems – even in the field of soundscape – as we shall see below.

The person behind the construction of colonial policies and the administrative apparatus was Benjamin von Kállay, a Hungarian who served as Joint Minister of Finance (Ger. *gemeinsamer Finanzminister*) in Vienna from 1882 to 1903, and was, as minister, responsible for the administration of Bosnia. Because of the nature of the occupation, the Habsburg authorities sought to restructure the relations between the Bosnian religious communities, whose clerical and political authorities were all based abroad. Thus they created new Catholic, Muslim and Serbian Orthodox religious hierarchies and rearranged the existing ones, thereby weakening the ties between Bosnian groups and outside authorities, on the one hand, and securing a dominant influence over the Bosnian confessional hierarchies, on the other.[6]

The authority of the Emperor-King constituted the most vital rallying point of the Austro-Hungarian Empire, which was why the colonial administration imported the cult of Franz Joseph from Vienna to Bosnia soon after the occupation. Various imperial holidays celebrated the major events in Franz Joseph's – and subsequently in his successor, Karl's – life: the Emperor's birthday, coronation day, saint's name day and wedding anniversary were major feasts in Bosnia.[7] The invented traditions of the emperor cult – illuminations, flags, tattoos, performances of the Imperial Hymn, congratulatory receptions, church bell ringing and salutes – served as models for other neo-traditions the regime created to support its colonial policies in Bosnia. As Eric Hobsbawm has famously noted, invented traditions are sets of practices which aim to establish certain values and norms of behaviour through repetition, and this recurrence automatically implies continuity with the past, even though the continuity is largely fictitious.[8]

A considerable part of the Kállay regime's religious policies sought to curry the sympathy and trust of the colonized population by creating various invented traditions connected with their religion. These neo-traditions included the ceremonial participation of high-ranking colonial officials in Christian religious ceremonies and receptions on the religious holidays of all confessions. Military brass bands playing confession-specific music and the Imperial Hymn *Kaiserhymne* (also known as the *Volkshymne*) by Joseph Haydn (1732–1809) as well as salutes and cannon signals were integral elements of these sets of neo-traditions for the holidays. Often the administration superimposed the invented traditions onto established religious rituals, thus introducing the symbols of state protection alongside various symbols of power.

Kállay had two special motives for his policies towards Bosnian Muslim traditions. Firstly, the motives were related to domestic policy: the administration attempted to win the support of the majority among the Bosnian population, and after Catholics, Muslims were logically the second most important ethno-religious group for this purpose. Besides, since Vienna had undertaken to honour Bosnian Islamic customs and traditions in the 1879 Novi Pazar Convention with the Porte, it was important not to disappoint Bosnian Muslims. In confidential documents, however, Kállay speculated on the possibility of Muslim apostasy with Muslims converting to Catholicism.[9] Due to Austro-Hungarian colonial policies, the occupation of Bosnia led to a different development from that in post-1878 Sofia in the Principality of Bulgaria, where almost all mosques were demolished by the late 1890s and church bell ringing had completely replaced calls to prayer.[10]

Secondly, based on its long-term foreign policy strategy, the regime strived to pave the way for Austro-Hungarian future expansion in the Balkans by showing the Muslims of the Sandžak of Novi Pazar, Kosovo and Albania how good its relations were with Bosnian Muslims. The policy was somewhat successful: an Austrian consul reported in 1896 how cannon salutes for the *Bajram* feast marking the culmination of the *Ramazan* month of fasting in Sarajevo impressed an Albanian leader and his fellow tribesmen.[11]

The beginnings of the policies date from the early years of the occupation: as soon as the situation in the occupied twin provinces settled down, the administration introduced the Austro-Hungarian system of bureaucratic control of music, musicians and the soundscape in general. The system sought, among other things, to prevent disturbances of night-time peace (Ger. *Nachtruhe*) by regulating the official closing times of restaurants, public houses and cafés and by taxing noise (Ger. *Spektakelgebühr*). Those public houses which sought to extend their opening hours had to apply for it and pay a fee.[12] As we shall see, however, the definition of a disturbing noise was not unambiguous. The relationship between sound and power illustrates the problem of defining noise in a concrete manner. R. Murray

Schaffer uses the term 'Sacred Noise' to denote noise associated with such strong social power and dominance that its authority enables it to make noise with impunity. Schafer notes that the sphere of 'Sacred Noise' expanded from religious sounds to those of profane rulers and even, during the Industrial Revolution, to industrial noise.[13] Such Noise is always linked with sacred or profane power, *inter alia* in Habsburg Bosnia, in Islamic calls to prayer, Christian church bell ringing, cannon salutes and signals in the early hours during the Serbian Orthodox Easter Sunday procession of the cross and the Islamic fast month of *Ramazan* and cannon salutes during the early-morning tattoos of the Emperor's birthday. As for industrial noise, the sounds of technology certainly disturbed the public although available sources rarely mention it. However, a German-language Sarajevo newspaper did write that 'the steam exhaust pipe of the city electricity plant made a constant noise equal to the roll of a million drums or the trumpets of Jericho'.[14]

Church Bells and the Soundscape

Prior to the mid-nineteenth century, publicly audible metal bells were very few in Bosnia. Most towns had an Islamic clock tower *sahat kula* (< Turk. *saat kulesi*) with a dial and chiming bell indicating the lunar hours for the daily prayers and, during the month of *Ramazan*, the hours of the fast. The Ottoman authorities restricted the use of Christian metal church bells but allowed wooden clappers (Gr. *semandron*; Serb. *klepalo*) in monasteries and small hand bells in religious services. In Bosnia, Orthodox Christian folk tradition favoured hand bell ringing if there was the threat of a storm and around the house on the Annunciation (Serb. *Blagovijest*) to ward off snakes.[15]

The change of the religious soundscape in Bosnia emerged during the period of the Ottoman *Tanzîmât* (Reorganization) reforms (1839–76) which improved the position of non-Muslims in the Empire. In 1856, Sultan Abdülmecid I (1839–61) announced *Islahat Fermanı*, an influential decree, which sought to protect the rights of all Ottoman religious communities. It was not always well received and enforced in the provinces. For instance, to avoid provoking public disorder, many local authorities declined Christian demands for permission to ring metal church bells rather than the traditionally allowed dull wooden ones.[16] Regardless of the problems, the reform resulted in the gradual dissolution of the monopoly of *ezan*, the Islamic call to prayer, and the chimes of the Islamic clock towers in the religious soundscape of religiously mixed areas.

Finally, observing the 1856 decree in 1860, the Ottoman authorities in Bosnia permitted the building of a bell tower for the Franciscan monastery church of St. John the Baptist in Kraljeva Sutjeska and the ringing of the new metal bells. In Sarajevo, construction of the new Orthodox cathedral was completed in 1872. The final phases of the construction project met fierce resistance from a group of conservative Muslim clergy which opposed the tall belfry and the bells of the cathedral as well as the installation of a small bell in the old Orthodox church. Both churches received their bells, however.[17] After analysing these discords in his 1924 dissertation, the future Yugoslav diplomat and Nobel Prize winning writer Ivo Andrić (1892–1975) claimed that, due to their conservatism, Bosnian Muslims had preserved the abhorrence of church bells typical of the Bosnian pre-Ottoman Bogomil sect.[18]

The Austrian occupation of Bosnia led to a boom in the construction of new Catholic and Orthodox churches with belfries. The demand for bells created a new market for them, which became obvious as early as 1880: Hungarian foundries, such as Anton Novotni's Bell Foundry in Temesvár (nowadays Timişoara in Bánát, Romania) and Stevan Bota's Bell Foundry in Versecz (nowadays Vršac in Vojvodina, Serbia) in Temes County, Hungary, advertised their products in a Sarajevo newspaper.[19]

The new policy of religious equality was instrumental in the realization of the colonial administration's political objectives. Take, for instance, the dedication of six church bells, religious books and other artefacts which the Emperor had donated to the Serbian Orthodox church in Gacko, Herzegovina, on the Orthodox New Year on 13 January 1880. A newspaper article describes how a crowd of 1,000 believers cheered for Franz Joseph and his representative. The tenor of the text is clearly propagandistic; the ringing of the new church bells symbolizes the freedom which the occupation has brought about:

> The crowd gathered already at dawn that day to wait to hear the happy and sweet sound of the bells, glory and freedom, which these oppressed people have lacked for centuries. ... After the bells and other precious imperial presents were dedicated, a divine service was celebrated in the presence of the imperial representative and his retinue [i.e., the tenth *Jaeger* battalion]. The bishop personally rang the bells for the first time, and the battalion fired fusillades. The soldiers paid homage in accordance with regulations, and the roar of 101 cannon salutes from the rampart of the town shook the *karst* area of Gatačko Polje.[20]

The dedication of the Catholic cathedral on 14 September 1889, which included a military parade and a procession with a military band, cannon salutes, bell ringing and a mass, marked a turning point in the religious soundscape of Sarajevo. Actually, the bells rang for the first time on the evening of 3 July, but, strangely enough, the press apparently did not mention the test. Another interesting detail is that although the peal of six could play triads as well as melodies in the Italian manner, no known source mentions melodies being played.[21] Besides bell ringing, another novelty in the cathedral changed the soundscape: the clocks in the belfries chimed the hours. One can only wonder whether this new practice gave rise to protests from Muslims since it disrupted the chiming from the Islamic clock tower.

Contrary to other Habsburg lands, church bells seem to have played a somewhat limited role in Bosnia since no sources mention their ringing as an alarm signal for fire or other emergency in the twin provinces, where the bells served two main functions. Firstly, they rang for holy services, christenings, weddings and funerals, religious holidays and inaugurations of Christian religious leaders. Secondly, the bells marked non-religious events usually connected with the court.

For example, when Crown Prince Rudolf and his wife Princess Stéphanie visited Banja Luka in June 1888, church bells rang and cannons fired salutes. The ringing procedure for Bosnian Catholic churches during the fifteenth anniversary of Franz Joseph's coronation in December 1898 was published in a newspaper. The archbishop's administration ordered the ringing of all bells in the evening of the day before the celebration and twice – at noon and in the evening – at the actual celebration for half an hour, respectively.[22]

The Great War brought about another change in the religious soundscape. The confiscation of large church bells for the war effort in 1916 almost restored the late Ottoman religious soundscape in Bosnia. Once again, the *ezan* became a more dominant soundmark than the bell ringing. The confiscation measures must have disquieted the Christian spheres since war propaganda touched on the issue. To silence critical voices, the essay 'The Bell and the Cannon' (*Zvono i top*) by the unidentified writer S.R., writing in the Entertainment Supplement of the government newspaper *Sarajevski list*, portrayed the pendulum-like transformation of copper from church bells into cannons and back again as an ancient and totally normal phenomenon.[23]

Cannon Signals and Religious Salutes

Despite the triumphal march of the church bells, the colonial administration maintained the fundamental characteristics of the pre-1878 soundscape, including celebratory gunfire and various cannon signals and salutes for Islamic holidays, often absorbing them into colonial policies. The Ottomans had marked the Islamic religious holidays of *Bajram*, *Kurban bajram* and *Mevlud* with cannon salvos throughout the nineteenth century. The Ottoman authorities also announced the departure of pilgrims to Mecca as well as the hours of fasting and eating during *Ramazan* by firing cannon salvos.[24] The Kállay regime further developed the system of Islamic salutes and extended the salvos to Christian holidays.

Signals and salutes were not always official cannon shots; it was customary for Bosnian males, regardless of ethno-religious background, to use celebratory gunfire, an old Near Eastern tradition. Such gunfire occurred during feasts such as weddings and the Orthodox *krsna slava* (Patron Saint's Day) as well as on *teferiči* (picnics) during the folk holidays of *Đurđevdan* (St George's Day) and *Ilindan* (St Elias' Day). After the outbreak of the Bosnian annexation crisis, the authorities banned the old custom of celebratory gunfire and made it punishable with a fine in December 1908.[25]

Very soon after the occupation, the Austro-Hungarian administration adopted the Ottoman tradition of firing the noon gun every day in Bosnian towns. This practice was by no means restricted to the Ottoman Empire since it was common practically everywhere in Europe. For instance, since 1735, a cannon has fired at noon in the Naryshkin Bastion of the Peter and Paul Fortress in St Petersburg. In Sarajevo, the artillery fired the midday gun and other signals as well as various salutes from the Yellow Bastion, an Ottoman fortification in the old part of the city.[26]

In Sarajevo and Mostar – probably in other urban areas as well – the artillery used cannon signals for marking the whereabouts of a fire. The Sarajevo system of signalling was as follows: two cannon shots marked a fire in the Fortress, three cannon shots marked a fire on the right bank of the Miljacka River, four cannon shots marked a fire on the left bank and three cannon shots during total darkness or thick mist meant that the fire was on either riverbank.[27] In addition, the press also mentions the use of the fire brigade's bugle call in Sarajevo and Mostar and the foghorn of the railway station and an alarm bell on the fire watchtower in the former city for signalling fires.[28] According to Hamdija Kreševljaković, the municipal fire brigade introduced an electric fire alarm system on 1 June 1913. At first, the new system may not have functioned as intended; newspapers continued to report on cannon signals for fire into early October 1913.[29]

Fire was not the only element Sarajevo had to fight against: yet another alarm signal in Sarajevo consisted of six consecutive cannon shots, which warned of high water levels or flood. At the same time, the fire brigade's bugler played signals while walking along the bank of the Miljacka River.[30]

According to one newspaper source, the Joint Ministry of Finance established the practice of firing cannon salutes regularly during the most important religious holidays in Bosnia in 1885 when it began supplying the artillery with blank cartridges of two different calibres. The salvos were fired from fortresses or bastions in Sarajevo, Mostar, Banjaluka, Travnik, Zvornik and Višegrad.[31]

The following list of cannon salutes for religious holidays in Sarajevo is based mainly on data from the *Bosnischer Bote* yearbook for 1901.[32] In addition, newspaper sources contain invaluable information on the details and variations of the salutes. Even though the *Bosnischer Bote* published the list for Sarajevo, the procedure for each holiday seems to have been the same in all Bosnian towns capable of firing cannon salvos. One should note, however, that the list excludes cannon signals for the *Ramazan* fast month.

The beginning and end of *Ramazan* were marked by seven cannon shots, the former during the evening prayer and the latter during the afternoon prayer. Three cannon shots guided believers during each night of *Ramazan*: the first cannon shot signalled the break of the fast through the *iftar* meal immediately after sunset, the second cannon shot indicated the hour of the *sehur* (< Turk. *sahur*) morning meal and the third cannon signal marked the beginning of daytime fasting at the hour of *imsak* slightly before sunrise.[33]

In river valley cities and towns, such as Sarajevo and Mostar, the echoing cannon booms must have had a strong impact on the public. An English traveller, observing an unfamiliar soundscape, noted the phenomenon in Sarajevo during a *Ramazan* evening:

> As the sun disappears in its bed of fire one hears the sharp boom of the cannon, and, almost before the hills have ceased throwing and multiplying the sound, a hundred different voices strike the ear – the plaintive call to prayer.[34]

Along with the cannon shot, another tradition signalled the time for the *sehur* meal: Roma musicians played in the streets and roamed from house to house to wake people up. The most common musical instruments for *sehur* music were the *zurna* folk oboe and the double-headed *bubanj* drum, both of Middle Eastern origin. The musicians could also sing a special waking song in front of each Muslim house.[35]

Clashing strongly with the Central European concept of *Nachtruhe*, *Ramazan* tended to reverse the normal rhythm of life and the soundscape of the day and night. Due to the nightly activity, the Bosnian Muslim community would return to its conventional daytime routine late in the *Ramazan* afternoons. Accordingly, the traditional *Ramazan* night-time lights, loud conversation, *teferič* picnics and noisy music led to social conflicts between Muslims and non-Muslims and occasionally to official complaints.[36]

The only known archival document protesting against *Ramazan* traditions in Bosnia is by the Czech apothecary and *Handelskammerrat* Gustav Proche who submitted an official complaint about *sehur* music in June 1918 suggesting that the authorities should restrict the nightly noise exclusively to Muslim neighbourhoods in his home town of Brčko. Proche suspected that musicians intentionally disturbed the *Nachtruhe* of non-Muslims in mixed neighbourhoods. The authorities replied that fasting belongs to the religious duties of Muslims and that Christians should therefore tolerate the nightly wake-up signals. Furthermore, the authorities asked Proche to produce more precise evidence of the malicious disturbance.[37] Thus, the authorities classified *sehur* music as 'Sacred Noise'.

During the three-day *Bajram* after *Ramazan*, the cannons fired 104 salutes.[38] The actual number was probably 105 since, according to the press, seven salutes accompanied each of the five daily prayers between sunrise and sunset during the three days.[39] The *Bosnischer Bote* notes that the festive cannon fire for the *Kurban bajram* consisted of 139 shots in total over a period of four days. Yet again, this number is one shot short if seven shots indeed accompanied each of the five daily prayers between sunrise and sunset during the four days, which totals 140. In addition, the feast began with one cannon shot on the eve of the feast. That said, a source from 1907 notes that the feast lasted for three days only and began with seven salutes in the afternoon. On the following day, 11 cannon salutes solemnized the festive prayer at eight o'clock in the morning.[40] According to the *Bosnischer Bote*, salvos for *Mevlud* (the Prophet's birthday) consisted of 35 cannon shots, which would logically imply seven shots during each of the five daily prayers. Nevertheless, newspapers mention seven shots on the *Mevlud* eve and either an unspecified number of shots the next morning or seven shots during the afternoon prayer *ikindija* (< Turk. *ikindi*). The number of salvos for *Mevlud*, it seems, varied considerably over the years.[41]

Salutes during Christian religious holidays combined cannon shots with fusillades from infantry rifles outside a church or next to a spring. Other simultaneous or subsequent components of such events were church bell ringing and a military brass band's performance of the *Kaiserhymne*. For fusillades during Catholic services, the infantry battalion was drawn from the Catholic areas of the Empire, whilst Serbian Orthodox services, whenever possible, enlisted a

Bosnian-Herzegovinian infantry battalion. Since the main Christian holidays (apart from Christmas) gave rise to a military church service, they included a military church parade (Ger. *Kirchenparade*).

Following the occupation, reports of Catholic Easter celebrations for Holy Saturday (Ger. *Karsamstag*, Cro. *Velika subota*) would regularly mention how cannon shots accompanied the procession's departure from the church in the afternoon, while during the prayers after the return of the procession to the church the infantry shot fusillades outside the church. Twenty years later, however, the *Bosnischer Bote* yearbook mentions three rifle fusillades and 20 cannon shots.[42] The order of the gun shots in the latter text may be incorrect.

The salutes for the Catholic feast of *Corpus Christi* (Ger. *Fronleichnamsfest*, Cro. *Tijelovo*, *Brašančevo*) would consist of five rifle fusillades, each followed by four cannon shots (20 cannon shots in total). An early source from 1881 mentions two separate occurrences of the alternating fusillades and cannon shots in the course of the procession, whereas a source from 1888 specifies that the *Kaiserhymne*, four sets of fusillades, cannon shots and church bell ringing accompanied the reading of the Gospels.[43]

The celebration of the Serbian Orthodox Easter Sunday (Serb. *Velika nedjelja*) included two sets of salvos: ten shots at three in the morning and four in the afternoon during the two Easter processions of the cross, respectively, for a total of 20 cannon shots. According to a newspaper from 1881, the early salutes consisted of 33 cannon shots at half past two in the morning, the number apparently symbolizing the lifespan of Jesus. The afternoon procession of the cross around the church was combined with a military parade.[44]

In Sarajevo, the Serbian Orthodox Theophany (Serb. *Bogojavljenja*) or the Great Blessing of Waters, took place in January at the Pjenkovac Spring in the Bendbaša neighbourhood. An infantry battalion fired five rifle fusillades during the blessing ritual and the cannons of the Yellow Bastion fired 101 times during the procession of the cross. Apparently the cannon salutes accompanied the route of the procession of the cross from the church to the spring and back to the church. Various newspapers mention that the cannons fired 26 times when the procession of the cross left the church, and that the infantry fired four fusillades alternating with cannon salvos during the blessing of the spring.[45]

The colonial administration was eager to exploit the policy of religious equality as the following excerpt from a propagandist report on the Serbian Orthodox Easter in Derventa in 1880 shows. The anonymous writer for the government newspaper links cannon salutes with civilization and modernity. The adjective 'surprised' in the text seems an understatement in describing the auditory perception of nocturnal cannon salvos, a newly introduced practice:

> On Sunday (that is, the Orthodox Easter) cannons began roaring at three o'clock in the morning on the hill above the Orthodox church, from which the sound spread throughout the town and its surroundings. Some townspeople were particularly surprised, as they had never before heard the roar of the cannon … The divine service ended with the roar of cannons and the fusillade

> fired by soldiers, and every moment of the shooting made our people's hearts dance with joy, all thanking our Governor on his word and most kindly our Emperor, Franz Joseph I, for his liberation and wise administration as well as for the freedom, which each religion enjoys according to its wish and desire.[46]

The writer interprets the new custom of saluting the main Christian holidays as a sign of freedom, in contrast to Ottoman rule and its exclusively Islamic saluting. Naturally, the report portrays the colonized population as utterly satisfied with the Austro-Hungarian administration and the rule of Franz Joseph. This propaganda was far from true: two years after the feast of the sanctification of the Gacko church, a revolt against the occupation began in Herzegovina.[47]

Imperial Holidays: Sound and Power

As Darin Stephanov observes in his analysis of Russian and late Ottoman festivities, celebrations in honour of the monarch, military performances and religious rituals often intertwined.[48] Also in Habsburg Bosnia, the alliance of throne, military and altar produced a mixture of salutes, guards of honour, religious services and processions, church bell ringing and tattoos (Ger. *Zapfenstreich, Retraite*). One can question whether the Emperor or the state took over religious feasts; the symbiosis of three powerful institutions – two secular, the third religious – is a convincing interpretation. Islam was a very different issue, however. Due to the status and traditions of Islam in Bosnia, co-operation between the colonial administration and the Islamic religious authorities hardly ever reached the high level of fluency which existed between the state and Christianity, especially Catholicism.

Let us begin by examining the hierarchy of sound and power from the very top. In Germany and Austria-Hungary, the 101-gun salute was primarily reserved for the Emperor's birthday, and during the first birthday celebrations of Franz Joseph on 18 August 1879 in Sarajevo, the cannons of the Yellow Bastion fired that specific salute. However, the 101-gun salute could occasionally solemnize other major festivities of the German cultural area, such as the centenary of the birth of Friedrich von Schiller in 1859.[49] Other instances include the salutes after the battle of Sarajevo on 19 August 1878 and the annexation of Bosnia in October 1908, both of which marked a significant territorial expansion. In addition, the grand salute apparently belonged to other occasions directly linked to the Emperor: as described above, cannons fired 101 salutes on 13 January 1880 in Gacko to solemnize the sanctification of new church bells and other gifts which the Emperor had donated to the local Orthodox church.[50]

That said, the standard procedure for the imperial birthday salvos in Sarajevo was as follows: a set of 24 cannon shots, one mixed set of rifle fusillades and 24 cannon salutes – after every third rifle fusillade, eight cannon shots – and finally a 24-gun salute. The first set of salutes took place at five o'clock in the morning during the flying of the imperial flag, the changing of the guard and the morning tattoo; the second took place after the Christian military church services and a military parade; and the final one took place in the afternoon when the official guests of the Provincial Governor (Ger. *Landeschef*) were toasting the Emperor and listening to the performance of the *Kaiserhymne* at the festive meal in Konak, the Governor's residence.[51]

The festivities for Franz Joseph's name day (3 October) included three sets of 24 cannon shots: the first occurred during the evening tattoo on the eve of the holiday, the second during the changing of the guard in the morning and the third during the morning festive service at the Catholic church.[52] The last set must have been a combination of rifle fusillades and cannon shots. The ceremonials for the Emperor's coronation date (2 December for Franz Joseph) resembled those for his saint's name day, and the same applied to his wedding anniversary (24 April). After the outbreak of the Great War, and especially after the enthronement of Franz Joseph's successor, Karl, on 21 November 1916, imperial festivities tended to be more modest.

One may wonder why the number of salutes for imperial holidays varied between 21, 24 and 101 in Bosnia and elsewhere in the Empire. This variation presumably reflected Franz Joseph's various positions as a ruler and the differences between the salute systems of the navy and the ground forces. Actually, the 21-gun salvo as the salute for kings and queens originates from the British Royal Navy which was founded in the sixteenth century during the reign of Henry VIII.[53]

The hierarchy of authority was also apparent in salutes for other notables who visited Sarajevo. When Archduke Wilhelm, Inspector-General of Artillery, left Sarajevo in March 1889, the cannons of the Yellow Bastion fired the imperial three sets of 24-gun salutes: firstly when the Archduke was leaving the Hotel Europe, secondly when his entourage passed the Tobacco Factory and thirdly as his train was leaving the station. Lesser dignitaries received fewer salvos. For instance, the Joint Minister of Finance, Count István Burián, was greeted with a 21-gun salute in May 1904. Religious authorities received fewer honours again: the newly inaugurated Serbian Orthodox Metropolitan, Nikola Mandić, was greeted with an unspecified number of cannon salutes, a mixture of rifle fusillades and cannon salutes and church bell ringing, whilst the newly inaugurated Reis-ul-Ulema, Džemaludin Čaušević, received 15 cannon salvos.[54] The differences between Christian and Islamic traditions are also evident in these two ceremonials.

Salutes were also a matter of acoustic composition. For instance, in eighteenth-century St Petersburg, cannon batteries and other sources of festive sounds were located considerably far away from each other in order to allow the acoustic composition to cover a large area of the urban space.[55] In Sarajevo, after the army had constructed new fortifications on the hills and mountains around the city, the artillery was able to shoot spatially more varied salutes than earlier. The army certainly utilized acoustic composition for the imperial birthday celebrations in 1913, but the first extensive use of sound sources considerably far away from each other may have occurred during Franz Joseph's arrival in Sarajevo on 30 May 1910. The *Sarajevski list* mentions cannon salutes from the fortresses on Mount Hum and Mount Vraca on opposite sides of the Miljacka Valley as the imperial train entered the city; simultaneously church bells pealed out a welcome. Somewhat later, as Franz Joseph rode in his carriage along the promenade by the Miljacka River, the cannons of the Yellow Bastion began firing salutes.[56]

A correspondent of the Vienna newspaper *Neue Freie Presse* described the soundscape of the Emperor's arrival in Sarajevo in greater detail – or perhaps more poetically – than did the local reporter of the *Sarajevski list*. As the city was waiting for the imperial trains, he wrote, the boom of the noon gun sounded endlessly long: louder and more festive than usual; it felt as though the gun produced a hundred-fold echo. As the imperial train approached Sarajevo, the artillery began firing salutes from the fortresses on mountains and hills at the four cardinal points around Sarajevo. This narrative is metaphorically more satisfying than that of the *Sarajevski list*, since the cannon roar from the cardinal points symbolizes Franz Joseph's power over the lands he was visiting. The *Neue Freie Presse* also mentions the performance of a military brass band and the solemn silence that fell after the music. According to the reporter, the next salutes came from a mountain fortress opposite the railway station. When the train arrived at the beginning of the promenade, the fortifications took turns firing salvos.[57]

The last significant visit of Habsburg royalty to Sarajevo took place when the heir to the crown, Archduke Franz Ferdinand and his wife, Duchess Sophie, arrived in the city on 28 June 1914. A large-scale programme of imperial salutes greeted the couple. As the imperial train left the near-by spa resort Ilidža for Sarajevo, the cannons of the Vraca fortress fired 24 salutes. The fortress repeated the salutes when the train was approaching the city and fired them for the third time after the train stopped at the station. Later in the afternoon, according to the programme, when the Archduke was to leave Konak, the Yellow Bastion was to fire the customary imperial three sets of 24-gun salutes.[58] This part of the visit never took place, of course, due to the fatal shots from Gavrilo Princip's pistol.

Instead, the cannons of the Yellow Bastion fired the next 24 salutes on the evening of 29 June as the hearse carrying the Archduke's and Duchess's bodies and the accompanying procession left Konak. Simultaneously the bells of all the churches began tolling. Later on, the imperial train left Sarajevo railway station as the *Kaiserhymne* played and the cannons thundered from the Mount Vraca fortress.[59] The sources fail to specify whether the farewell volley consisted of three sets of 24-gun salutes.

Conclusion

Various sounds marking the Bosnian urban soundscape during Habsburg rule were part and parcel of Habsburg colonial policies. The salutes and performances of military bands superimposed on religious rituals and holidays symbolized state protection of the three main religions in Bosnia. The authorities utilized aural symbolism in both domestic and foreign policies, particularly towards the Muslim population in Bosnia and elsewhere in the Balkans. Especially in the early years of the occupation, descriptions of soundscapes mingled with colonial pro-Habsburg propaganda in the press. Furthermore, artillery salvos upon the arrival and departure of dignitaries demonstrated the position of the guest in the power hierarchy of the Empire. It is remarkable that the combination of Ottoman and Habsburg traditions and the idea of religious equality made the quantity of ceremonial and signalling cannon fire multiply in post-1878 Bosnia. The number of annual cannon signals and salvos in the dual provinces was certainly the highest in the Habsburg Empire and possibly the highest in Europe as well. The soundscape in Bosnia, then, changed considerably in the 40 years covered by this chapter: firstly, modernization brought about industrial noise and made the use of cannon signals as fire alarms unnecessary; secondly, domestic and international political crises – and especially the Great War – directly affected localized Bosnian soundscapes; the ban on private celebratory gunfire during the annexation crisis and the confiscation of large church bells during the Great War are telling examples. Habsburg Bosnia's urban soundscape constituted a complex semiotic system, therefore, in which Islamic calls to prayers, church bells and especially rifles and cannons intermingled, not only disseminating information but simultaneously serving as a political tool through which to affirm Habsburg legitimacy.[60] The hierarchical messages of church bell ringing and salutes were embodied in the following simple basic principle: the higher the rank of the person or the more important the occasion, the more noise for the longer time in a spatially wider area was to be applied. Intermingling sonic symbols from religion, the military and the throne – and superimposing these sonic signifiers in multifarious ways onto religious festivities and civic, nationalistic and imperial occasions – the soundworld of urban Habsburg Bosnia was punctuated by these complex sonic expressions of colonial policy and the sounding out of its imperialist power.

Notes

1 Henceforth, I refer to Bosnia-Herzegovina as 'Bosnia'.

2 A soundmark is 'a community sound which is unique or possesses qualities which make it specially regarded or noticed by the people in that community'. See R. Murray Schafer, *The Soundscape: Our Sonic Environment and the Tuning of the World* (Rochester, VE: Destiny Books, 1994 [1977]), p. 274.

3 Cf. Bruce R. Smith, *The Acoustic World of Early Modern England: Attending to the O-Factor* (Chicago: Chicago University Press, 1999), pp. 58–9.

4 David Garrioch, 'Sounds of the City: The Soundscape of Early Modern European Towns', *Urban History*, 30/1 (2003): 5–25, at p. 6.

5 Robert J. Donia, *Islam under the Double Eagle: The Muslims of Bosnia and Hercegovina, 1878–1914* (Boulder: East European Monographs, 1981), p. 14.

6 Ibid., pp. 18–21.

7 Robert J. Donia, *Sarajevo: A Biography* (London: Hurst & Co., 2006), pp. 80–81.

8 Eric Hobsbawm, 'Introduction: Inventing Tradition', in Eric Hobsbawm and Terence Ranger (eds), *The Invention of Tradition* (Cambridge: Cambridge University Press, 1983), pp. 1–14, at p. 1.

9 Robin Okey, *Taming Balkan Nationalism: The Habsburg 'Civilizing Mission' in Bosnia, 1878–1914* (Oxford: Oxford University Press, 2007), p. 60.

10 See Božidar Jezernik, *Wild Europe: The Balkans in the Gaze of Western Travellers* (London: Saqi, 2004), pp. 214–15.

11 Tomislav Kraljačić, *Kalajev režim u Bosni i Hercegovini (1882–1903)* (Sarajevo: Veselin Masleša, 1987), p. 365.

12 Risto Pekka Pennanen, 'Controlling Sound and Music: Aspects of Musical Life in Sarajevo under Austro-Hungarian Rule (1878–1918)', in Tamara Karača and Senad Kazić (eds), *Collection of Papers: 4th International Symposium 'Music in Society', Sarajevo, October 28–30, 2004* (Sarajevo: Musicological Society of the FBiH, 2005), pp. 114–25.

13 Schafer, *Soundscape*, p. 76; Alain Corbin, *Village Bells: Sound and Meaning in the Nineteenth-Century French Countryside* (New York: Columbia University Press, 1998), p. 210.
14 *Sarajevoer Tagblatt*, 17 August 1912.
15 *Bosanska vila*, 11–12 (1890): 181.
16 M. Şükrü Hanioğlu, *A Brief History of the Late Ottoman Empire* (Princeton: Princeton University Press, 2008), p. 85.
17 Maude M. Holbach, *Bosnia and Herzegovina: Some Wayside Wanderings* (London: J. Lane, 1910), pp. 238–9; Noel Malcolm, *Bosnia: A Short History* (London: Macmillan, 1994), pp. 130–31; Donia, *Sarajevo*, pp. 33–4. In 1914, the government newspaper *Sarajevski list* claimed that the bells were installed in the belfry of the new Orthodox church only after the occupation, which does not hold true (*Sarajevski list*, 27 June 1914).
18 John F. Loud, 'Andrić on Bosnia: The 1924 Dissertation', in Wayne S. Vucinich (ed.), *Ivo Andrić Revisited: The Bridge Still Stands* (Berkeley, CA: University of California, 1995), pp. 187–99, at pp. 195–6.
19 *Bosansko-hercegovačke novine*, 19 September 1880; Ibid., 23 December 1880.
20 *Bosansko-hercegovačke novine*, 25 January 1880. This translation, and the following translations in this chapter, are my own unless otherwise stated.
21 A description of the *Corpus Christi* celebrations in May 1888, however, implies that the new cathedral already had a temporary bell at that time (*Sarajevski list*, 1 June 1888); *Sarajevski list*, 15 September 1889; Ibid., 25 September 1889.
22 *Sarajevski list*, 13 June 1888; *Bosanska pošta*, 29 November 1898.
23 *Sarajevski list*, 4 August 1917. For the collection of copper in Austria-Hungary during the First World War, see J. Robert Wegs, 'The Marshaling of Copper: An Index of Austro-Hungarian Economic Mobilization during World War I', *Austrian History Yearbook*, 12–13/1 (1976): 189–202.
24 Darin Stephanov, 'Cannon Salvos for the Monarch: Notes on the Ceremonial Usage of Artillery at Nineteenth-Century Ottoman and Russian Accession, Coronation and Other Public Dynastic Festivities', unpublished conference paper presented at the Bosphorus University's Ottoman Material Culture Workshop (Istanbul, Turkey, September 2006).
25 Bosnian Muslims call the inter-religious feasts of *Đurđevdan* and *Ilindan* respectively *Jurjevo* and *Aliđun*. For celebratory gunfire at Cretan weddings and baptisms, see Michael Herzfeld, *The Poetics of Manhood: Contest and Identity in a Cretan Mountain Village* (Princeton, NJ: Princeton University Press, 1985), pp. 89, 68; *Bošnjak*, 11 December 1908.
26 Adolf Walny (ed.), *Bosnischer Bote. Universal-Hand- und Adreßbuch für Bosnien und die Hercegovina* (Sarajevo: Self-published, 1901), p. 276; Irina A. Chudinova, 'Song, Bells, Ritual: The Topography of St. Petersburg's Ecclesiastical Musical Culture', *Russian Studies in History*, 45/3 (Winter, 2006–07): 31–59, at p. 42. For the midday gun in the Ottoman Empire, see Loius Gardey, *Voyage du sultan Abd-ul-Aziz de Stamboul au Caire* (Paris: E. Dentu, 1865), pp. 64, 250.
27 *Sarajevski list*, 29 December 1883.
28 *Bošnjak*, 17 July 1902; *Bosnische Post*, 23 January 1907; Ibid., 2 December 1912; Ibid., 12 March 1913.
29 Hamdija Kreševljaković, *Sarajevo za vrijeme austro-ugarske uprave, 1878–1918* (Sarajevo: Arhiv grada Sarajeva, 1969), p. 35; *Sarajevski list*, 1 October 1913.
30 *Bošnjak*, 17 July 1902.
31 *Bosnische Post*, 27 November 1884.
32 Walny (ed.), *Bosnischer Bote*, p. 276.
33 *Bosnische Post*, 2 January 1900; Ibid., 1 February 1900; Ibid., 25 August 1911.
34 Roy Trevor, *My Balkan Tour: An Account of Some Journeyings and Adventures in the Near East, Together with a Descriptive and Historical Account of Bosnia & Herzegovina, Dalmatia, Croatia & the Kingdom of Montenegro* (London: John Lane, 1911), p. 102.

35 Enver Mulahalilović, *Vjerski običaji Muslimana u Bosni i Hercegovini.* 2nd enlarged edition (Sarajevo: El-Kalem, 1989), p. 149.
36 Antun Hangi, *Život i običaji Muslimana,* 2nd enlarged edition (Sarajevo: Daniel A. Kajon, 1906), p. 35. Roy Trevor (*My Balkan Tour*, p. 89) describes how the *muezzin* of the Ferhadija Mosque woke his Scottish companion several times during a *Ramazan* night when a tourist group was staying at the Hotel Europe in Sarajevo in 1909. The gallery of the minaret was outside the companion's open window.
37 *Arhiv Bosne i Hercegovine*, ZVS, 22–22/11/1918: the complaint of Gustav Proche, 4 July.
38 Walny (ed.), *Bosnischer Bote*, p. 276.
39 *Bošnjak*, 12 April 1894.
40 Walny (ed.), *Bosnischer Bote*, p. 276; *Bosnische Post*, 17 January 1884; Ibid., 24 January 1907.
41 Walny (ed.), *Bosnischer Bote*, p. 276; *Sarajevski list*, 16 October 1891; *Bošnjak*, 13 September 1894.
42 *Bosanska pošta*, 12 June 1897; *Sarajevski list*, 27 March 1883; Walny (ed.), *Bosnischer Bote*, p. 276.
43 *Bosansko-hercegovačke novine*, 19 June 1881; *Sarajevski list*, 1 June 1888.
44 Walny (ed.), *Bosnischer Bote*, p. 276; *Bosansko-hercegovačke novine*, 28 April 1881; *Sarajevski list*, 11 April 1890.
45 *Bosnische Post*, 20 January 1884; *Sarajevski list*, 21 January 1900; *Bosansko-hercegovačke novine*, 20 January 1881; *Bosnische Post*, 19 January 1914.
46 *Bosansko-hercegovačke novine*, 13 May 1880.
47 See Okey, *Taming*, pp. 52–3.
48 Stephanov, 'Cannon Salvos'.
49 *Die Presse*, 19 August 1879; Bruce Duncan, 'Remembering Schiller: The Centenary of 1859', *Seminar: A Journal of Germanic Studies*, 35/1 (1999): 1–22, at p. 1; After the introduction of the *Tanzîmât* reforms, the Ottoman Empire observed the European system of salutes as well: during Sultan Murad's short reign in 1876 and festivities for the accession to the throne, the Ottoman fleet fired 101 cannon salutes in the Bosphorus (Stephanov, 'Cannon Salvos'). In Serbia, during the coronation of King Alexander I, which took place immediately after the 500th anniversary of the Battle of Kosovo on 4 July 1889 in Žiča monastery near Kraljevo, the artillery shot 101 cannon salutes in front of the monastery. Church bells were simultaneously pealing throughout the country (*Bosanska vila*, 12 [1889]: 188).
50 Josef Koetschet, *Aus Bosniens letzter Türkenzeit* (Vienna: A. Hartleben, 1905), p. 109; *Pester Lloyd*, 8 October 1908; *Bosansko-hercegovačke novine*, 25 January 1880.
51 *Sarajevski list*, 21 August 1898; *Sarajevoer Tagblatt*, 15 August 1912.
52 *Sarajevoer Tagblatt*, 2 October 1913.
53 *Wiener Zeitung*, 25 April 1879; *Die Presse*, 19 August 1879; *Meyers großes Konversations-Lexikon: Ein Nachschlagewerk des allgemeinen Wissens*. 6th edition. (Leipzig: Verlag des Bibliographischen Instituts, 1907), see the entry for 'Ehrenbezeigungen'; N.A.M. Rodger, *The Safeguard of the Sea: A Naval History of Britain, 660–1649* (London: HarperCollins, 1997), pp. 221–37.
54 *Bosnische Post*, 30 March 1889; *Bošnjak*, 5 May 1904; *Bosanska pošta*, 22 December 1896; *Sarajevoer Tagblatt*, 27 March 1914. The 15-gun salute was reserved for envoys and resident ministers (*Meyers großes*, see entry for 'Ehrenbezeigungen').
55 Chudinova, 'Song, Bells, Ritual': 46–7.
56 *Bosnische Post*, 2 December 1913; *Sarajevski list*, 31 May 1910.
57 *Neue Freie Presse*, 31 May 1910; Corbin, *Village Bells*, p. 210.
58 *Sarajevski list*, 27 June 1914.
59 Ibid., 30 June 1914; *Neue Freie Presse*, 30 June 1914.
60 Garrioch, 'Sounds of the City': 18.

Section III

Urban Soundscapes of Europe

Introduction

Ian Biddle and Kirsten Gibson

The chapters in this section all focus on urban soundscapes of Europe's past. They are, as in the previous sections, geographically and historically dispersed, ranging from the cities of Maximilian I's empire at the turn of the sixteenth century, early modern Italian cities and late sixteenth-century London to nineteenth-century Lyon and *fin-de-siècle* Lisbon. While each chapter is focused on the sonorities of particular urban centres, they deploy a range of approaches and methodologies, from various disciplines – musicology, literary studies and architectural studies – to consider past experiences of, and engagements with, these soundscapes. From explorations of everyday civic musical life and religious festivities that utilized the physical and sonic fabric of particular urban spaces to studies of the aural consequences of developments in architectural design and town planning, the introduction of new technologies and new modes of sociability and leisure activity, these chapters attempt not simply to evoke these lost soundscapes, but to ask how, for people living in these urban centres – at these particular times – they were experienced aurally. The underpinning principle, as in the previous two sections, is, then, the recovery of the semantic agency of sonority for subjects past, mediated here through the changing urban environments of Europe from the late fifteenth to the early twentieth centuries.

Chapter 8, by Helen Coffey, draws on civic council records and contemporaneous accounts to consider uses of music in the everyday lives of the citizens of the cities of Maximilian I's domains. Coffey's examination of civic music presents a complex urban soundscape in which musicians employed by the municipality – the *Stadtpfeifer*, pipers and drummers, trumpeters and watchmen – intermingled with the cities' amateur musicians to varying degrees at a range of civic occasions and social gatherings. In particular, Coffey's account draws a detailed picture not only of the types of music commonly heard and used in such towns and cities, and the various uses of music in everyday urban life, but also the ways in which music was controlled by the cities' patricians, and was used to demarcate social and class differences amongst the citizens alongside aurally defining the urban spaces associated with differing social groups. In the towns and cities examined here, the diverse urban sounds encountered – especially those of musical performances – reflect the underlying social and political structures that determined all aspects of civic life.

In Chapter 9, Daniele V. Filippi explores the ways in which Biblical notions of 'otherworldly' sound – both heavenly and hellish – were echoed in the Cathedrals, churches, squares and streets of early modern Italian cities, giving their citizens a sensory, aural foretaste of the afterlife. Drawing on the writings of Luis de Granada and Jeremias Drexel, Filippi outlines the aural dimensions of early modern understandings of the afterlife, and, in particular, the status of music in the heavenly sphere. These ideals, as Filippi shows, were drawn on and evoked in earthly urban life, in both liturgical and secular art music practices, in the aural dimensions of the daily cycle of worship, in the architectural planning of sacred spaces and in the ceremonies and processions of urban feast days. Aural evocations of hellish soundscapes could also be heard in the streets of Italian towns and cities, though to a lesser extent than the heavenly, but, as Filippi observes, such evocations were to be found on the stage, playing an important role in some of the earliest operas at the beginning of the seventeenth century. That the evocation of infernal soundscapes would continue both as a source of interest and meaning for composers, librettists, performers and theatre audiences has been shown in Chapter 6, in which the Orpheus myth was adopted in eighteenth-century Paris as a significant site for both acting out, and resisting, the 'politics of silence'.

Everard Guilpin's 1598 satiric representation of London, *Skialetheia*, forms the focus of Chapter 10 by Adam Hansen. As Hansen notes, sound plays a significant role in Guilpin's evocation of the city, both in terms of references to the sounds of the city, and the poetic devices he employs – alliteration, onomatopoeia, rhythm, homophones – in its illustration. Drawing on scholarship on classical and early modern satire, on the satire of Juvenal and recent historical sound studies scholarship, Hansen considers the ways in which Guilpin reproduces the sounds of the city poetically – in order to capture for the reader/listener 'the effects of immediacy'[1] – in order, at the same time, to derogate them. Yet, simultaneously, his poetic persona and the subjects he addresses are inextricably implicated in the sounds of the city, as subject to and producers of urban noise. The city and its sounds, for the poetic speaker, are represented as something to deplore and resist, as something that can seduce and as something in which the listener is always already inevitably implicated. Through his close reading of 'Satire V' from Guilpin's *Skialetheia* Hansen asks, ultimately, 'How did the city's sounds constitute and compromise urban and satiric subjectivities?'

The final two chapters in this section deal with the aural implications of material changes to the urban fabric. Chapter 11, by Olivier Balaÿ, shifts the focus from poetics to the materiality of the city and its sounds. Focusing on nineteenth-century Lyon, Balaÿ examines the ways in which urban development – renovations, the widening of streets and the removal of overhangs, corbelled constructions and awnings – in the mid-nineteenth century altered the conditions under which sounds were produced and promulgated. Balaÿ's research utilizes city maps and illustrations, urban planning reports and by-laws, alongside sonometer readings, to map out the aural consequences of changes to the fabric of the city. Drawing on 'the sound effect concept' currently used at Cresson (Centre de recherche sur l'espace sonore et l'environnement urbain), Balaÿ is able to trace the past aural terrain of the city, and to chart the changing quality of sounds from the Lyon of the *Ancien Régime* to its 'Haussmannian' counterpart.

Chapter 12, by João Silva, similarly focuses on the aural implications of changes to the urban fabric of a city, in this case Lisbon, from the second half of the nineteenth to the turn of the twentieth century. It considers the coming of modernity to Lisbon, influenced, like many European cities, by Georges-Eugène Haussmann's reconfiguration of Paris, while at the same time exploring the interrelated implications for the urban soundscape of the development of sound reproduction technologies and emergent leisure markets. Drawing on the work of Walter Benjamin, Silva considers the aural porosity of Lisbon, through, firstly, an examination of how the modernization of the city – in terms of the rationalization of public and private spaces, the implementation of a modern transport infrastructure and, in particular, the development of entertainment venues across the city – changed dominant modes of sociability and altered its sonic landscape. As Silva's study demonstrates, changes to the city's entertainment venues had implications beyond the stage, not only as the burgeoning of theatres in the city brought new theatrical genres to the city's audiences, but also because the music first heard on the stage frequently seeped out onto the city's streets and into the parlours of the city's bourgeois residents. For those living in the older neighbourhoods and on the fringes of the city sounds – the performance of fado, street cries, church bells and the music of *banda* ensembles – likewise permeated the public and private spheres of everyday life.

While each chapter presents a localized case study, taken together they sketch out the overarching narrative of the changing fabric – and aurality – of European urban spaces from the late fifteenth through to the early twentieth century. These chapters, therefore, bear witness to the gradual shift from the sonorities of the narrow, winding streets and official civic and religious spaces of pre- and early modern towns and cities – through the process of 'Haussmannization' in the nineteenth century – to the beginnings of the rationalized, sprawling modern urban centres we know today. Yet, in their varied approaches, they also tease out myriad interrelationships between the acoustic consequences of the changing materiality of urban space and other kinds of human interventions in these spaces. Changes in these various aspects of the physical and aural dimensions of urban space, of course, moved at differing rates, at different times; the long-standing ubiquity of street cries, for instance – as witnessed in the chapters by Hansen, Laurance and Silva across the collection as a whole – provides a certain kind of continuity in the sonic character of European urban centres, even as, at the same time, the acoustic make-up of the towns and cities in which distinctive street cries reverberated was altered in the processes of architectural modernization. These chapters, then, draw out the ever-shifting, interwoven narratives between the changing physicality of urban centres and changing attitudes towards particular types of sound and the ways in which it was managed in the sounding of European towns and cities from the Middle Ages to the cusp of the twentieth century.

Note

1 Lawrence Manley, *Literature and Culture in Early Modern London* (Cambridge: Cambridge University Press, 1995), p. 375.

8 City Life and Music for Secular Entertainment during the Reign of Maximilian I[1]

Helen Coffey

By the end of the fifteenth century, performances by instrumentalists were an intrinsic feature of life within the many cities that were dotted across the German lands of the Holy Roman Empire.[2] Their music could be heard from all corners of these municipalities: from the towers that overlooked each urban space, to the specially constructed dance halls, the musicians' balconies at the town halls and even on the rivers and streets. The importance of instrumental music to the cities is testified by the frequency and regularity of references to musicians in civic payment lists, which exist for the fourteenth century onwards.[3] Other kinds of civic documents corroborate this perception of the significance of civic instrumentalists; official correspondence, council records and contemporary chronicles all contribute to the formation of a rich and lucid picture of the musical life of German cities at this time.[4] Previously scholars have drawn on these invaluable sources to highlight the significance of these cities to musical developments of the period: the continual patronage of instrumental ensembles by the presiding city councils; the pre-eminence of German instrumentalists within the network of musical personnel that connected the cities and courts of Europe; and the emergence of many of the cities – most notably Nuremberg – as centres for the production and trade of instruments, have all been addressed in exemplary studies.[5] German cities were indeed important musical centres at this time. In addition to the civic patronage of instrumentalists and instrument makers mentioned above, Maximilian I's frequent visits to these municipalities during his relentless travels through the *Reich* and the regular use of the cities as venues for the political gatherings of his reign meant that resident civic musicians were often exposed to external influences, conveyed by the musical retinues of the visiting monarch and nobility. This chapter will present the activities of the instrumentalists of a number of Imperial or 'free' cities (the independent municipalities answerable directly to the Emperor) as entwined in the daily lives of the citizens as well as in the many secular festivities that occurred throughout the year. The performances of these resident minstrels were integral to the celebrations of weddings, carnivals and the myriad dances that were a customary and regular feature of city life. Through consideration of council records and other contemporary accounts, insights can be gained into musical performances for these and other civic customs, which were all defined by the political, social and cultural circumstances of the time.

Life within the cities presented a number of employment opportunities to secular instrumentalists. The most esteemed of these was a position within the civic wind ensembles – the *Stadtpfeifer* (town pipers) – employed by the city councils to provide appropriate musical performances for the entertainment of the citizens and their guests, thereby acting as the cities' official musical representatives. Adorned accordingly with the shield of the city and a livery of their patron town's colours,[6] the *Stadtpfeifer* were bound to the service of their council through the declaration of an official oath. In the early sixteenth century, the council of Schwäbisch Hall demanded that their musicians 'will swear to be obedient, loyal, ready and willing, to the mayor and council here, for whatever or wherever the council wishes to employ them, be this in the city or outside, in the region, far or near'.[7] The extent of the patronage of *Stadtpfeifer* ensembles was a clear reflection of social and financial conditions within each municipality. This was the case throughout the Empire, where different degrees of wealth and stature were expressed through performances of the musical retinues of the monarch, nobility and cities – from the grandeur of Maximilian I's court with his substantial trumpeter corps, acclaimed *Hofkapelle* and numerous secular instrumentalists, to the lesser ensembles of minor bishops and nobles.[8] Musical patronage in the cities, on the lowest tier of the Empire's social hierarchy, was subject to the decisions and actions of the ruling classes and of course the level of prosperity of each region – the cities varied considerably in their size and wealth and this obviously impacted on the degree to which they could subsidize musicians. The size of *Stadtpfeifer* ensembles therefore ranged greatly. The largest and most affluent cities, such as Nuremberg and Cologne – rich through trade – could support an ensemble of five musicians, while their prosperous neighbours Augsburg and Ulm persisted in their employment of an ensemble of four instrumentalists. Smaller cities of lesser wealth, such as Memmingen, Nördlingen and Schwäbisch Hall, were restricted to ensembles of three musicians, leaving the most meagre municipalities – Bad Windsheim, Dinkelsbühl – with modest ensembles of just two *Pfeifer*.[9]

The instruments of the *Stadtpfeifer* ensembles followed the standard ensemble configuration employed in most civic and court institutions from the fourteenth century. This was the *alta capella* – a wind ensemble that consisted of two or three performers on reed instruments (i.e. shawm, bombard) – commonly denoted *Pfeifer* in council records, and a *Trompeter*, playing by now the slide brass instrument and precursor to the modern-day trombone.[10] The addition of other wind instruments such as the crumhorn and cornett had followed by the start of the sixteenth century as both civic and court ensembles developed.[11] Although the use of the generic terms *Pfeifer* and *Trompeter* often makes it difficult to determine the exact pattern of expansion of the *Stadtpfeifer* ensembles, there are frequent references to the 'pusauner' ('*trombonists*') who played within these groups,[12] and the records of Nuremberg reveal details of the purchase of an array of instruments for the ensemble there. 'Ein pfeyffen als genant ein pumhart'[13] (*a pipe called a bombard*) was acquired by the council in 1478, followed by '8 krum pfeiffen und zwo schalmeyen'[14] (*8 crumhorns and 2 shawms*) in 1503 and '8 flötn pfeyffen unsern Stat pfeyffern'[15] (*8 recorders for our Stadtpfeifer*) in 1512.

Nuremberg's patronage of instrumentalists was exceptional, not only possessing a *Stadtpfeifer* ensemble, but also a lute-and-portative duo for most of the fifteenth century.[16] This was a significant addition to the musical resources of the city as such prestigious *bas* instruments were customarily only afforded by the wealthiest (usually royal or noble) patrons.

In addition to the pre-eminent *Stadtpfeifer*, the political strife of the period and the vulnerable position of the cities demanded the employment by the councils of a number of signal instrumentalists. The watchmen sounded the hour, or raised any alarms on a brass or wind instrument, bell or drum, from one of the many towers within the cities, when the threat of fire, or indeed an enemy, arose.[17] Brass and wind instrumentalists were also employed in times of war, when the cities were frequently required to provide troops. Trumpeters, fifers and drummers were all enlisted in this capacity, from the greatest civic powers, such as Augsburg, to the more meagre, like Schwäbisch Hall.[18] Nuremberg employed a piper-and-drummer duo continuously from 1509,[19] whose services were also often called on for the entertainment of the citizens. In all the cities, the music of those employed by the councils was enriched by the presence and practices of numerous freelance and amateur instrumentalists. In addition to their arbitrary performances amidst the everyday activities of the citizens, these were also often required to participate in the many civic celebrations and social gatherings. Together, the performances of these various strands of civic music – the *Stadtpfeifer*, pipers and drummers, trumpeters, watchmen and freelance or amateur musicians – all combined to form a rich civic soundscape.

The city councils controlled all aspects of civic life – from jurisdiction, business and trade, to social customs, cultural activities and musical practices. Despite the diversity in size and stature of the German cities, the oligarchic power of the presiding councils was common to all, as was the social elite of patricians, nobles or leading merchants who predominated over the tradesmen, craftsmen and others of lower status. This council control and social structure is clearly expressed in civic musical performances of the time, for which venues and instrumentalists were often stipulated by the ruling classes, according to the participants' social status. The regulation of social conduct and musical practices is notably exhibited in the city authorities' restrictions on performances in the open – the potential for improper behaviour resulting in frequent council intervention. In 1492, the council of Nuremberg attempted to assert some sort of control over the evening performances of their *Stadtpfeifer*, instructing that they should no longer play on the streets after midnight,[20] an order that was repeated in 1494: 'to inform … the Stadtpfeifer that they have been playing too late at night on the streets, and to forbid them from doing the same'.[21] As the official musicians of the Nuremberg council, the performances of the *Stadtpfeifer* would have been monitored with particular care so that they behaved with all propriety and did not cause any disturbance to the community.

Yet it was not only the official musicians of the cities who were subject to the scrutiny of the councils – other outdoor performances received similar treatment as the authorities attempted to maintain decorum in their towns. The council of

Schwäbisch Gmünd announced that the following pastimes were to be forbidden: 'shameful singing and shouting, as has been exhibited on the streets at night, together with the playing of pipes, trumpets, lutes, fiddles and other string instruments, with no exceptions'.[22] An act administered by the council of Regensburg in 1509 also reflects the extent of this regulation of civic revels; here, restrictions on the performance of music during sleigh rides were pronounced:

> Regarding the sleigh, and how one should travel with a drummer, the mayor and council accept that the nobility and citizens or other respectable individuals like to travel by sleigh, yet wish that no bawdiness nor shouting, and therefore no commotion nor other mischief arises, occurs or is caused by this. One may also make appropriate use of honourable amusements such as singing and playing on bells, pipe and drum. Yet this should not be permitted to the craftsmen or similar unsuitable persons.[23]

In addition to the council's attempt here to assert some degree of control over the apparently frivolous nature of the sleigh ride, the social divide between the nobility or other 'respectable individuals' and the craftsmen, for whom such activities were forbidden, is also clearly defined. This determination of entertainments and musical activities according to social status, and the town councils' repeated endeavours to control the behaviour of the community, extended to all kinds of musical practices. This chapter will consider musical performances for civic weddings, dances and carnivals, to demonstrate how various forms of urban entertainment were determined by these council constraints which reflected the social division within civic communities at this time.

Weddings

The social and musical demarcation of civic amusements is clearly expressed in the regulation and restriction of wedding festivities. As one of the most important celebrations in civic life, these happy events combined all kinds of entertainments including dances and banquets. Numerous entries in the records of the Nuremberg council convey the privileged position of the social elite on these occasions. Citizens of higher status were permitted the use of the rooms of the town hall together with the official musicians of the city, as was granted for a wedding of a member of the patrician family Haller in 1491: 'to lend the town hall, rooms and musicians to Wolf Haller, Jobst Haller's son, with Anthon Koburger's daughter for their intended wedding on St. Sebastian's eve'.[24] The *Stadtpfeifer* are also known to have performed at the weddings of visiting nobility, sometimes together with musicians of their courts. For the nuptials of Margrave Casimir of Brandenburg and Susanna of Bavaria, which occurred in Augsburg amidst the proceedings of the Imperial Diet there in 1518, the *Stadtpfeifer* of the city received payment 'for the dance when the lady of Bavaria held her wedding here with the Margrave or he with her'.[25] Other records reveal details of the performances of visiting royal and noble musicians at various times on the wedding day.[26]

Unsurprisingly, the potential for wedding-day musical performances by both resident and visiting musicians often resulted in intervention by the local authorities. The fastidious regulations asserted by city councils regarding the particulars of wedding celebrations, as dictated in the many *Hochzeitsordnungen* of the period, attest to the regular contributions of various resident and visiting musicians to these events.[27] These documents emphasise the social standing of the bride, groom and their guests and the consequently appropriate musical accompaniment for their festivities.[28] A Nuremberg instruction of 1485 reveals this stringent council control and the hierarchy of various wedding parties and their musicians. Here the preference for official city and court musicians is patently clear:

> Item no minstrels or individuals should be admitted to a wedding nor to the meal nor be permitted to eat there, other than those who accompany the bridegroom or bride from the country or wear the city shield, together with the *Hegelein*, who lead the dance. And if there are any princes here that day, whose musicians come to the wedding, they are permitted to eat without objection. If however, an individual does not wish or cannot have the musicians who bear the city shield, they may have and appoint one, two or three others in their stead, without objection.[29]

This ordinance also goes on to detail the conditions for use of the city's *Stadtpfeifer* for evening performances, as ever, every element of their work overseen by their watchful council employers:

> Henceforth, if an individual wishes to employ the *Stadtpfeifer* to play in the evenings, they should award them one guilder, and they should then be committed to perform before eight houses or areas upon the wishes of the same individual … If someone wishes to have them serenade for longer or at more venues, the same individual should be obliged to give the *Stadtpfeifer* half an *ort* of a guilder for every venue or before every house.[30]

The stipulations of the councils were not limited to instructions regarding the appointment and payment of musicians. This same Nuremberg ordinance proceeds to comment on the dance repertoire musicians were to perform on these occasions. This was to be confined to more traditional pieces, the disapproval of unusual or new styles being clearly stated. Instrumentalists were instructed to: 'not pipe, play nor perform any such dances other than the customary dance, of old traditions'.[31]

A Frankfurt order of 1489 is more socially discriminatory in the restrictions applied to the use of musicians:

> Item a bridegroom of higher status may have six musicians for his wedding and one of lower status, three musicians and no more … if he brings princes' or other cities' musicians, namely trumpeters and pipers, he may only give each one no more than two guilders and any that live here in the city, no more

> than one guilder, but a visiting lutenist may also be given one guilder and a resident lutenist, also a fiddler, drummer or bagpiper, half a guilder and no more … .[32]

Here, not only is the number of musicians dependent on the status of the bridegroom, but a hierarchy among the instrumentalists employed is also exposed. Visiting *haut* instrumentalists received the highest monetary amount, yet the resident *Stadtpfeifer* would only be awarded up to half as much. Lutenists of an external patron were entitled to the same level of payment as local wind instrumentalists, therefore demonstrating the lute's prestige among *bas* instruments, while those of lower status – the fiddlers, drummers and bagpipers – received a mere nominal fee. This instruction again expresses how social hierarchy and musical representation were closely interlinked, while also conveying the wishes of the Frankfurt council to maintain a level of musical skill and decorum through their restrictions on instrumentalists who were permitted to perform.

Dances

The impact of the overbearing control of the town councils and the pervasive social divide on civic festivities is clearly articulated in the dance practices of the period, which formed part of most urban celebrations. Two dances in particular illustrate the difference between the practices of the social elite and the rest of the community. That identified with the upper classes was the slow, processional dance, stemming from the *basse danse* tradition of the Burgundian court, which was ostentatiously performed by couples and maintained an air of dignity and refinement. The livelier *Reigen*, or round dance, could involve participants from any social group, who performed in an open or connected circle or chain, either in single or double file, with the dancers' movements ranging from vibrant twists and turns to slow, more elegant steps.[33]

The venues for dances were suited to the range of movement of each genre, as well as the status of the participants. Many of the cities, such as Nördlingen, Augsburg and Frankfurt, boasted dance halls or rooms within the town hall, which, thanks to their great length, could accommodate the processional dances of the elite. To facilitate the performance of music for these occasions, balconies were often built at these venues, from which the accompanying musicians could observe the dancers and provide appropriate music. As noted earlier with regard to wedding celebrations, the use of the town hall and the *Stadtpfeifer* for dances was generally only granted to those of appropriate social standing and required prior approval by the council. Nuremberg records include numerous references to the consent granted for these elitist events, as was agreed for a social gathering of a member of the Haller family in 1486: 'Jobst Haller the Younger is to have the town hall, the rooms and pipers for his intended dance today'.[34] Payments to the councils' musicians for these events are numerous, as are references to food, wine and other essential items. In 1485 the Augsburg council made payment for 'candles and for wine to the pipers for three dances and of "four gilders" to the four pipers for three dances.'[35] There are also several entries in Augsburg financial accounts relating to the involvement of the city's *Pfeifer* in dances for the visiting royalty and nobility: in 1489 a payment was recorded to the 'three pipers for the dance when the Duke of Braunschweig was here'[36] and two years later, another was made to 'the 3 pipers for 2 dances when the Princes of Saxony were here'.[37]

While the elite were therefore permitted the town hall or dance halls for these refined activities, the general populace entered with verve into the more liberated round dance, which was generally performed in the open air to cater for the large number of participants and great amount of movement. The informality of these public dances is reflected in the fact that there are far fewer references to these gatherings than to the dances officially approved by the council. It is therefore often difficult to establish who danced and played music for these diversions, yet the *Stadtpfeifer* were in all likelihood involved on occasion. As noted above, we know that they regularly performed in the open, late at night (to the chagrin of their council employers), and that their performances were therefore not restricted to the entertainment of the elite: in 1489, the town council of Augsburg made a rare payment for 'wine for the pipers for the citizens' dance'.[38]

Easier to establish is the participation of other (most probably freelance or amateur) musicians – another indication of the informality of these pastimes. These played on various combinations of wind, string and keyboard instruments, including the ever-popular pipe and tabor; and it was also customary that the participants performed songs in a round while they danced, sometimes without instrumental accompaniment.[39] These practices are confirmed in a number of administrative documents, which again convey the repeated attempts of the oligarchic town councils to assert some kind of control on the citizens' dances despite their ostensibly informal nature.[40] In Basle in 1492, a decree was issued forbidding 'the dances which have until now been pursued on the open streets with the playing of pipes, lutes and other string instruments'.[41] Likewise, the songs performed during round dances in Strasbourg were under the constant watch of the authorities, who in 1533 ordered: 'The round dances are permitted yet not the singing of shameful songs therewith'.[42]

Music for Shrovetide

Dances as well as banquets, plays and, most notably, carnivals all formed part of the extravagant civic revels for Shrovetide, which occurred in the days preceding Ash Wednesday, and offered a final opportunity for celebration before the temperate season of Lent. During the carnival period, the usual conventions of civic life were seemingly overthrown, as was expressed in the costume parades, masked balls, satirical plays and the generally unrestrained behaviour that engulfed all corners of the cities and their communities. Yet, although these revelries exhibited the greatest amount of freedom and flamboyance, the constraints of the social hierarchy and ubiquitous council control still lay at the heart of the celebrations. In Regensburg, the elite still maintained an exclusive and dignified circle at the town hall, again employing the city's official musicians for the accompaniment of their dances. Accounts of the council include numerous payments to the *Stadtpfeifer* of the city for their participation in these annual celebrations: 'when on the three nights of Shrovetide, dances were held at the town hall, we had wine and confectionery and other things … we gave the pipers for the same dances, the three, vj ß xvj d …'.[43] Carl Theodor Gemeiner, in his study of the city's history, reported of the Shrovetide celebrations at Regensburg's town hall which appear to have been granted to the patricians and visiting nobility:

> The most celebrated was the Shrovetide dance, that was held for four days in the rooms of the town hall or in the drinking hall … It seems that the festivity was held through shared costs, in such a way that both the ecclesiastical prelates [the Bishop and Abbot of St. Emmeram] supplied the pastries, the council provided wine and music … For these sumptuous nights, during which performances and figured dances were also pursued, the municipal funds incurred not inconsiderable costs.[44]

The 'figured dances' referred to here were most probably the carefully ordered processional dances of the patricians – the *basse danse* or German equivalent, which were, as was customary, accompanied by the council's musicians, the *Stadtpfeifer*. Maximilian I himself is also known to have participated in the Shrovetide festivities of the civic elite: during the 1491 Imperial Diet in Nuremberg, he attended a dance at the town hall, and in Constance in 1507, he was present at a banquet and dance convened for the more affluent citizens of the town.[45] Although maintaining an air of refinement, these celebrations also involved an element of the costumed carnivals experienced on the streets: in 1522, a report of a Shrovetide celebration at Nuremberg town hall makes mention of mummers who interrupted a dance of the visiting nobility.[46]

Yet it would appear that the Shrovetide festivities of the elite were not restricted to the town hall. The carnival of Nuremberg – the so-called *Schembartlauf*[47] – provides an interesting example of a public parade on the streets, which, despite drawing together various elements of the populace and their musicians, was strongly defined by the rulings of the city council and the involvement of the city's patricians. The *Schembartlauf* is particularly significant in its representation of Nuremberg to onlookers – a city where the authority of the council was unrivalled and where the elite could shape the parade to suit their interests.[48]

Recorded in a number of like manuscripts created after the carnival's demise, the *Schembartlauf* occurred annually from 1449 to 1524 with a brief revival in 1539.[49] The parade stemmed from a long-standing tradition enacted by the butchers of the city, who had been granted permission to perform an annual dance as a result of their loyalty to the city council amidst the guild uprisings of 1348/9.[50] The butchers thus performed a round dance every year, with the participants' hands linked by holding leather rings that supposedly represented sausages. City accounts during this period reflect the continuing support of the council for the butchers' dance, as there are regular, annual references to the *Stadtpfeifer*'s musical accompaniment of this custom: 'Item to the craft of the butchers for their

Shrovetide dance, to be loaned the Stadtpfeifer … as according to ancient tradition'.[51] As was customary with round dances, it is of course possible that the butchers performed songs while they danced, in addition to the musical accompaniment provided by the *Stadtpfeifer*. The loan of the *Stadtpfeifer* by the council is especially significant here – the use of the official civic musicians on these occasions was a clear expression of the privileged position of the butchers.[52]

From around 1449, a parade was introduced, which preceded the butchers through the streets so as to clear the way for their dance. These *Läufer* ('runners') acted as a masked guard for the butchers and although they had initially comprised both patricians and craftsmen of the city, from 1468 the group primarily consisted of sons of the elite. Due to their lavish costumes, embellished with motifs such as flowers, flames and stars – unaffordable by the common *Bürger* – the parade, instead of the butchers' dance, came to dominate the Shrovetide festivities.[53] By the early sixteenth century, a second group appeared alongside the elegant runners, of grotesquely costumed individuals dressed as giants and wild men. These represented the carnival's reversal of the usually ordered world of the city, albeit under the supervision of the council – in 1469 the council forbade unauthorized citizens from wearing masks and similar costumes, or from exhibiting inappropriate behaviour in their celebrations.[54]

The *Läufer* customarily performed a 'springender Tanz' typically accompanied by a piper and drummer, which therefore contrasted with the music of the esteemed *Stadtpfeifer* allowed to the butchers. In 1460, the *Läufer* were attired in costumes 'divided in red and white with small red and white flames; they had no drummer but instead a bagpiper, who was clothed in the same colours'.[55] The regular use of the official piper and drummer of the city for this dance is reflected in periodical payments to these musicians, as recorded in 1510: 'to Wegerhan and his fellow the drummer for the evening of Shrovetide ½ guilder …'.[56]

From 1475, the patrician 'Läufer' introduced a pageant to the parade – the *Hölle* (hell) – which they pulled through the streets, and was then traditionally stormed by the people and set alight in front of the town hall.[57] These tableaux typically presented secular themes, ranging from gardens, castles and boats to mythical creatures such as dragons, basilisks and giants. As the grand finale to the Shrovetide celebrations, this pageant also drew together diverse musical resources of the city. Several, slightly contrasting descriptions of the *Hölle* of 1518 confirm this integral musical presence, one of which reports:

> The Hölle was a Venus Mountain … 20 people were appointed in the *Schembart* colours … consisting of several goldsmith fellows, who had their own drummer and piper and … an appointed trumpeter … furthermore, a trumpeter of the honourable council, Neuschel, together with his fellows.[58]

The 'trumpeter of the honourable council, Neuschel' was Hans Neuschel – the esteemed brass instrument maker and city trombonist,[59] who here played together with his fellow *Stadtpfeifer*, as well as an additional drummer, piper and brass instrumentalist on a float involving the goldsmiths of the city. The combination of these musical resources reflects the grandeur and apparent freedom of the *Schembartlauf* on the whole, which enfolded different parts of the city in its path, as well as various elements of civic society. The social divide of the Nuremberg community at first appears to have been transgressed on this occasion: artisans and patricians all participated in the same carnival, as did various musicians of the city. Yet, in reality the dominance of the social elite remained: the carefree world of the carnival was in fact strictly controlled by the civic authorities.

Conclusions

Life within German cities during the reign of Maximilian I was entirely defined by the social hierarchy enforced by the oligarchic municipal authorities. All elements of the daily lives of the citizens reflected the social division between the elite and the common folk, and the different sounds – and musical performances – which arose from their everyday affairs were an audible representation of the complexities of urban society. From the city watchmen and military musicians whose signals met the territorial and political demands of the period, to the travels of the citizens by sleigh or by boat, scenes in the cities' streets, squares and public buildings, and private matters in the home – these all contributed to a multifaceted civic soundscape. Events that required performances by the cities' various musicians – such as for weddings, dances and Shrovetide – clearly exemplify this musical demarcation of the social structure and the control of the city councils, who prescribed or precluded elements of each celebration, including the instrumentalists to be used. Together, the diverse sounds of the urban space not only reflected the affairs and affluence of each city, but also the political and social structures which underpinned all aspects of civic life, and musical practices therefore contributed to the social decorum that the town councils wished to create.

Archival Documents

Augsburg Stadtarchiv, Baumeisterbücher: 78–113 (1485–1519)
Constance Stadtarchiv, Hs. A I/ 8, vol. 2.
Nuremberg Staatsarchiv, Handschriften 438
Nuremberg Staatsarchiv, Rep. 52b, Nr. 240: Nürnberg Hochzeitsordnung von 1503. Mit Nachträgen bis 1510
Nuremberg Staatsarchiv, Rep. 54/18–25, Kleines Register (1478–1520)
Nuremberg Staatsarchiv, Rep. 54/181 and Rep. 54/182, Grosse Stadtrechnung (1486–1536)
Nuremberg Staatsarchiv, Rep. 60a, Verlässe des Inneren Rats (1484/5–1530/1)
Nuremberg Staatsarchiv, Rep. 60b/4–11, Ratsbücher (1483–1521)
Regensburg Stadtarchiv, Cameralia 17–25a (1472–1535)
Regensburg Stadtarchiv, Literalien 438a: Ambtpuech 1507
Regensburg Stadtarchiv, Bürgerbuch 3 (1500–1520)
Schwäbisch Hall Stadtarchiv, Rep. 4/198, Eidbuch
Schwäbisch Hall Stadtarchiv, Steuerrechnungen, Rep. 4/a11–a18 (1484–1521)

Notes

1 The research for this study was carried out as part of my DPhil dissertation 'Stadtpfeifer and Varende Lewte: Secular Musical Patronage in the Imperial Cities of Germany during the Reign of Maximilian I (1486–1519)' (University of Oxford, 2006), which was funded by an AHRC Postgraduate Award.

2 This chapter will focus on the period of Maximilian I's reign as King of the Romans (under his father Emperor Frederick III), then Holy Roman Emperor, from 1486 until his death in 1519. Although Germany did not exist in its present state during this period, the term 'German' will be used in this study to refer to the same geographical area as the present country, together with cities such as Basle and Strasbourg, which belonged to the Empire at that time (Basle would in fact renounce the Empire to join the Swiss confederacy in 1501).

3 For the gradual acceptance of instrumentalists as court and civic employees, and the origins of civic and court ensembles, see Walter Salmen, *Der fahrende Musiker im europäischen Mittelalter* (Kassel: Johann Philipp Hinnenthal Verlag, 1960) and Keith Polk, 'The Trombone, the Slide Trumpet and the Ensemble Tradition of the Early Renaissance', *Early Music*, 17 (1989): 389–97.

4 I would like to thank the civic archives of Augsburg, Bad Windsheim, Constance, Dinkelsbühl, Nuremberg, Nördlingen, Regensburg and Schwäbisch Hall for allowing access to relevant documents.

5 See, for example, Keith Polk, *German Instrumental Music of the Late Middle Ages: Players, Patrons and Performance Practice* (Cambridge: Cambridge University Press, 1992) and Keith Polk, 'The Schubingers of Augsburg: Innovation in Renaissance Instrumental Music', in Friedhelm Brusniak and Horst Leuchtmann (eds), *Quaestiones in musica. Festschrift für Franz Krautwurst* (Tutzing: Schneider, 1989), pp. 495–503; Ekkehart Nickel, *Der Holzblasinstrumentenbau in der freien Reichsstadt Nürnberg* (Schriften zur Musik 8; Munich: Emil Katzbichler, 1971); Fritz Jahn, 'Die Nürnberger Trompeten- und Posaunenmacher im 16. Jahrhundert', *Archiv für Musikwissenschaft*, 7 (1925): 23–52.

6 There are numerous references to the shields and liveries bestowed by the councils on members of the *Stadtpfeifer* ensembles. In Regensburg in 1507, a 'Caspar hofman goldsmid' (*Caspar Hofman goldsmith*) was paid for the repair 'von funff schilltn fur die pfeiffer vnd trometter' (*of five shields for the pipers and trumpeters*) (Regensburg Stadtarchiv, Literalien 438a, fol. 3^r) and around the same time, payments were registered there for 'rote Nurnberger tuech' (*red Nuremberg cloth*) given to the 'zwaien pfeiffern' (*two pipers*) and the 'pusawner' (*trombonist*) for their liveries. See Regensburg Stadtarchiv, Bürgerbuch 3 (1500–1520), fols 132^v–133^r. All translations are my own.

7 '… werden schweeren, das sie ainem Stetmaister vnd Rat, alhie, gehorsam, getrew, gewertig, vnd willig sein, warzu, oder wahin sie ain Rat gebrawchen werde. Es seÿ in der Stat oder awsserhalb In das veld, verr oder nahet' (Schwäbisch Hall Stadtarchiv, Rep.4/198, Eidbuch, fols 53^r–53^v). Further oaths exist for the *Stadtpfeifer* of Cologne, Ulm and Worms, as well as appointment documents for members of the ensembles of Nördlingen, Nuremberg and Regensburg, which express their sworn duty to the council.

8 For the range of musical patronage in German cities and courts during this period, see Polk, *German Instrumental Music*.

9 For the great range of civic ensembles, see Polk, *German Instrumental Music*, pp. 237–9. See also Helen Green (now Coffey), 'Defining the City "Trumpeter": German Civic Identity and the Employment of Brass Instrumentalists *c.*1500', *Journal of the Royal Musical Association*, 136 (2011): 1–31, at pp. 18–20.

10 Lorenz Welker, '"Alta Capella": Zur Ensemblepraxis der Blasinstrumente im 15. Jahrhundert', *Basler Jahrbuch für historische Musikpraxis*, vii (1983): 119–65. There has been much discussion and contention regarding the form of the brass instrument

used in the *alta capella*. See, for example, Peter Downey, 'The Renaissance Slide Trumpet: Fact or Fiction?', *Early Music*, 12 (1984): 26–33; Polk, 'The Trombone'; Ross W. Duffin, 'The "*trompette des menestrels*" in the 15th-century "*alta capella*"', *Early Music*, 17 (1989): 397–402; Herbert W. Myers, 'Slide Trumpet Madness: Fact or Fiction?', *Early Music*, 17 (1989): 383–9 and Trevor Herbert, *The Trombone* (New Haven and London: Yale University Press, 2006), especially pp. 52–60.

11 Polk, *German Instrumental Music*, pp. 112–13. Instruments such as crumhorns and recorders tended to be played in complete consorts of instruments rather than joining the basic shawm band.

12 See Polk, *German Instrumental Music*, pp. 60–71.

13 Nuremberg Staatsarchiv, Rep.54/18, fol. 96^{r}.

14 Nuremberg Staatsarchiv, Rep.54/181, fol. 332^{v}.

15 Ibid., fol. 519^{r}.

16 See Keith Polk, 'Voices and Instruments: Soloists and Ensembles in the 15th Century', *Early Music*, 18 (1990): 179–98, at p. 182.

17 Several extant oaths and chronicles document the work of the city watchmen. For details of their duties, see Edmund A. Bowles, 'Tower Musicians in the Middle Ages', *Brass Quarterly*, 5 (1961–62): 91–103 and Green, 'Defining the City "Trumpeter"'.

18 In 1488, payments for Augsburg's campaign to Flanders included one made to the 'pfeyffern vnd Trumbslager' (*pipers and drummer*) (Augsburg Stadtarchiv, Baumeisterbücher Nr.81, fol. 85^{r}). 'Hanns Trumeter' is listed amongst the soldiers employed by the council of Schwäbisch Hall during the final decades of the fifteenth century (Schwäbisch Hall, Stadtarchiv, Rep. 4/a11–a13).

19 See, for example, Nuremberg Staatsarchiv, Rep.54/181, fols 466^{v,} 484^{v}, 505^{r}, 523^{v}, 547^{r}, 568^{v}, 594^{r} and 615^{v}.

20 Nuremberg Staatsarchiv, Rep. 60b/5, fol. 223^{v}.

21 'die Stat pfeiffern … zesagen, daz Sie nachts so spat auff den gassen hofieren gangen sind, vnd Inen zu verpoten, solichs vnd der gleichen zethunde'. Nuremberg Staatsarchiv, Rep. 60b/6, fol. 70^{r}.

22 'das schanntlich singen und schreyen, so des nachts uff der gassen beschickt, dazu das pfeyffen, trumpen, lautenschlageren, geygen und ander Sayttenspil, keine ausgenommen'. Eberhard Stiefel, 'Musikgeschichte der ehemaligen Reichsstadt Schwäbisch Gmünd', PhD dissertation (University of Tübingen, 1949), p. 120.

23 'Der Schlitten halb, und wie man mit der Pauken soll fahren, ist durch Hauptmann und Rath vertragen, daß Edel und Burger oder andere der Erbarkeit mit dem Schlitten wohl fahren mögen, doch daß kein Unzucht und solch Geschrey, deshalb ein Aufruhr oder ander Unrath daraus erwachsen möcht, entstehe und getrieben werde. Man mag sich auch daneben eines erbarn Freudenspiels als Singen, Lauten, Pfeiffen und Trummen ziemlich gebrauchen. Doch den Handwerkgesellen und dergleichen unstatthaften Personen soll es nit erlaubt seyn'. Rathsprotokoll, fol. 148, cited in Carl Theodor Gemeiner, *Regensburgische Chronik*, ed. Heinz Angermeier (Munich: C.H. Beck, repr. 1987), iv, pp. 156–7.

24 'wolfen haller Jobst hallers Son mit Anthoni Koburgs tochter zu Ir vorhabend hohzeit vff Sant Sebastians abent, das Rathaws die Stuben, vnd die hofierer ze leihen' Nuremberg Staatsarchiv, Rep.60a, Nr.259, fol. 5^{v}.

25 'vom tanntz als das frolin von Bayrn mit dem Marggrafe oder er mit Ir hie hochzeit het' Augsburg Stadtarchiv, Baumeisterbücher, Nr.112, fol. 67^{v}.

26 Gerhard Pietzsch, 'Die Beschreibungen deutscher Fürstenhochzeiten von der Mitte des 15. bis zum Beginn des 17. Jahrhunderts als musikgeschichtliche Quellen' *Anuario Musicale* 15 (1960): 21–62, at pp. 35–6.

27 Records of the Nuremberg council not only refer to the participation of the *Stadtpfeifer* in wedding celebrations, but also of the official fifer and drummer of the city. See Nuremberg Staatsarchiv, Rep.60a, Nr.525, fol. 18^{r}.

28 For an overview of the councils' control of such celebrations see Neithard Bulst, 'Feste und Feiern unter Auflagen. Mittelalterliche Tauf-, Hochzeits-, und Begräbnisordnungen in Deutschland und Frankreich', in Detlef Altenburg, Jörg Janut and Hans-Hugo Steinhoff (eds), *Feste und Feiern im Mittelalter. Paderborner Symposium Mediävistenverbandes* (Sigmaringen: Thorbecke, 1991), pp. 39–51.

29 'Item man sol auch ainicherlay spillewt oder lotter zu ainicher hochzeit nit herein, noch zu der malzeyt laden. noch da essen lassen, ausgenomen die, die mit ainem preutigam oder praut von lannd herein komen, oder die der statt schylt trügen, mitsambt dem hegelein, der zum tanntz ledt. Unnd ob auf denselben tag ainich fürst hie were, und des spilleut auf die hochzeyt kemen, die mocht man da essen lassen, ungeverlich. Ob aber yemandt solicher spilleut, die der statt schilt trügen, zu haben nit vermöcht oder haben wölt, der mocht wol an derselben statt ainen, zwen oder drey annder haben, und bestellen, ungeverlich'. Cited in Rainer Gstrein, 'Tanzmusik-Ensembles zur Zeit und am Hofe Kaiser Maximilians I' *Jaarboek van het Vlaamse Centrum voor Oude Muziek*, III (1987): 79–96, at p. 81.

30 'So hinfuro yemant die statpfeiffer nachts zuhofieren gebrauchen will, der soll zur (Ine) ein guldin zugeben, vnnd sie darumb pflichtig sein, auff begere desselben, vor acht heusern, oder orten zuhofirn ... Wolt aber yemant daruber, Weyter oder an mer orten lassen hofirn, derselb solt von einem yeden ort vnnd vor yedem hawse zu hofiern ein halb ort eines guldins den statpfeiffern zugeben, schuldig sein'. Nuremberg Staatsarchiv, Rep. 52b, Nr.240, fol. 18$^{v.}$

31 'eynichen sollichen tanntz annders dann was gewonlicher tenntz, die von alter herkomen seindt, nit pfeyffen, schlahen noch machen'. Cited in Gstrein, 'Tanzmusik-Ensembles': 81.

32 'Item eß mag auch eyn brudgam im merern stande sehs und im niddern stande dry spielelute zu syner hochzyt und nit daruber haben ... wo er der fursten oder sunst anderer stete spielute, nemlich trumpter und pfiffer her inne brechte, der eynem sol er nit uber zwene gulden und eynem, der inne der stat hie wonet, nit uber eynen gulden, aber eynem frembden lutensleger auch eynen gulden und eynem heymschen, auch eym gyger, buckeler und sackepfiffer eynen halben gulden und nit dar uber ...'. Cited in Gstrein, 'Tanzmusik-Ensembles': 81.

33 Walter Salmen, 'Dances and Dance Music *c.* 1300–1530', in Reinhard Strohm and Bonnie J. Blackburn (eds), *The New Oxford History of Music* iii, Part 1: *Music as Concept and Practice in the Late Middle Ages* (2nd edn., Oxford: Oxford University Press, 2001), pp. 162–90, at p. 173–6.

34 'Jobsten Haller dem Jungn Ist zu seine vorhabend gesellentantz vff hewt zehaben das Rathauß die Stuben vnd pfeiffer' Nuremberg Staatsarchiv, Rep.60a, Nr 193, fol. 3^{v}.

35 'kertzen vnd vmb Wein den pfeyffern bey dreyen Tantzen; den vier pfeyffern von dreyn Tantzen'. Augsburg Stadtarchiv, Baumeisterbücher Nr.78, fol. 31^{r}.

36 'dreyen pfeiffern vom Tantz als der Hertzog von Prawnschweigk hie was'. Augsburg Stadtarchiv, Baumeisterbücher Nr.82, fol. 36^{v}.

37 'den 3 pfeiffern von 2 Tanntzen als die Fürsten von Sachssen da warn'. Augsburg Stadtarchiv, Baumeisterbücher Nr.85, fol. 43^{v}.

38 'wein den pfeyffern zum Burgers Tanntz'. Augsburg Stadtarchiv, Baumeisterbücher Nr.82, fol. 38^{r}.

39 Salmen, 'Dances', pp. 184 and 187.

40 For an overview of the restrictions see Wolfgang Brunner, 'Städtisches Tanzen und das Tanzhaus im 16. Jahrhundert', in Alfred Kohler and Heinrich Lutz (eds), *Alltag im 16. Jahrhundert: Studien zu Lebensformen in mitteleuropäischen Städten* (Wiener Geschichte der Neuzeit. Bd.14; Wien: Verlag für Geschichte und Politik, 1987), pp. 45–64.

41 'die tantz, so uff offener gassen mit pfiffen, lutenslahen und andern seitenspilen bisshar gepflegen sind' Fritz Ernst, 'Die Spielleute im Dienste der Stadt Basel im ausgehenden Mittelalter (bis 1550)', *Basler Zeitschrift für Geschichte und Altertumskunde*, 44

(Basel, 1945): 79–236, at p. 203. In Schwäbisch Gmünd, public dances were also forbidden, although it is questionable whether the citizens conformed to this command – the celebration of dances 'uff der gassen' was repeatedly precluded. See Eberhard Stiefel, 'Musikgeschichte', p. 135.

42 'Die rundtaenze seien zugelassen doch nicht schaendliche Lieder daran zu singen', Martin Vogeleis, *Quellen und Bausteine zu einer Geschichte der Musik und des Theaters im Elsass, 500–1800* (Geneva: Minkoff, 1979), p. 228.

43 'als man die drej nacht zu vasnacht auf dem Rathaws Tannzt habn wir vns wein Confeckt vnd anders … wir gabn den pfewffern von denselbn Tännzn ir dreyen vj ß xvj d'. Regensburg Stadtarchiv, Cameralia 17, fol. 92[v]. There are also frequent entries in the Augsburg accounts which refer to the employment of the *Stadtpfeifer* for dances for Shrovetide, although the venue and participants of the dances are not noted. In 1487, payments were made 'vmb 2 tantz den pfeiffn zu der faßnacht' (*for 2 dances to the pipers for Shrovetide*) and in 1502 'den Stattpfeiffern vom Tanntz zü vasnacht' (*to the Stadtpfeifer for the dance at Shrovetide*). See Augsburg Stadtarchiv, Baumeisterbücher Nr.80, fol. 31[r] and Nr. 96, fol. 35[v].

44 'Um feyerlichsten war der Fastnachttanz, der die vier Tage auf dem Rathhaussaale oder auf der Trinkstube gehalten worden war … .Es scheint, die Lustbarkeit war auf gemeinsame Kosten veranstaltet worden, solchergestalt, daß die beiden geistlichen Prälaten für die Krapfen, der Rath für Wein und Musik gesorgt hatte … Es hatten diese schwelgerischen Nächte, während welcher auch Vorstellungen und figurirte Tänze gegeben zu werden pflegten, der Stadtkammer nicht unbeträchtliche Kosten veranlaßt'. Gemeiner, *Regensburgische Chronik*, iv, pp. 157–8.

45 See: 'Etliche Geschichten 1488–1491', in Historische Commission bei der Königlichen Akademie der Wissenschaften, ed. Carl Hegel, *Die Chroniken der fränkischen Städte. Nürnberg*, vol. 5, Die Chroniken der deutschen Städte, vol. 11 (Leipzig: S. Hirzel, 1874), pp. 715–33, at p. 732 and Constance Stadtarchiv, Hs. A I/ 8, ii, p. 224½.

46 Nuremberg Staatsarchiv, Handschriften 438, fols.23[r]–24[r].

47 It is not possible to establish the exact meaning of the name, *Schembart*. It has previously been hypothesized that it derives from the word 'Schoenbart' in reference to the grotesque masks worn by many of the participants. See Hans-Ulrich Roller, *Der Nürnberger Schembartlauf. Studien zum Fest- und Maskenwesen des späten Mittelalters* (Magstadt: Horst Bissinger, 1965), pp. 48–9.

48 Samuel Kinser, 'Why Is Carnival So Wild?', in Konrad Eisenbichler and Wim Husken (eds), *Carnival and the Carnivalesque: The Fool, the Reformer, the Wildman, and Others in Early Modern Theatre* (Amsterdam: Rodopi, 1999), pp. 43–87, at pp. 58–9.

49 Roller, *Der Nürnberger Schembartlauf*, p. 8.

50 These uprisings resulted in the council's abolition of all guilds in the city.

51 'Item dem hantwerck der fleischhacker zu Irem Vasnachttantz die Stat Pfeiffer … ze leihen wie von alter herkomen ist'. Nuremberg Staatsarchiv, Rep. 60b/6, fol. 2[r].

52 This dance, together with the *Stadtpfeifer*, is depicted in a number of the manuscripts that commemorate the carnival. See for example Stadtbibliothek Nürnberg, Nor.K.444, fol. 1[r].

53 Samuel L. Sumberg, *The Nuremberg Schembart Carnival* (New York: Columbia University Press, 1941), pp. 72–5; Kinser, 'Why Is Carnival So Wild?', p. 62.

54 Sumberg, *The Nuremberg Schembart Carnival*, p. 97 and Samuel Kinser, 'Presentation and Representation: Carnival at Nuremberg, 1450–1550', *Representations*, 13 (Winter 1986): 1–41, at pp. 4–5.

55 'in Rodt und Weyß abgethailt mit Rodt vnd Weÿsen Flemlein, hetten Kein Trumel sondern einen Sackpfeiffer, Jn gleiche farb geklaidet': Sumberg, *The Nuremberg Schembart Carnival*, p. 44.

56 'dem Wegerhan und seinen gesellen drumelschlaher czu nacht an der faßnacht ½ gulden …'. Wilhelm Loose (ed.), *Anton Tuchers Haushaltbuch (1507 bis 1517)* (Tübingen: H. Laupp, 1877), p. 77. Similar payments to the fifer and drummer of the city include to 'Hannß Schober Pauckher' and 'Hanns Behaim Pfeiffer' in 1518 and again in 1539, to a 'Thrumel schlager' and 'Pfeifer'. See Sumberg, *The Nuremberg Schembart Carnival*, p. 88.

57 Again, the derivation of the term 'Hölle' is unclear, yet is supposed to have represented a world of sin to the citizens before the commencement of Lent. See Therese Bruggisser-Lanker, 'Die Stadtpfeifer von Nürnberg im 16. Jahrhundert. Ikonographische und quellenkritische Hinweise zur Aufführungspraxis', *Schweizer Jahrbuch für Musikwissenschaft*, n.s.10 (1990): 43–72, at p. 46–7.

58 'Die heell was ein Fenusperg … Inn den schemparttsfarben seintt 20 personen verordnett gewessen … aus ettlichn golttschmidt gesellen, haben ihrer aigene drumell Vnd pfeiffen gehabt vnd … ein verordneter drumeter … Mehr eines Erbarn raths, Drumetter der Neuschel samptt seinen gesellenn …'. In Roller, *Der Nürnberger Schembartlauf*, p. 215. Further descriptions confirm this combination of instrumentalists while another refers instead to the presence of string instrumentalists. See Sumberg, *The Nuremberg Schembart Carnival*, pp. 164–5 and Roller, *Der Nürnberger Schembartlauf*, p. 216.

59 For details of Neuschel's work as brass instrument maker see Martin Kirnbauer, 'Die Nürnberger Trompeten- und Posaunenmacher vor 1500 im Spiegel Nürnberger Quellen', in Walter Salmen (ed.), *Musik und Tanz zur Zeit Kaiser Maximilian I: Bericht über die am 21. und 22. Oktober 1989 in Innsbruck abgehaltene Fachtagung* (Innsbruck: Helbling, 1992), pp. 131–42.

9 Sonic Afterworld

Mapping the Soundscape of Heaven and Hell in Early Modern Cities[1]

Daniele V. Filippi

Introduction

Early modern Europe inherited the far from obvious idea of a sonic afterworld from earlier Christian traditions. It is not possible here to trace the complex history of this idea or even to précis it;[2] rather, in the first part of this chapter I will give an account of the notion of 'otherworldly' soundscapes in the sixteenth and seventeenth centuries through the works of two leading spiritual authors of the time: the Dominican, Luis de Granada (1504–88), and the Jesuit, Jeremias Drexel (1581–1638). Their works were read widely and can be said both to have reflected standard ideas on the subject and to have contributed to their shaping. In the second part of the chapter, I will explore the interaction between celestial and terrestrial soundscapes in early modern cities and, finally, I will briefly discuss the negative ideal of the infernal soundscape.

The main features of these 'otherworldly' soundscapes are to be found primarily in the Bible (though integrating with other, sometimes earlier, sources) and are mentioned and discussed by ancient and mediaeval ecclesiastical authors with remarkable insistence and continuity,[3] within descriptions of the afterlife as a whole. During the Late Antiquity and Middle Ages, the constant interest in the life to come found expression in a series of highly imaginative literary works – such as the famous *Visio Tnugdali* – which were conceived as reports of travels in the realms of Heaven and Hell. These works, which had an immense diffusion all over Europe, significantly developed the sonic characteristics of the afterworld, merging ideas of biblical or ecclesiastical-liturgical origin with elements of apocryphal or Greek-Roman derivation while also adding new features.[4] Dante's *Comedy* can be seen as the supreme apex of this tradition.

It is important to note that in magisterial documents and official theological-doctrinal writings we usually find very sober descriptions, particularly of Heaven, in which, moreover, the *visual* aspects prevail (beatitude being commonly characterized as a *visio beatifica*, in accordance with a complex constellation of theological and philosophical beliefs). While they provide the fundamental doctrinal premises for the conception of a sonic afterworld (the most important premise being that the five senses will be active in the blessed and in the damned),[5] it is in other genres that this theme is described with richness of detail: in the 'travel reports' already mentioned; in mystical and *lato sensu* spiritual literature; in sermons; in liturgical and non-liturgical poetry; and, in hagiographic writings. These sources exerted great influence on believers and inspired figurative artists, architects and musicians.

The idea that Heaven possesses a relevant sonic dimension gave music a unique status among human *artes*, as we read in Aurelian of Réôme's (fl. 840–50) *Musica disciplina*:

> [Praise of the discipline of music] What more? We read that even among the celestial citizens the splendour of this art is celebrated, as in the Apocalypse: 'Having the harps of God ...' [Apocalypse, 15: 2] and elsewhere ... And what more? *The art of music surpasses all the arts*. No reader of the Apocalypse can doubt that also the angels in the heavenly citadel praise God after the manner of this discipline.[6]

Proud words of this sort recur frequently in mediaeval treatises and other music-related writings, often in formulaic patterns. One of the most bombastic examples of these encomia is to be found in Amerus's *Practica artis musice* of 1271:

> She is certainly the excellent queen of all the liberal arts, *the only one who leaves the other six arts below and flies to the tribunal of the Omnipotent*, as in the eighteenth chapter of the Apocalypse [*recte* Apocalypse, 14: 3]: 'And they sung as it were a new canticle, before the throne and before the four living creatures'. She is the supreme lady, who *serves God in the Church militant and triumphant with her sweet sounds* ... Every clergyman who aspires to a position or any duty in the Church of God should know that; as Isidore says in the book of Etymologies: 'not to know how to sing is not less shameful than not to know letters' [cf. Isidore of Seville, *Etymologies*, b. 3, ch. XVI]. Because the saints in heaven, with the angels and archangels, the thrones and dominations, and all the troops of the heavenly host never stop singing an hymn every day: 'Holy, holy', and so on. Also on earth the angels sang: 'Glory to God in the highest' [Luke, 2: 14] – *and angels neither are interested in literature, nor engage in logical disputations*.[7]

Similar thoughts are echoed by early modern theorists such as Andreas Ornithoparchus (even though in a humanistic, classicizing context):

> And only [music], leaving earth, flies to the tribunal of the Supreme Judge, where the instruments of the saints play continuously: where Angels and Archangels indefatigably sing an hymn to God, where Cherubim and Seraphim cry 'Sanctus' with unceasing voice. Moreover, no discipline can be perfect without music [cf. Isidore of Seville, *Etymologies*, b. 3, ch. XVII].[8]

The Soundscape of the Afterworld According to Granada and Drexel

In the course of the meditations for each day of the week included in his best-seller manual *Libro de la oración y meditación* (Salamanca, 1554; I will quote from Richard Hopkins's English translation of 1582),[9] Louis de Granada invites his devout readers to reflect on the Four Last Things: Death, Judgement, Hell and Heaven. Leaving aside the first, as still pertaining to this world, let us briefly consider the sonic elements of his description of the other three.

The first classic sonic marker of the Day of the Judgement comes directly from the Old Testament (Zephania, 1: 16): it will be '*a daie of the trompette and alarom against the stronge cities, and against the highe towers*'.[10] Men will wander in great anguish and fear of death 'hearinge the terrible roringes of the Sea'.[11] According to the Sibyls, Granada informs us, 'the beastes shall goe bellowinge and roringe throughe the feildes, and cities'.[12] After 'that dreadfull sownde of the trompett, wich the Archangell shall blowe, to sommon and calle all the generations of the worlde to assemble together in one place' for the judgement, 'Our Lorde shall come like a tempest, and furious whyrlewinde'.[13] Among the signs preceding the general judgement, Granada mentions earthquakes and sea storms: 'Such as dwell by the Sea side shalbe in great dread and terror by reason of the great rysinge of waters: and such as dwell further of shalbe wounderfullie afraide of the horrible roringes, and noyces of it, wich shalbe extremelie outragious, that they shalbe hearde manie myles of'; men will be astonished, confounded, 'bereud of their senses, of their speech, and of their tast of all thinges'.[14] In sum, 'all thinges shalbe tossed and turmoiled upside downe, and be full of terror, and confusion': but the good will raise their heads and sing joyfully to God.[15] The soundscape of the Judgement, according to Granada, foreshadows the sonic and conceptual oppositions we will find in fuller form in Heaven and Hell: order versus disorder; harmony versus noise; human song versus bestial cries and loud, alarming noise.

Regarding Heaven, Granada fully adopts the classic doctrine: the angels 'doe singe lawdes and prayses unto almightie God' and the blessed exultantly join in.[16] 'There is the place where that songe is alwaes Songe, that S. Iohn hearde in his revelations, which songe he tearmeth, *Quasi canticum novum: As it were a newe songe*. For that althowghe the songe be alwayes after one maner … yet it is alwaies newe, as concerninge the taste and delighte it hath'.[17] In Heaven, 'Each one of [the senses] shall there have his owne proper delighte and glorie … The eares shall alwaies heare that wounderfull musicke, which is so exceedinge sweite, and pleasant, that one onely sownde of it', Granada concludes, 'were able to bringe all the hartes of this worlde a-sleepe'.[18]

On this curious characterization of the exceeding sweetness of celestial music, we leave Granada and turn to Drexel.[19] The Jesuit tries, in typical Seicento fashion, to give his readers an idea of the sonic delight of Heaven by means of hyperbolic descriptions ('the music of many thousands Orpheus will charm the ears. Whose pleasure will be twofold: [the pleasure] of conversations and of music')[20] and imaginative comparisons. If a music lover, he writes, could hear the wonderful

Psalms performed by just 20 angels (six string instrument players, six singers, six 'citharoedi' and two organists), he would undoubtedly exclaim: 'Lord Jesus, if I had one hundred ears, if my whole body were an ear, I would listen to these singers day and night, and I would not be satiated with listening; neither food nor sleep would take me away from this honey for the ears!'[21]

According to Drexel, amazing aural experiences await us in Heaven. On this earth, one cannot perceive all the most beautiful musical passages simultaneously: 'Now one hears sweet melodic curves; now elegant cadences; after a while, again sweet melodies; then a gracefully ornamented figure; here and there some things meddle in, which are not as pleasing for the ears'.[22] But in Heaven 'every joy is all the pleasure, and in every moment one perceives the same pleasure as at his first entrance. "You shall make them drink of the torrent of your pleasure" [Psalms, 36: 9]'.[23] Moreover, the blessed will enjoy – indeed, the dream of every solitary singer! – the gift of a polyphonic voice: 'in heaven each saint generates an individual harmony of sounds … so that a single blessed man can produce an harmonious concord of sounds [*symphoniam et concentum*]. How, then, will the exulting song [*jubilum*] of such a multitude of singers be?'[24]

The heavenly soundscape is, thus, characterized by the perpetual liturgical music sung and played by the angels and the blessed, fulfilling the vision of the Apocalypse. The extreme sweetness of this music, causing unspeakable delight, the incalculable number of celestial performers and the extraordinary sonic experiences made possible by their supernatural skill, are, at the same time, elements of a transfiguration of earthly music and of a longed-for archetype.

As in Heaven the blessed are delighted through their senses, in Hell the damned will suffer the *poena sensus* (beside the *poena damni*): 'there shalbe no sense, neither within nor without a man, but that it shall suffer his proper tormente … The eares, with the confusion of such horrible cryes, and lamentations as shall there be harde', writes Granada.[25] He suggests to his readers that Hell should be imagined as 'a horrible and darcke Cittie wholie bourninge with terrible flames of fire, in which none other noise were there to be harde, but onely the furious raginge of hellysh tormentors, and ruethfull lamentations of damned persones tormented with continual weapinge, and wailinge, and gnashinge of teeth'.[26] Further on, the Dominican specifies:

> The eares shalbe torme[n]ted with hearinge of perpetual horrible cries, clamours, lamentations, and blasphemies … For like as in heaven there shalbe none other sounde harde, but onely a continual Alleluia, and praises of almightie God: even so shall there none other sounde be harde in this infernall howse of tormentors, but onely blasphemies, cursinges, and banninges of almightie God, and a disordered horrible melodie of infinite iarringe noises, roringe, cruinge, sueakinge, and howlynge, at the terrible sounde of the hammers, and strookes of the hellishe tormentors, wherein shalbe … confusion, and varietie of noyses, such great howlinges, and lamentations, emonge all that miserable pryson … And that thou mayst conceive somewhat of this horrible paine, imagin with thy selfe that thou diddest passe by a verie greate deepe valley, that were full of an infinite nomber of prisoners, some hurte,

> some wounded, and some sicke, and that they were all cryenge, roringe, and howlinge … These are the mattins which there are songe. This is the miserable chappell of the prince of darknes. These be his musitions, and singinge men.[27]

Besides similar remarks in his *Infernus damnatorum carcer*,[28] Drexel underscores the inhuman, bestial factor of this fearful soundscape: 'The damned … they rage, they roar, and fill the air with horrid outcries … All of them fill their Denns with confused howlings, as beasts do, which are presently to be lead away to the slaughter. Out, alas! what musick is this, which will never cease for all eternity? … as if so many Oxen were roasted alive, or so many mad dogs strugled to break their chains, but could not'.[29] In sum, the main traits of the infernal soundscape consist of a 'disordered melodie' of loud, disturbing noises emitted by the instruments of torture and the perversion of the human voice, now able only to utter blasphemies, to weep and howl, when not annihilated into the evangelical 'gnashing of teeth' or bestialized.

Early modern Christian authors, living in a comparatively urbanized culture, tended to favour urban models over more 'rural' visions in their conception and imagination of the afterworld. As to our specific theme, two opposite paradigms loom: the City of God, whose soundscape is characterized by a perpetual, universal, exultant liturgy (strongly reminiscent of the Apocalypse), and the 'horrible and darcke Cittie' of Hell, where all human sonic experiences are perverted. These models in part reflect available urban soundscapes, and in part help to define the sonic archetypes to which earthly cities should or should not correspond: an urban space marked by splendid worship places, resonating with manifold voices and instruments praising God, versus a hideous environment invaded by disorder and noise. In the next section, I will show how the first otherworldly model interacted with early modern urban soundscapes, exploring the threefold relationship between aural experiences, conceptions about the celestial soundscape and strategies adopted to foreshadow it on earth.

Celestial Soundscape on Earth

Circular Interaction

It is striking to notice how often music-related texts of the sixteenth and seventeenth centuries mention the music of Heaven.[30] This forms the subject of countless references, metaphors and comparisons: performers are invariably compared to angels; performing spaces transform themselves into Heaven; and composers express the hope that their music will arouse in listeners and performers the desire of hearing the choirs of angels in the life to come. C. Antegnati, for example, states:

> I consider that these *Salmi* – like all the compositions I sent to the press – reach their aim when they are sung and listened to by people, who, relishing this loud sound of air compressed and put into vibration among most humble bodies, elevate their desire to the eternal and most perfect concerts of the angelic choirs in Heaven.[31]

The connection of liturgical music to Heaven is almost obvious ('Catholic liturgy, in particular, is impregnated with paradisiacal themes', observes Jean Delumeau,[32] referring to elements of both the daily and annual liturgical cycles). But the importance of this theme becomes evident when one considers that also sacred non-liturgical and even secular performances are frequently likened to heavenly music. Gino Stefani quotes, for instance, an anonymous letter of 1611 regarding the famous *virtuosa* Adriana Basile: 'they remained there listening to the music of an Angel [Adriana Basile]', whose 'singing kept everybody's soul in suspense, or even enraptured to Heaven, for two hours'; her 'melodious sighs … left them in doubt, whether that was by chance an earthly place, or the happy seat of Paradise'.[33] The ubiquitous nature of this topos (examples could be endlessly multiplied) reveals how strict and almost automatic was the association between heavenly and earthly music in the culture of the time. On the one hand, this seems to be the natural development of a long-standing tradition, but, on the other, specific cultural trends need to be taken into account.

The rich and varied aural experiences available after the spectacular 'rise of European music' (to quote the title of a famous book by Reinhard Strohm) in the fourteenth and fifteenth centuries undoubtedly influenced the developing sense of importance of the sonic dimension in conceptions of heavenly beatitude. According to Delumeau, the effect of the rise of music in European culture (prompted by the growth of polyphony and the multifarious expansion of the instrumental families) is unmistakably reflected in the iconography of Heaven during the period *c.*1350–*c.*1550:

> The iconography that adorned heaven with angelic orchestras was in fact the expression of the moment in which European sensibility, exploring the riches and the extension of the universe of music, accorded music a new place among the constituents of culture.[34]

A circular interaction arose: as music was becoming more varied, complex and captivating, 'a synonym of happiness … and the essential partner of love'[35] on earth, it was being given a more prominent place among the *gaudia Paradisi* by spiritual authors, preachers and painters; the other way round, the fascination of heavenly music became so strong, that musicians and music lovers could not help taking it (more or less explicitly) as their main point of reference.

A Foretaste of Heaven

The contemplation of Heaven as the ultimate horizon of life played, for early modern Catholics, a far more important role than we usually tend to think. Therefore, its representation, providing a foretaste of its joys, played an essential role in religious festivals, and constituted a catechetical tool no less useful than the vivid depiction of the torments of Hell.

Music, in its double nature of sensual and intellectual delight, could contribute in a unique way to this foretaste. The aural experience of music goes beyond the expressible and beyond the visible, as clearly stated by Grazioso Uberti in his *Contrasto musico* (1630):

> it is not possible to explain in words the glory of Heaven. It is too little to speak of that jubilation, that joy, that sweetness, that happiness of the Blessed. People do not understand it enough, especially the simple-minded. In order to make men love it more, in order to move their soul, we paint choirs of angels in the act of singing and playing various music instruments. But since painting is mute and does not offer any melody to the ears, we add the singing of voices and the sound of instruments, to give a more vivid aural representation of a sweet harmony: and judging from it, it becomes possible to argue, in conversation, how much greater is the sweetness of celestial music, how much greater the melodiousness of the heavenly concert.[36]

The sung word depicts the state of beatitude much better than the spoken word. Music functions as the only possible sonic representation of Heaven:[37] 'And who ignores that man, when he savours the sweet concord of Music, has the impression of enjoying Paradise …?'[38] Early modern churches and other sacral spaces are consistently conceived, constructed and decorated as images of Heaven – 'a church is an image of Heaven [*Templum est imago coeli*]' (Bellarmino)[39] – windows on Paradise and symbols of the celestial harmony. Architectural elements (especially domes and cupolas), frescoes, trompe-l'oeils, retablos, the richest possible materials and decorations, are all designed to represent a *different* reality.[40] Music can interact with all these complex symbolic constructions in a peculiar, multidimensional, 'surround' way: 'Thus it seemed to happen that all our more refined senses – in part hearing melodious songs and very sweet sounds, in part smelling various fragrant scents, in part, lastly, gazing at every kind of admirable ornaments – rejoiced in many ways'.[41] This possibility of tasting in advance the joys of paradise through synaesthesia is particularly effective in cases like the Sanctuary of Santa Maria delle Grazie in Saronno (Lombardy). According to recent studies, in the once centrally planned church (the nave was a later addition) musicians played and sang from tribunes situated in the drum below the dome famously frescoed by Gaudenzio Ferrari with an extraordinary large angelic orchestra.[42] Thus, during solemn liturgies, the faithful standing under the dome could experience the amazing vision of the angelic choirs while surrounded by music. Artistic and musical means were employed to fulfil a theological vision and a precise pastoral programme.

The Angelic Ideal of Singers

An almost obvious practical translation of the archetype of heavenly music was the development of a specific ideal of singers modelled on angels as 'heavenly performers'. Seventeenth-century manuals for ecclesiastical singers revived and reworked the mediaeval tradition on this topic, shaping a choral spirituality that, in the words of Stefani, combined the monastic ideal of angelic imperturbability with the paradigm of the Heavenly Jerusalem in the image of 'a well-regulated Choir of Angels in human guise' (Matteo Coferati, 1682).[43] The frequent comparison of female convents and male monasteries to Heaven invariably recalls this concept: 'the convent of the nuns is a starry sky, where the flashing stars make sweet harmony around the Sun. The enclosure of the sacred virgins is a Paradise, where the fervent seraphs make a pleasant concert for the ears of the eternal God'.[44] The principle that singers should model themselves on their angelic counterparts helps to explain, in part, the kind of disciplinary measures taken by Church reformers to exclude married singers, or at least to compel them to wear ecclesiastical robes (consider famous cases such as Carlo Borromeo's efforts in Milan or the vicissitudes of Palestrina in the Papal chapel). A far more relevant consequence of this ideal, from the aural point of view, is the vast utilization of boy singers in (mediaeval and) early modern musical practice. Prepubescent voices were easily associated with angels' voices, and *pueri cantores* offered the best possible aural and visual image of the conjunction of earthly and heavenly praise of God. This unmistakable association is sometimes made explicit also in architectonic terms: Willem Elders reports that many mediaeval churches 'had a so-called *chorus angelicus* or *chorus angelorum* in an elevated part of the building ... In some cathedrals we even find a small gallery built high against one of the inner walls', sometimes 'richly decorated with angelic musicians ... From this gallery, a number of choristers sang antiphonally with the choir in procession below'.[45]

Outside liturgical music – and outside the immanent logic of the ecclesiastical recruitment system – boyish-angelic voices marked the soundscape of early modern Italian cities on many festive occasions: the female and male pupils of the Scuole della dottrina cristiana (one of the main educational agencies of the sixteenth century) were usually trained to sing *laudi* and hymns, and frequently participated in processions and ceremonies.

In a characteristic mixture of concrete 'urban' details and metaphysical flashes, bishop Agostino Valier describes the success of the Scuole in Borromaic Milan, where angelic voices singing God's praises filled the whole city with joy on festal days, so that Milan could be said to be a corner of the heavenly home ('coelestis patriae quidam angulus'). 'Sometimes, on festive days, there burst the countless voices of boys and girls, sweetly singing and praising God, in such a way that the choirs of the angels seemed to be made visible [Milan, *c.* 1580]'.[46] Often, the singing boys literally represented angels:

> All the Schools of Christian doctrine, an infinite number of boys, well adorned and dressed as angels; they sang devoutly, and they too had lighted candles in their hands [Bologna, 1586].
>
> At the 'baldachin' of the Virgin 'sat a choir of boys dressed as angels, which sang most sweetly' [Rome, 1625].[47]

There was clearly a continuous interplay of references between the figurative and performing arts, between visual and aural stimuli and between symbolic spaces and urban contexts. Another sixteenth-century chronicle describes a figurative procession on the subject 'The Church Militant and the Church Triumphant', enlivened by an ensemble of a hundred musicians ('un'armonia di cento musici') on carriages. A float with the Virgin Mary Assumed into Heaven was followed by another carriage carrying 'seraphs, cherubs, angels, archangels, evangelists, all musicians, who sang beautiful *laudi*, with the accompaniment of string and wind instruments [Florence, 1577]'.[48] By means of such multimedia strategies, the angelic ideal of singers and players was presented in the most persuasive way.

Scala paradisi

The possibility of experiencing heavenly music on earth was explored also by means of compositional strategies. A musical work could indeed be described and function as a *scala paradisi* – an image obviously reminiscent of Jacob's ladder, a well-known mystical symbol. 'This work ... like a ladder, can give you sometimes the opportunity of raising your mind to Heaven, and envisaging the harmony made by the angels in Paradise', wrote Cremonese composer Bernardo Corsi in the preface to his *Concerti ... opera quinta* (Venice, 1613).[49] If, in relation to our theme, plainchant did represent perfectly some aspects of heavenly music, and especially what Hammerstein gathered under the term *una voce* ('gemeinsam-einhelliger Lobpreis', 'collective-unanimous praise'),[50] polyphony could prefigure other aspects of it, such as its overwhelming *dulcedo* and the unity in variety of multiple voices. Elders identified some markers useful to recognize Renaissance polyphonic compositions purposely designed to represent the music of Heaven: for instance, the unusual scoring for nine voices, or the subdivision into nine parts (where the number 9 refers to the nine angelic choirs); the combination of canonic and ostinato voices allowing for endless repeatability; the presence of texts insisting on praise; and a pervasive 'sweetness'.[51]

While this has mainly a symbolic rather than sonic relevance, other phenomena in late Renaissance music deserve our attention as aural translations of the heavenly soundscape. The polychoral technique enjoyed widespread popularity in the late sixteenth and seventeenth century. Outside a few examples and authors (chiefly Giovanni Gabrieli and other Venetians), this vast repertoire is still largely understudied. Its origins are obviously deep-rooted in technical and purely musical matters, but every technique has a spiritual underpinning. Thus, I believe that the extraordinary success and dissemination of polychorality may find an explanation, at least in part, precisely in this order of ideas. Polychorality not only enhanced the magnificence of the musical apparatus, it was also perfectly suited to depict some essential aspects of heavenly music. The antiphonal exchanges, echo-like answers, layered superimposition and occasional *tutti* of polychoral music seem the best possible sonic representation of the ineffable performances of celestial choirs, the mysterious Seraphim calling *alter ad alterum* of Isaiah, 6: 3. As Michael Praetorius observes in the

Vorrede to his *Urania* (1613): '*Musica per choros caelestia canens* or heavenly choir music ... because the correct way of singing varying the choirs [*per choros*] is truly the correct, heavenly way of making music'.[52] The solution given by Athanasius Kircher to an enigmatic canon by P.F. Valentini (96 voices in 24 choirs, or 512 voices in 128 choirs, or 128,000 in 32,000, 256,000 in 64,000 and so on)[53] takes to the extreme the symbolic potential of the multiplication of choirs and demonstrates, notwithstanding its pedantic eccentricity, the proximity of the polychoral technique to the ideal of *musica perennis* and to the 'new song' of the 'hundred and forty-four thousand who had been redeemed' in Apocalypse, 14.

Polychorality maximized some of the most effective paradise-evoking aural experiences, through its own sonic physiognomy, its typical, and sometimes deliberately illusionistic, effects of spatialized sound and 'surround',[54] and the synaesthesic interaction with the figurative and architectural context (even if its genesis from the disposition of the singers' tribune in Saint Mark's is now a worn-out fable).

The ever-present ideal of heavenly music had other consequences for the actual musical experience of the period. The promotion of *lauda* singing and other devotional practices involving music in the post-Tridentine era can be understood from this perspective. The Oratorian poet-musician and polemicist Giovenale Ancina considered praiseworthy the work of authors who wrote

> those kinds of rhymes and music, in which one may not only find delight, but also raise oneself to the pure and fervent love of God ... and fall in love with the divine beauties, and, touched by unwonted devotion, feel in oneself a wonderful and very sweet taste of Paradise [1599].[55]

And later writers agreed: 'Music is necessary also in private oratories, because ... through sensible harmony it is suitably manifested, how greater is the harmony and sweetness of the famous music of Heaven [1665]'.[56]

The City and the Heavenly Jerusalem

To identify the earthly city with the Heavenly Jerusalem was more than an inherited archetype in the early modern period: it was a deliberately chosen pedagogical strategy. I will explore, by means of examples pertaining to Italian cities, some implications of this idea on aural experience.

In the ceremonies held on major urban feast days, a representation of Heaven often took place in one of the key-points of the urban geography; and this representation invariably involved the aural dimension. In Siena, for instance, on the occasion of the canonizations of saint Bernardine (1450) and saint Catherine (1461) a 'paradise', or heavenly stage set, was constructed in Piazza del Campo. Young people 'dressed as saints' ('vestiti in fighura di santi') stood on the stage and the saint's ascent to Heaven was saluted with 'very beautiful songs' ('bellissimi chanti').[57] There were also other important events, when the self-representation of the city implied this identification with its heavenly counterpart: at Pope Pius II's arrival in Siena, on January 31, 1460,

> the entire route was decorated between the Duomo and the Porta Camollia … where a beautiful *apparato* was prepared which imitated a choir of angels, or Paradise itself [*a guisa d'un coro d'angeli, ovvero un Paradiso*]. And when Pius reached that place, an angel descended from that angelic host and sang some verses, and then turned from the image of the Virgin Mary to the pope, and commended Siena, Her city and ours, to him.[58]

In the *Commentaries* of the same Pius II we find similar descriptions regarding other Papal ceremonies. On Corpus Christi Day 1462, the Pope was in Viterbo, where numerous representations were staged along the procession route, involving angel musicians. In the cathedral square, after Mass, the assumption of the Virgin Mary was represented:

> Thereafter, full of joy, she was taken up into heaven, *sweetly singing*. And her Son, at the same time her Father and Lord, went to meet her, and kissed his mother on the forehead, and having brought her before the eternal Father, he placed her at his right. *Then the hosts of heavenly spirits sang, played music instruments, rejoiced, exulted, and everything in heaven was merry.*[59]

Through aural and visual means (which would require a more detailed description) the Viterbese succeeded in transfiguring the Pope's itinerary through their city into a journey across the Heavenly City:

> Whoever entered Viterbo that day [Pius concludes] and looked upon so many marvellous things and such a pomp and order walking through the city, thought he had entered without any doubt the abode of gods, not the dwelling place of men, and said he had seen the heavenly home while still in his flesh.[60]

While a transfigured soundscape, filled with concerts of angels, an unusual amount of voices and instruments resonating in the streets and squares and other festive sonic markers, is the typical result of ephemeral self-representations of the city as the Heavenly Jerusalem, in some cases the aural prefiguration of the City of God is rooted in long-standing structures of urban identity.

A seminal contribution for the comprehension of these phenomena has been given by Robert Kendrick. He writes: 'Milan itself prefigured the Heavenly Jerusalem, immanently in the 132 sanctoral bodies, 119 saintly heads, and 2,880 other relics that Morigia dutifully tallied among the churches' holdings. The polyphonic repertory flowered only in a profoundly sacral place'.[61] A famous map by Nunzio Galizia (*Pianta di Milano liberata dalla peste*, 1578) gives a visual demonstration of the correspondence between the two cities: in its upper part it shows the heavens opened, with Christ and his saints, while in the lower part we find a planimetric and volumetric representation of Milan, with its main religious sites well in evidence.[62]

It is important to underline the almost 'biunivocal' correspondence between polyphonic music and religious sites in the Milanese soundscape: polyphonic music resonated almost exclusively in the city's main sacral (or sacralized, in the case of urban processions, *entrate* and *trionfi*) places. Crossing the threshold of one of the city's main churches, the faithful enter into a reality visually very different from the environment of their daily life, incomparably richer and more harmonious: the same happens from the sonic point of view. If the churches act as 'windows on Paradise', as already discussed, by means of symbolic and synaesthesic-experiential stimuli, the fact that they convey peculiar musical experiences turns them into the main points of contact between earthly and celestial soundscapes.

Mapping the 'geography of [Milanese] urban polyphony', Kendrick concludes: 'On the largest (and most optimistic) scale … the city's sounds attempted to reproduce the musical complement of Paradise, a topos of similarity invoked again and again'.[63] The use of music as a tool to configure and enact Milan's ideal association with the Heavenly Jerusalem played a role in the pastoral programme of both the Borromeos: if Carlo fostered the Scuole della dottrina cristiana and their devotional singing (see above), Federigo encouraged the development of musical activities in the convents. From the late sixteenth century on, these became more and more important musical places in the city, gaining international renown: the nuns' singing 'served to mark their status as earthly angels, whose music sonically and symbolically represented the celestial choirs, not least by the extraordinarily high timbres of their vocal ensembles'.[64]

Carlo Borromeo was canonized in 1610. During an urban procession made in 1638 to obtain the Saint's intercession *pro pace*, his body was carried through the city in its crystal sarcophagus. 'The presence of musicians … attempted to make the city portray Heaven':[65]

> After the collegiate [clergy], there came in front of the cathedral a numerous and well-organized squad of musicians, *which opened an air-hole to heaven for the ear* with their sweet and harmonious concert, and this [Milanese] church, an important member of the Church Militant, became a simulacrum of the Church Triumphant, while that portable paradise was appearing under a majestic baldachin.[66]

The body of Milan's co-patron was seen as a 'portable Paradise'. Its journey through the city was accompanied by music. Thus the conjunct presence of the most venerated relics and of sweet and harmonious music transfigured the procession of the Milanese Church militant into a simulacrum of the heavenly triumph. It is hard to find a more convincing example of how the experiential association of celestial and earthly soundscapes could be linked to central elements of a city's identity: its streets, its patron, its sense of community, its political concerns, its religious and civic traditions.

It is impossible to conclude this short survey without mentioning Rome, the city that for many decades in the Early Modernity – despite the *leyenda negra* of the Counter-Reformation – was the theatre of the most innovative spiritual and creative experimentations. Regarding the role of the innumerable churches in its urban sonic geography as aural windows on Paradise, suffice it to cite a famous passage by the English priest Gregory Martin:

> It is the most blessed varietie in the world where a man may goe to so many Churches in one day, chose where he wil, so heavenly served, with such musike, such voices, such instrumentes, al ful of gravitie and majestie, al moving to devotion and ravishing a mans hart to the meditation of melodie of Angels and Saintes in heaven [1581].[67]

For the sake of space, I will briefly consider only one Roman example, among the many possible, documenting another aural implication of the reading of the city as a prefiguration of the Heavenly Jerusalem. Filippo Neri (1515–95) was the founder of the Oratorian congregation, one of the most influential and ground-breaking institutions of early modern Catholicism. As is well-known, music was one of his favourite evangelization tools,[68] and the reflection on the Four Last Things was a prominent theme in his spirituality. The musical productions of his entourage and the writings of his disciples and followers (above all Giovenale Ancina and Agostino Manni) would provide many useful materials for a history of heavenly music on earth. But let us consider here an account of the Saint's favourite devotional itinerary, the pilgrimage to the *Sette Chiese*, usually undertaken during Carnival. Many people (to the hundreds and thousands) followed Filippo in this daylong visit to the seven main Roman basilicas, and singers and wind players (usually among the best musicians in Rome) were also in the procession. Sermons were given in every church, and Mass was sung in San Sebastiano; many participants received the Holy Communion:

> On the way they always sang psalms and hymns, and there was such a joy, that *it seemed they were in Heaven.* Many people were converted during this pilgrimage of the Seven Churches, and all this was very fruitful for the souls. The midday meal took place in various places, at the vineyard of the Mattei, at St. Stephen in the Round, at the Holy Cross, or elsewhere; and there was such a silence, that everybody was astonished … and during the meal there was a music of voices, cornetts and flutes: and going back home, people *had the impression of having been in heaven.*[69]

The route touching the most sacred shrines of the Holy City, the singing of *laudi* and hymns along the street, the Eucharistic banquet, the sober meal thereafter in a garden to the sound of 'voices, cornetts and flutes': no wonder if, going back home, the participants thought they had experienced Heaven. In opposition to the secular interpretation of the city given by the Carnival, the Oratorian devotional itinerary read Rome as a map of its heavenly counterpart. Once more, music played a key role in substantiating the association of the two cities through their soundscapes.

Infernal Soundscape on Earth

As we have seen above, the soundscape of Hell was traditionally characterized by sonic disorder: loud noise, cries and howlings.[70] Given its obvious nature of anti-model, we do not expect to find it deliberately reconstructed in real life, if not in the context of fictional representations. It is worth noting, however, that fragments of the hellish soundscape surface in chronicles influenced by the current way of thinking: in an early biography of Carlo Borromeo, for instance, we read that during his solemn entrance into Milan in 1565 possessed people bellowed and howled and screeched. And during his funeral (1584), the shrieking devils, 'tormented by the presence of his holy body', could be heard 'all the way long, and even in the Duomo … so that it almost seemed the end of the world'.[71]

For its great evocative power, the hellish theme and its sonic attributes had been common currency in mediaeval and early Renaissance sacred representations and mystery plays.[72] Interestingly, the opposition between Hell/Hades and Heaven plays an important role in some of the first operas of the early seventeenth century, notably in Emilio de' Cavalieri's *Rappresentatione di Anima, & di Corpo* (1600) and in Claudio Monteverdi's *Orfeo* (1607).

In Cavalieri's third act, the allegorical characters Intelletto and Consiglio ask the damned and blessed souls (appearing on different stage levels, representing Hell and Heaven) about their respective conditions: the damned sing their answers with minor harmonies, either through dull homophonic declamations in low registers, or short solos marked by dissonances and sobbing rests, while the blessed answer with major harmonies and beautifully ornamented sequential lines in high registers. During Orfeo's katabasis, Monteverdi assigns the underworld a louder orchestra with cornets, trombones, bass gamba and the snarling timbre of the regals, while the gentler tones of soft strings and the harp in particular are associated with Orpheus and his *cetra*.[73]

Both solutions are, however, only stylized and softened renditions of the appalling soundscape of Hell, that the contemporaneous German composer and theorist Michael Praetorius evokes, in the *Vorrede* to his *Polyhymnia Caduceatrix* (1619), describing the 'whinnying and barking, caterwauling and yelling, clattering of teeth, howling and baying, the woeful, appalling music of the cries of pain of the burning hellish chapel' led by the Devil.[74]

Conclusions

The idea of afterlife occupied a central place in the early modern mentality. And in the Christian tradition, afterlife had an intriguing feature: it possessed a sonic dimension. In an epoch during which the art of music was undergoing an unprecedented expansion and was acquiring a new role among the constituents of European culture, this idea could not remain without consequences. In fact, it was adopted and developed by countless authors, in many different contexts, and became part of the great pedagogical effort of the Catholic Church, especially in the post-Tridentine era.

If, on the one hand, the conception of a celestial soundscape animated by voices and instruments was modelled, at least in part, on aural experiences in this world, on the other hand it affected in turn the soundscape-building policies of early modern cities, serving as an archetype and a symbolic referent. The processes resulting from this interaction influenced many aspects of early modern sonic experience: for instance, the configuration of sacral spaces, devotional itineraries and ceremonies; the promotion of spiritual music; the preference given to some performance options and compositional techniques. In many ways music, as a privileged part of the soundscape, played a key role in substantiating the association of terrestrial cities with their celestial counterpart, the Heavenly Jerusalem. The evidence concerning this dynamic is manifold. It is to be hoped that, despite its limits, this first attempt at a comprehensive description will foster a new understanding of a largely underestimated (and only apparently abstruse) phenomenon of Western sonic culture.

Notes

1 This chapter was originally written during my research project 'A History of Sonic Experience in the Renaissance', funded by a 2010/2011 grant from the University of Pavia (Department of Musicology, Cremona). Although I have meanwhile made some revisions, I did not attempt to update the bibliography.

2 For a history-of-ideas approach to the Christian afterworld, besides the entries in standard reference works such as the *Dictionnaire de spiritualité*, *Dictionnaire de théologie catholique* and similar dictionaries, see for instance Jean Delumeau, *Une histoire du paradis – 3. Que reste-t-il du paradis?* (Paris: Fayard, 2000); Colleen McDannel and Bernard Lang, *Heaven: A History* (New Haven: Yale University Press, 1988); Jeffrey Burton Russell, *A History of Heaven: The Singing Silence* (Princeton: Princeton University Press, 1997); Herbert Vorgrimler, *Geschichte der Hölle* (Munich: Fink, 1993). On the specifically musical aspects, see Reinhold Hammerstein, *Die Musik der Engel. Untersuchungen zur Musikanschauung des Mittelalters* (Bern: Francke, 1990). I will quote these works in notes only on specific issues, but they were constantly used as reference tools during the preparation of this chapter.

3 Delumeau, *Une histoire du paradis*, p. 197.

4 Hammerstein, *Die Musik der Engel*, p. 70.

5 Indeed all the senses in the blessed will be rewarded and in the wicked will be punished: see for instance *Summa Theologica*, Suppl., qq. 82 and 97; L. Lessius, *De summo bono* (Antwerp: Ex officina plantiniana, 1616), b. III, ch. 8; R. Bellarmino, *De aeterna felicitate sanctorum* (Antwerp: Ex officina plantiniana, 1617), b. IV.

6 Aurelian of Réôme, *Musica disciplina*, ed. L. Gushee ([Rome]: American Institute of Musicology, 1975), pp. 58–9 and p. 132. The latin text reads: '[De laude musicae disciplinae] Quid plura? Etiam apud supernos cives legimus huius artis insignia celebrari, ut in Apocalipsi: Habentes citharas Dei, et alibi … Et quid plura? *Musica ars omnes exsuperat artes*. Angeli quoque, quod Deo laudes more huiusce discipline in arce referunt sidereo, lector Apocalipsi nemo qui dubitet'. Emphasis added, here and in all the following citations. All translations are mine unless otherwise indicated.

7 Amerus, *Practica artis musice*, ed. C. Ruini ([Rome]: American Institute of Musicology, 1977), pp. 19–20: 'Hec est illa regina omnium liberalium scilicet artium excellentissima que ceteris sex artibus dimissis in inferioribus, *sola volat ante tribunal omnipotentis* ut in Apocalypsis capitulo octavodecimo: "Et cantabant quasi canticum novum ante sedem Dei et ante quatuor animalia". Hec est illa altissima domina que *in*

ecclesia triumphanti et militanti Deo servit dulcisona … Hoc ergo cognoscens omnis clericus qui vel dignitate[m] vel aliquod auxilium in ecclesia Dei affectat, sicut dicit Ysidorus in libro Ethimologiarum quod "non minus est dedecus nescire cantare quam litteras ignorare". Quoniam sancti in celis cum angelis et archangelis, thronis et dominationibus, cumque omni milicia celestis exercitus non cessant hympnum canere quotidie: Sanctus sanctus et cetera. Et in terra ab angelis cantabatur: Gloria in excelsis Deo, *a quibus nec de literatura queritur nec de logica disputatur*'.

8 Andreas Ornithoparchus, *Musice active micrologus* (Leipzig: Valentin Schumann, 1517; facs. ed., New York: Dover, 1973), b. II, ch. 1, 'De artis huius Utilitate ac laudibus': '[Musica] Que et sola, relictis terris, volat ante tribunal summi iudicis, ubi sonant iugiter organa sanctorum: ubi angeli et archangeli hymnum deo infatigabiliter decantant, ubi Cherubin et Seraphin, sanctus, incessabili voce proclamant. Adde quod nulla disciplina sine musica possit esse perfecta'.

9 Louis de Granada, *Of Prayer and Meditation*, trans. Richard Hopkins (Paris, 1582). I have chosen to quote from early modern translations of Granada and Drexel, when available, not only because they document the reception of these works in the English-speaking world, but also because they are somehow 'nearer' to the style and the spiritual atmosphere of the originals.

10 Granada, *Of Prayer*, fol. 207^{v}. Italics as in the original.

11 Ibid., fol. 208^{r}.

12 Ibid.

13 Granada, *Of Prayer*, fol. 208^{v}.

14 Ibid., fol. 216^{v}.

15 Ibid., fol. 217^{v}.

16 Ibid., fol. 246^{v}.

17 Ibid., fol. 257^{r}.

18 Ibid., fol. 258^{r}.

19 Jeremias Drexel, *Caelum beatorum civitas* (Antwerp: J. Cnobbaert, 1635). The subsequent quotations are taken from b. II, ch. 5, 'Quintum in caelo Gaudium, Voluptas Aurium'.

20 '… tot mille Orpheorum Musica permulcebit aures. Quarum gemina erit voluptas; a colloquiis & a Musica'.

21 'Domine Iesu, si mihi sint centum aures, si totus sim auris, perdius & pernox hos cantores audiam, nec satiabor audiendo; non cibus, non somnus ab hoc aurium melle me avellet!'

22 'Jam dulces cantus flexiones, jam rotundae clausulae, paulo post iterum suaves moduli, jam chroma nobile auditur, quaedam hinc illinc miscentur, quae non adeo grata sunt auribus'.

23 '… gaudia omnia omnis voluptas simul, & omnibus momentis percipitur, quae primo ingressu caeli percepta est. Torrente voluptatis tuae potabis eos'.

24 '… in caelo Sanctorum quilibet singularem constituit symphoniam … sic unicus beatorum hominum symphoniam & concentum potest edere. Quale igitur jubilum erit in caelo a tot myriadibus cantorum?'

25 Granada, *Of Prayer*, fols 227^{v}–8^{r}.

26 Ibid., fol. 227^{v}.

27 Granada, *Of Prayer*, fols 234^{v}–5^{r}.

28 Jeremias Drexel, *Infernus damnatorum carcer* (Munich: C. Leysser, 1631); I quote from the English translation *A Pleasant and Profitable Treatise of Hell*, trans. not given (London, 1668).

29 Ibid., pp. 44, 46.

30 See for instance Gino Stefani, *Musica barocca 2. Angeli e Sirene* (Milan: Bompiani, 1988).

31 C. Antegnati, *Salmi a otto voci* (Venice, 1592), cited in Stefani, *Musica barocca*, p. 128: '… questi Salmi come tutte le altre mie compositioni date alla stampa, allora io reputo

che abbiano conseguito il fin loro, quando sono cantate, ed udite da persone che, dal gusto di questo strepitoso rimbombo d'aria fra vilissimi corpi ristretta e percossa, ergono il desiderio agli eterni e perfettissimi concerti de' cori angelici nel Paradiso'.

32 Delumeau, *Une histoire du paradis*, p. 10.

33 Stefani, *Musica barocca*, p. 227: '"si trattennero in udir la musica d'un Angelo" cioè Adriana Basile, che "tenne per due hore continue gli animi di tutti cantando sospesi, anzi rapiti al Cielo"; i suoi "soavi sospiri … Lasciavan dubbio in loro se quello era per avventura terreno luogo, o pur il fortunato seggio del Paradiso"'.

34 Delumeau, *Une histoire du paradis*, p. 222. See also pp. 13, 209 and 218.

35 Ibid.

36 Grazioso Uberti, *Contrasto musico: opera dilettevole* (Rome: Grignani, 1630; facs. ed., Lucca: LIM, 1991), pp. 110–11: 'la gloria del Paradiso non si può esplicare con parole. È poco il dire quel giubilo, quel gaudio, quella dolcezza, quella soavità, quel contento de' Beati. Non si comprende a bastanza, e particolarmente dalle persone semplici. Per innamorarne maggiormente gli huomini, per commovere maggiormente gli animi, si dipingono i Chori degl'Angeli in atto di cantare, e di sonare varij stromenti Musicali. Ma perché la pittura è muta, e non porge all'orecchie melodia alcuna, si aggiunge il canto delle voci, & il suono degli stromenti, per rappresentare più vivamente all'udito soave armonia, dalla quale si possa discorrendo argomentare quanto maggiore sia la dolcezza di quella celeste Musica, quanto maggiore sia la soavità di quel concerto del Paradiso'.

37 See also Richard Rastall, *Music in Early English Religious Drama. 1. The Heaven Singing* (Cambridge: Brewer, 1996), p. 183.

38 Lodovico Casali, *Generale invito alle grandezze, e maraviglie della musica* (Modena, 1629), p. 99, quoted in Stefani, *Musica barocca*, p. 229: 'E chi non sa, che l'uomo gustando il soave concento della Musica, li pare propriamente di goder il Paradiso …?'

39 Roberto Bellarmino, *De controuersiis christianae fidei, aduersus huius temporis haereticos: tribus tomis comprehensae* (Ingolstadt, 1581), p. ii.

40 Delumeau, *Une histoire du paradis*, ch. XVIII.

41 From Giannozzo Manetti's famous description of the consecration of S. Maria del Fiore in Florence (1436), a text rich in references to heavenly music. See Caroline van Eck, 'Giannozzo Manetti on Architecture: the *Oratio de secularibus et pontificalibus pompis in consecratione basilicae Florentinae* of 1436', *Renaissance Studies*, 12/4 (December, 1998): 449–75, at pp. 474–5. The Latin text reads: 'Ita per hunc modum fieri videbatur, ut omnes paulo humaniores sensus nostri partim suavissimos cantus dulcissimosque sonos audientes, partim varios ac redolentes vapores olfacientes, partim denique admirabilia omnia ornamentorum genera conspicientes varie ilarescerent'.

42 See Andrea Di Lorenzo, 'Gaudenzio Ferrari e la cupola di Saronno', in Luigi Lazzaroni and Alfredo Fusetti (eds), *Il concerto degli angeli* (Milan: Silvana, 1994), pp. 24–33.

43 Stefani, *Musica barocca*, p. 153: 'un ben regolato Coro di Angioli sotto umana sembianza'.

44 Ibid., p. 112: 'il Convento delle Monache è un Ciel stellato, dove le fiammeggianti stelle fanno soave armonia intorno al Sole. La Clausura delle Vergini sacre è un Paradiso, dove gl'ardenti Serafini fanno grato concerto all'orecchie dell'Eterno Dio'.

45 Willem Elders, *Symbolic Scores* (Leiden, New York, Köln: Brill, 1994), p. 222.

46 A. Valier, *Vita Caroli Borromei* (Verona, 1586), p. 32: 'Sic enim interdum diebus festis innumerabiles puerorum et puellarum voces emittebantur, suaviter canentium et laudantium Deum, ut Angelorum chori representari viderentur'.

47 Stefani, *Musica barocca*, p. 23: 'Tutte le Scuole della dottrina Christiana, che era un infinito numero di fanciulli ben'ornati, & vestiti da Angioli, & cantavano divotamente, con lumi accesi ancor loro in mano'; 'Al "talamo" della Vergine "sedeva un Choro di fanciulli vestiti da Angeli, che cantavano suavissimamente"'.

48 Stefani, *Musica barocca*, p. 24: 'serafini, cherubini, angeli, arcangeli, evangelisti che, musici tutti, accompagnati da strumenti a corda ed a fiato, cantavano bellissime laudi'.

49 Stefani, *Musica barocca*, p. 223: 'quest'opera … a guisa di scala potrà tal volta porgerle occasione di levare la mente in Cielo, e imaginare l'Armonia che fanno gli Angeli in Paradiso'.
50 Hammerstein, *Die Musik der Engel*, p. 70.
51 Elders, *Symbolic Scores*, pp. 221–51.
52 Quoted in Hammerstein, *Die Musik der Engel*, p. 45: 'Musica per choros caelestia canens oder Himmlische Chormusik … dieweil die rechte Art per choros zu singen in Wahrheit die rechte himmlische Art zu musizieren ist'.
53 See Lorenzo Bianconi, *Il Seicento* (Turin: EDT, 1991), p. 56.
54 Compare other contemporary 'spatialized' techniques, such as Ignazio Donati's *cantar lontano*: see the preface to his *Sacri Concentus* (Venice: G. Vincenzi, 1612).
55 G. Ancina, *Tempio armonico della Beatissima Vergine* (Rome: N. Mutii, 1599), bass partbook, Dedication to f. G.M. da Brisighella: 'quelle sorti di rime e di musiche, dove altri possa non solo ugualmente dilettarsi, ma insieme col diletto innalzarsi al puro e fervente amor di Dio … ed invaghirsi delle divine bellezze, e tocco da inusitata divozione sentir dentro in se stesso un mirabil e soavissimo gusto di Paradiso'.
56 Carlo Pellegrini, *Museum Historico-legale bipartitum* (Rome, 1665), p. 95, cited in Stefani, *Musica barocca*, p. 223: 'In privatis oratoriis est quoque necessaria musica, quia … per harmoniam sensibilem magis apte repraesentatur, quanto maior sit harmonia, et dulcedo illius celebris musicae Paradisi'.
57 See Frank A. D'Accone, *The Civic Muse: Music and Musicians in Siena During the Middle Ages and the Renaissance* (Chicago and London: The University of Chicago Press, 1997), pp. 680–88.
58 Chronicle by A. Allegretti; I quote (with slight changes) the translation given by Fabrizio Nevola, *Siena: Constructing the Renaissance City* (New Haven: Yale University Press, 2007), p. 65.
59 Enea Silvio Piccolomini, *I commentarii*, ed. L. Totaro (Milan: Adelphi, 1984), p. 1608: 'Exin plena gaudio, *dulce canens* in coelum assumpta est. Cui occurrens Filius, idemque Pater et Dominus, in fronte osculatus matrem et oblatam aeterno Patri, ad dextram suam collocavit. *Tum canere coelestium spirituum agmina, musica instrumenta tangere, laetari, gestire et cuncta in coelo ridere*'.
60 Piccolomini, *I commentarii*, p. 1612: 'Quicunque ea die Viturvium intravere et tot miracula tantumque rerum ordinem et apparatum per urbem incendentes intuiti sunt, Deorum certissimum domicilium sese ingressos, non hominum habitacula, putaverunt et coelestem patriam vel vivos in carne sese spectasse dixerunt'.
61 Robert L. Kendrick, *The Sounds of Milan, 1585–1650* (New York: Oxford University Press, 2002), p. 21.
62 See Adele Buratti, 'L'azione pastorale dei Borromeo a Milano e la nuova sistemazione urbanistica della città', in Adele Buratti (ed.) *La città rituale. La città e lo Stato di Milano nell'età dei Borromeo* (Milan: Franco Angeli, 1982), pp. 9–56, at p. 11. A reproduction of the map appears in *Arcimboldo: artista milanese tra Leonardo e Caravaggio*, ed. by S. Ferino-Pagden (Milan: Skira, 2011), p. 292.
63 Kendrick, *The Sounds of Milan*, p. 87.
64 Robert L. Kendrick, *Celestial Sirens: Nuns and their Music in Early Modern Milan* (Oxford and New York: Oxford University Press, 1996), p. 166.
65 Kendrick, *The Sounds of Milan*, p. 161.
66 *Racconto della processione fatta in Milano l'anno 1638 il dì di San Carlo …* (Milan: F. Ghisolfi, 1638), quoted in Kendrick, *The Sounds of Milan*, p. 161 (I have slightly edited Kendrick's translation); the Italian text is given at page p. 481: 'Dopo le Collegiate veniva innanzi alla Cattedrale un numeroso, e ben ordinato drappello di Musica, che con dolci, ed armoniosi concerti apriva quasi uno spiraglio del Cielo all'udito, e questa Chiesa membro principale della militante, era divenuta un simulacro della Trionfante, comparendo a poc'a poco quel Paradiso portatile, sott'un maestosissimo baldacchino'.

67 Gregory Martin, *Roma sancta, 1581*, ed. George Bruner Parks (Rome: Edizioni di Storia e letteratura, 1969), p. 96.

68 See Daniele V. Filippi, *Selva armonica. La musica spirituale a Roma fra Cinque e Seicento* (Turnhout: Brepols, 2008).

69 From a testimony of 1595, quoted in Alberto Venturoli, *Visita alle Sette Chiese* (Rome: Città Nuova, 2006), p. 25: 'Et, quando andavano per strada sempre si cantavano salmi et hinni, et vi era un'allegrezza, che *pareva di stare in Paradiso*. Et, in questa andata alle Sette Chiese, si convertivano molte gente et si faceva un gran frutto per le anime. La refettione si faceva in vari luoghi, alla vigna de Mattei, a San Stefano rotondo, a Santa Croce et altri luoghi; et vi era un silenzio, che ognuno si meravigliava … et, mentre si mangiava, vi era musica di voce, cornette et flauti: che, nel tornare a casa, *la gente li pareva di esser stati in paradiso*'.

70 See also Hammerstein, *Die Musik der Engel*, pp. 102–5. This chapter was already completed when I discovered Nigel Wilkins's *La musique du diable* (Sprimont: Mardaga, 1999), which offers a wealth of information on infernal music from antiquity to the present.

71 See G.P. Giussano, *Vita di S. Carlo Borromeo* (Rome, 1610), pp. 38–9 and p. 496.

72 See for instance Martine Clouzot, *Images de musiciens (1350–1500). Typologie, figurations et pratiques sociales* (Turnhout and Tours: Brepols-CESR, 2007), pp. 226–9.

73 See F.W. Sternfeld, *The Birth of Opera* (Oxford: Clarendon Press, 1993), p. 135.

74 Quoted in Hammerstein, *Die Musik der Engel*, p. 106: 'Wiehern und Bellen, Myehern und Screyen, mit Zeen Klappen, heulen und bellen, die elende, schreckliche Music von Zettergeschrey der hellishen, heissen Capell'.

10 The Sounds of the City, 1598

Everard Guilpin's London in *Skialetheia*

Adam Hansen

Introduction

> But soft, whom haue we heare?[1]
> Nothing essential happens in the absence of noise.[2]

This chapter developed out of my teaching on a module called Early Modern Cultures. One of the topics we explore on this module is the early modern city, and one of the texts we use for that exploration is a brief section from Everard Guilpin's 1598 collection of poems, *Skialetheia, Or A Shadowe of Truth in Certain Epigrams and Satyres*. When preparing my teaching on this material, I was struck by the resonant ambiguities generated by Guilpin's use of literary devices to evoke London life, especially devices that had a disturbing acoustic power: alliteration, onomatopoeia, rhythm, homophones. In other words, I became fascinated by the ways Guilpin's poetry merges aurality and literacy, through realizing the sites and sounds of the city. How can a writer deplore such sounds without echoing them? How did the city's sounds constitute and compromise urban and satiric subjectivities?

What fascinates me is exemplified by the first epigraph above, from 'Satire V'. In the midst of what we will come to see as Guilpin's reproduction of the cacophony of the early modern city, his satiric persona appeals for quiet reflection, invites his audience to question the identity of auditors and noise-makers (including themselves) and creates a pun linking location and audition (here/'heare'). Yet in this expression, and not least in the pun that forms part of it, the persona makes their own contribution to the city's confusing sounds, validating the premise of my second epigraph, from Jacques Attali, a modern theorist of the social production and consumption of sound. Noise – whether literary or urban, musical or discordant or mingling all these – makes meaning happen.

How can we further theorize or historicize Guilpin's text? Existing scholarship on classical and early modern satire identified the mode as 'an urban art' and as being 'set in the city, particularly in the metropolis with a polyglot people'.[3] To enrich our comprehension of Guilpin's satires in such an urban, polyglot context, it is important to recognize how they consciously echoed satirists like Juvenal (and I try to do so). But one might also attend to the growing volume of criticism on sound in the early modern period. Most of this criticism begins with a basic premise, namely that there are 'fundamental differences' between modern and early modern soundscapes:

> Two inventions – electricity and the internal combustion engine – make it difficult for us even to imagine what life in early modern England would have sounded like.[4]

This does not mean modern concepts of audition have no bearing on our understanding of sound in earlier periods. However, whatever we say or think about sound in the period, we need to recognize sounds' production by, reception in and effects on individuals and communities. One way to do this is to localize and temporalize sounds in sites such as a city. For my purposes, that city was London in the 1580s and 90s.

The Sounds of the Early Modern City

Bruce R. Smith suggests that, like other early modern cities, London was 'full of sound': bells, traffic, animals, industry, crafts and trades, running water, musicians, conversation.[5] It featured significant 'soundmarks': the metal foundries in Lothbury, churches like St. Paul's, the 'genuinely polyglot' Royal Exchange, markets like Billingsgate and the entertainment zones around inns and playhouses.[6] If generated or heard 'in an inappropriate place, or at inappropriate times' sounds could also be perceived as noises in certain social contexts: ''Noise' was used to describe sounds that were musical or unmusical, pleasant or unpleasant, and could also be applied to quarrelling, strife or the spreading of rumours'.[7] Some of these sounds and noises were linguistic, and therefore diverse. Laura Wright notes that 'English as a second language was spoken in the capital', that 'continental languages and continental accents were heard daily' and that 'it is likely creole languages were spoken'.[8] She continues: 'Early modern Londoners heard foreign language systems spoken in their city, both from within and outside Britain, and forms of English that were informed by other substrate languages'.[9] Some literary texts record this. In John Eliot's 1593 series of humorously earthy dialogues, *The Parlement of Pratlers*, one character asks another where he'll be 'about twelue a clocke', to which his interlocutor responds:

> I will be below in the Change, either walking among the Italians, or troking with the French, or pratling amongst our English, or carousing with the Flemings at the Cardinals hat: And if I be not in these places, you shall finde me aboue in the Pawne, deuising with the faire Semsters.[10]

At the 'Pawne' or market, people, sounds and stuff of diverse style and provenance only proliferate. The capacity of markets and consumers complicated social, material and sonic demarcations: 'As a soundscape, early modern London was far too diffuse to be contained by parish boundaries. Within the acoustic horizon of a single parish, for example, might live several different speech communities'.[11] If the city's socio-economies compelled communication, then

London's buildings and streets amplified the possibility of acoustic interchange. Whether they belonged to the poorest or to artisans, residences were 'multifunctional' with 'bipermeable' rooms.[12] Complete privacy was a near impossibility. As David Garrioch notes: 'Given that many people did not have glass in their windows, but used paper to keep out the winter cold, external sounds readily penetrated'.[13] What were the effects of this on individuals and communities? Perhaps none. Otherwise brilliant scholars considering early modern subjectivity have tended to theorize 'the history of self-conceptions' and 'the epistemology of inwardness' in terms of a 'dialectic of vision and concealment', spectacle and display, surface and depth, thereby overlooking the significance of the audition of anything other than discourse – 'speech and coercion' – for the construction of selves.[14] As the musicologist Richard Leppert has suggested though, sound and sight have often been embodied in the form of the individual subject, meaning we do not need to privilege one over the other:

> The body, simultaneously site, sight, and possessing sight, is an object of tactile sensation and an aural phenomenon. The body sounds: it is *audible*; it hears. Sound constitutes the atmosphere supporting and comforting life on and in the terrain of the body. … For much of Western history, at the most fundamental levels of human perception, the sound *is* the sight, and the sight *is* the sound.[15]

Moreover, contemporary theorists of sound and cities, modern and early modern, have suggested that sound 'plays a crucial role' in the 'perceptual and cognitive processes' by which geographical and social spaces are mapped out and negotiated by individuals.[16] Sound made the early modern city comprehensible and intelligible. It was 'a vital element within an urban information system', and as such operated for individual subjects and for communities as a 'semiotic system … helping people to locate themselves in time and space': the ring of a church bell would be an example of this.[17]

Yet in places such as early modern London where boundaries between exteriors and interiors were so poorly defined and insulated, audition could induce sensations of *dis*location: people might feel out of it rather than at home. Though sounds let the early modern self be fixed, in this city of capacious, incontinent spaces and penetrative, alienating noises, 'sounds helped to create multiple identities, some local and others much broader'.[18] These issues of subjects being both located and dislocated by sounds become still more significant when we consider that the ability to produce and suppress sound in the early modern period was intrinsically linked to authority: 'Whoever controlled sound commanded a vital medium of communication and power'.[19] To early modern theorists like Helkiah Crooke, the physiology of audition embodied human subjection to and direction by a higher intelligence:

> Mark and observe also the art and curious workmanship appearing in the inward part of the ear, how exquisitely it is made and trimmed with little labyrinths, windings, little windows, a sounding tympan or timbrel; three small bones, a stirrup, an anvil and a hammer; the small muscles, the nerve

> or sinew of hearing, and the cartilaginous or gristle passage, prepared for conveying all sounds unto the sense … the infinite goodness and bounty of God shineth in this excellent workmanship.[20]

Yet because ears were ‘uncontrolled orifices’, continuous with the world beyond the self, they made subjectivity and agency vulnerable to contamination.[21] Certainly, even in this short passage from Crooke we can see how a series of reactions interior to the self and the body are conceptualized in terms of exterior places and phenomena: a noisy blacksmith’s, a saddler’s, ‘labyrinths’ and passages, all sites and sounds of the city.

Recording the Sound of the City

Before considering the implications of these concerns for a reading of Guilpin’s poetry, it might be instructive to offer a contrapuntal prelude by studying another contemporary reaction to the sounds of the city, evinced in the work of the composer Richard Dering. Dering was born in or about 1580. Now most famous for his *Cantiones sacrae* (1618), in later life he travelled to Venice where he assimilated Italian musical influences. However, as a much younger and evidently more secularly minded man, Dering wrote a piece entitled *The Cries of London*, which, though entered in the Stationer’s Register for 1599, existed only in manuscript form until modern publication. This text, and the songs of hawkers, traders, craftsmen and assorted others it records, is a remarkable hymn to sonic interpolation and acoustic liminality. As such, it conveys how sound travels, operates and is used and perceived in the city. In the most insightful recent analysis of this work, Bruce R. Smith relates Dering to other composers interested in incorporating London’s street-sounds into music. To Smith, Dering and his fellow composers ‘domesticate the cries’ and stage them ‘for gentle ears’, harmonizing them in an act of musical containment.[22] The form of Dering’s work also serves to contain the *city*, sequencing diverse catches into a diurnal pattern, regulating noise temporally. We might develop Smith’s analyses, however, to begin to apprehend the challenges artists like Guilpin faced when trying to accommodate themselves to the city’s soundscapes.

Certainly Dering’s songs *do* measure how the city’s appetites change through the day. In the earliest section, sea-and-river-food is traded for its freshness: ‘New mackerel, mackerel new, quick periwincles, quick, quick, quick’. By mid-morning to lunch time, the comestibles alter: ‘Hot pippin pies hot / Hot apple pies hot’. Dering’s verses reflexively admit this obsession with regularly marking time in the changing songs, as some of the songs themselves advertise devices for recording events: ‘Will ye buy a very fine Almanack?’ By the end of the day, the songs have changed again: ‘Lanthorn and candle-light, hang out, hang out maids’. The sequence ends in rhymes that lock words together, and that demarcate spaces and gendered subjects into discrete, continent functions: ‘Twelve o’clock, look well to your lock, / your fire and your light and so good night’.[23] This seems conclusive. Yet Dering’s text expresses daily change as much as it regulates it, and gives voice to consuming appetites and desires even as they are being made melodic. *Cries of London* also implies iterability: its diurnal structure signifies this happens again tomorrow. However they are organized, the sounds of the city are inexorable, unstoppable and uncontainable. And however much that organization is down to the composer's singular harmonizing sensibility, they reproduce others' songs, to be performed again.

These qualities are evoked by the songs themselves. The sequences of cries and melodies begin by both recording the stimulation of desire and stimulating that desire, evoking wants and lacks that the street-vendors and the songs Dering re-presents might appease: 'What do ye lack do ye buy sir, see what ye lack, / pins, points, garters, spanish gloves or silk ribbons?'[24] The consumables on offer and desired are intimate, close to the skin, to be taken into the household, the chamber and the body, whether they are foodstuffs or fancy goods: 'Will ye buy a very fine cabinet, a fair scarf, / or a rich girdle and hangers?' And as the voices and lines of melody interweave, so the sounds express skills and stuff that move between interiors and exteriors, domestic and public, national and international. Things brought from beyond the city and the household are consumed and internalized within them. Hence we catch sellers offering 'Fine Seville oranges', and we hear from coopers, shoe-blacking sellers, rat-catchers, all of whom trade on accessing nooks and crannies: 'Ha' ye any work for a tinker? / Ha' ye any old bowls or trays or bellows to mend? / … Will ye buy a mat, a mat for a bed?' It is not just a matter of stuff being brought *into* the household: household materials are recycled out of the home, as hawkers cry out for cast-offs: 'Doublets, doublets, old doublets, … ha' ye any old doublets?' This insistent, circular interpenetration inevitably and fittingly results in puns intimating activities that show Dering cheekily connect places, people and words and that promise to service gendered spaces with vigour:

> Sweep, chimney sweep, Sweep, chimney sweep,
> sweep, chimney mistress sweep with a hoop derry, derry, derry,
> sweep from the bottom to the top, sweep, chimney sweep.[25]

Along these lines, Dering suggests how one female seller bluntly offers fare that is both sexual and edible, playing on a term for a prostitute and a type of fruit: 'Medlers fine'.[26]

Dering celebrates and memorializes the city and its songs, but even as he tries to contain both, he lets the edited cries articulate that both are hard to contain, and rely upon an urban socio-geography where sonic and material containment is impossible. In comparison, as we will hear, Guilpin detests yet reproduces the sounds of the city, and can do little else as satirist of and yet subject to early modern London.

Satire, Self and Sound

Born on or around 1572, to a bureaucrat in the Court of Exchequer, Guilpin went to Highgate Grammar School (where his father was a governor), matriculated at Cambridge in 1588 and entered Gray's Inn in April 1591. Until recently, critical opinion tended to read his satires as 'a product of the limited world of the

Inns of Court'.[27] Yet our understanding of the place of satire like Guilpin's in a city like early modern London was redirected by Lawrence Manley. Manley suggests that 'verse satire at the end of the sixteenth century represents an alternative form of ethical innovation in response to the disorienting effects of urbanization'; it became '*the* medium for discrimination, for the moral and social judgement that could establish differences no longer given or apparent in contemporary social life'.[28] To Manley, the satirist's authority derived from this 'power of discrimination' – a force of ethical judgement that required and allowed them 'to prosecute the world extemporaneously, without reliance on traditional moral frameworks'.[29] In works creating 'the impression of a personal, uncompromising voice', the satirist therefore 'draws special force from the subjective angle of his moral vision'.[30] In turn, this 'discriminating subject' relied on performing a cultivated detachment from the mores and matter of the urban world.[31] Isolation and autonomy are both preconditions and evidence of moral power, and Guilpin certainly performs this type of satiric subjectivity.

However, as Manley suggests, for the satire to strike home in a timely, relevant fashion, authors like Guilpin also had to rehearse 'the effects of immediacy', recreating and animating the very scenes and figures they were critiquing.[32] One signifier of immediacy, not explored by Manley, is the sound of the city and its inhabitants: Guilpin reproduces this, if only to derogate it. Yet his works also convey how the sounds of the city problematized the isolation satiric subjectivity required, and in turn problematized the possibility of moral discrimination.

Though Manley does not discuss the sound of the satires, or the city, in these ways, he does discuss Guilpin's epigrams in ways that might be developed for such a project. Manley affirms that despite – or sometimes *because* of – their generic repetition, concision and precision, epigrams were inherently unstable: they marked an 'effort to define and thereby master a protean city', but 'the neat limitation and closure promised by the epigram would give way to endless confusion'.[33] Manley discusses Guilpin's 'Epigram 68: Of Caius' in these terms:

> Guilpin's Caius is neatly reduced and fixed by his own manic repetitiveness … But on the other hand, through Caius' undiscriminating use of the phrase 'oh rare', the city ceases to be an articulate space. The many varieties of experience collapse into the one potentially endless experience of perambulation. This experience is punctuated by sounds – the blackman's noise, the pippin-seller's cry, the cobbler's song – and Caius' phrase rings its changes in the manner of a ballad refrain. Indeed, these fluid, oral traits threaten to dissolve the epigram's lapidary neatness into balladeering chaos … where the soul is eroded by the environment.[34]

Manley goes on to expand his scope from Guilpin to others working in the same mode:

> [T]he epigrammatists' attempt to encode the city shows the strain of the material and social factors that contribute to the making of epigrams but cannot

> be contained by them. As its neat devices begin to soften and expand, the epigram admits to a difference between the order it desires and the alien, mysterious life that order would contain.[35]

Extending this approach into the satires themselves raises questions, though: what if that 'alien, mysterious life' provoking a desire to contain and order is *itself* desirable? What if the 'punctuation' produced by the sound of the city makes up the structure of the satire, and informs the subjectivity of the satirist? And what if the satirist's 'soul' is not simply 'eroded' *by* but also engaged *in* the urban 'environment'?

Guilpin foregrounds such concerns. He concludes his sequence with what seems to be a strong statement from or about a self that cannot be eroded: 'For let them praise them, or their praise deny, / My lines are still themselues, and so am I'.[36] The persona claims their identity is settled and unchanging ('still') however their lines are interpreted. Nonetheless, in the seemingly solid rhetorical parallelism of this phrase, there are curious grammatical quirks, initiated by the echoing of object and subject in the repeated 'them' (eliding into 'themselues'). This develops into a sense that while the singular subject ('I') agrees with the singular verb ('am') it also links back to the plural subject-verb form ('lines ... are ... themselues'): rather than being unitary, sure and unchanging, is the 'I' composed of multiple 'selves'? The form might demand this. Satire involves assuming different identities, some of which accrue more disapprobation, and require more exculpation, than others:

> Excuse me (Reader) though I now and than,
> In some light lines doe shew my selfe a man,
> Nor be so sowre, some wanton words to blame,
> They are the language of an Epigrame.[37]

Moreover, as we will hear, satire also audibly involves reproducing sounds that both confirm and complicate the satirist's ethical and urban position.

The Sound of Satire

Deplorable urban noises resound in Guilpin's work, breaking bounds of privacy and propriety. Echoing Juvenal's lines about belching and flatulent patrons in his *Satire* 3, one of Guilpin's poems anticipates its subject will 'let a fart'.[38] The unruliness inherent to apparently transitory and fickle urban relations is characterized as noise, with verbal arguments preluding public disorder:

> Friends and good fellowes all now, by and by
> Thei'le be by the eares, vie stabs, exchange disgraces,
> And bandie daggers at each others faces.[39]

Even the sound of something harmonious like music merely evokes something base like licentiousness. Again, Guilpin follows Juvenal's lead. In *Satire* 6 Juvenal described how musical pipes excited loins, and how a woman was fond of handling instruments.[40] For Guilpin, comparably, one man's 'talke's' simply composed of crude puns about 'crotchets and quauers'; 'twixt' one woman's legs, she 'holds her instrument'; people learn 'a prick-song-lesson without ditty'.[41] In a plaguey, venal and venereal city like early modern London, these kinds of somatic and linguistic intercoursings might prove fatal, as the breath from bodies and tongues communicates 'infection'.[42] Words themselves are corruptible, evoke corruption and corrupt. And in this degenerate, undiscerning environment, decent writers – like Thomas Deloney, to whom Guilpin directs one poem – are muffled by the hubbub of idiots: 'At euery streets end *Fuscus* rimes are read, / And thine in silence must be buried'.[43] Instead of decency, people hear poetic 'fiddlers' with 'oyle-buttred lines', 'whimpring Sonnets, puling Elegies'. These are '*Lydian* tunes which blunt our sprights / And turne our gallants to *Hermaphrodites*'. Poets toast 'A health to Circes', live in 'hogsties' and are thus themselves 'transformd to Goates lasciuiously', as they fill 'chast ears with theyr pens Gonorrhey'.[44] Yet sound travels in the early modern city, beyond the contagious noise of the capital's literary marketplace. Indiscrimination contaminates not only aesthetics, but also the ethics and forms of bodies politic, economic and personal. '*Opinion*' notes one Satire, '*is as various as light change ... the Eccho of inconstancie*'.[45] Even – or especially – smugly self-professed aesthetes cannot discriminate between one noise and another:

> As *Caius* walks the streets, if he but heare
> A blackman grunt his note, he cries *oh rare*!
> He cries *oh rare*, to heare the *Irishmen*
> Cry pippe, fine pippe, with a shrill accent …
> He'le cry, *oh rare*, at a Gongfarmers cart,
> *Oh rare* to heare a ballad or a fart.[46]

Appropriately, intellectual errancy is persistently figured through metaphors of '*noyse*':

> *It's but the hisse of Geese ...*
> *The tongue of humours, and phantasticke voice.*[47]

And in sound of such cacophony, masculine and national identities audibly diminish:

> Of cap and courtship complements, and new
> Antike salutes, I care not for th'embrace,
> The Spanish shrug …
> Pronounc'd with lisping, and affected graces, …
> But this I wonder, that th'art so estranged,
> And thy old English looks to outlandish chang'd,
> Howsoe're thy selfe by English birth art freed,

> Thou hast neede to haue thy looks endenized: …
> Th'art quite turn'd Dutch, or some outlandish lowt,
> Thou hast cleane forgot thine English tong.[48]

Here the English adopt foreign fashions so completely that it is hard to discriminate if they are free, indigenous citizens, or denizens and so *not* naturalized citizens, but merely individuals allowed by royal letters patent to hold and acquire land.[49] Such lines reveal a xenophobic (and homophobic) reaction to 'outlandish' innovations in language, poetry and deportment. They also evince the sense that the anxieties provoking such reactions might have been acute for some living in late Elizabethan London, a time and place where, given mortality rates, it has been estimated that between 3,750 and 6,000 immigrants were required each year to sustain the levels of population growth.[50] These levels of immigration were not constant, or constantly rising: proportionally fewer strangers lived in the capital in the 1590s than at Elizabeth's accession. All the same, *perceptions* of large numbers of incomers abounded, and these were sometimes negative.

So Guilpin's persona betrays an anti-urban repugnance, composed around the sound of the city and who makes it. Yet counterpointed with this are the noises Guilpin knows he and his personae make. In this nationalistic xenophobia, Guilpin deliberately and ironically mimics a non-English satirist, repeating Juvenal's concerns in *Satire* 3 about foreign influences in *his* city: 'I cannot stand a Greekified Rome'.[51] This echo undercuts Guilpin's insularity. The ironies resound to intensify the compromises afflicting his distinction. In an early epigram, Guilpin's persona terms his work 'these pedler rimes'.[52] This humility topos implies they are of little worth, comparable to the mean stuff sold by street-vendors. Yet this also connects Guilpin's satire to the sounds and stuff of the city. Indeed, a later poem suggests this connection between words and world is no accident, but a fundamental premise for the satire to have a curative effect:

> I know they are passing filthy, scuruey lines,
> I know they are rude, harsh, and vnsauory rimes:
> Fit to wrap playsters, and odd vnguents in.[53]

Invoking rudeness is of course more false modesty – these are classically cultured 'rimes'. However, the scabrousness, and, crucially, the *volume* of their attack does constitute 'vnsauory' immodesty. Thus 'vnmannerly', says the persona, these poems' '*Muse*' can 'beray our folly-soyled age', by 'Bringing a foule-mouth Iester who might sing / To rogues'.[54] In other words, the city-sounds this urban satire makes are vital to its impact, and those satiric sounds both deplore and echo noises on the streets. If Guilpin seeks to criticize the soundscapes of his city, he has to reproduce them: 'A soundscape consists, therefore, not just of the environment that the listener attends to but of the listener-*in*-the-environment'.[55] Guilpin reflexively acknowledges this. His satires are aware that to offer discerning discrimination, they can only do so by creating their own noise, and thereby adding to the cacophony. This poetry repeatedly signals how it disturbs any specious peace to vociferously question the noises other poets make:

> Shall I still mych in silence and giue ayme,
> To other wits which make court to bright fame?
> A schoole boy still, shall I lend eare to other,
> And myne owne priuate Muses musick smother?[56]

Breaking 'silence', Guilpin's satires promise and offer 'To sound an Antheme of … folly'.[57] Noise combats noise and sound beats sound, as the militant tenor of these lines makes clear, with reference to another 'outlandish' foreign voice that might usefully be incorporated into this polyglot city:

> Oh that whip of fooles, great *Aretine*,
> Whose words were squibs, and crackers euery line,
> Liu'd in our dayes, to scourge these hypocrites.[58]

'Satire V'

The ambiguities of this struggle for sonic hegemony and subjective poetic autonomy in a city like early modern London reach their crescendo in 'Satire V'.[59] This satire begins with the persona wanting isolation, particularly from the depraved urban environment around them: 'Let me alone I prethee in thys Cell, / Entice me not into the Citties hell;' According to the persona, they have no need to 'conuerse' with anyone other than 'those diuiner spirits' contained within 'bookes and papers'. When in 'meditation' with such company, as 'graue … auditor' to Plato for example, the persona says, 'I am not alone'. They certainly seem to have no need or desire to stop being 'encoffin'd' in their room, with 'calme content', when the alternative is 'all the Citties lushious vanity'. In fact, 'this *Eden*' becomes a substitute for the city, and privately consumed texts allow the persona to recreate the urban space as divisible, controllable, traversable and escapable:

> (For this my study is indeede m'Exchange)
> Heere may I sit, yet walke to *Westminster*
> And heare Fitzherbert, Plowden, Brooke, and Dier
> Canuas a law-case: or if my dispose
> Perswade me to a play …
> If my desire doth rather wish the fields,
> Some speaking Painter, some Poet straitway yeelds
> A flower bespangled walk …
> Thus my books little case,
> My study, is mine All, mine euery place.

Urban inhabitants in later periods would develop 'quiet spaces in their homes to concentrate on their books'.[60] And later works, such as Ben Jonson's 1609

Epicoene, would of course stage noise as an analogue to disease, against which characters hope in vain for 'domestic prophylaxis'.[61] Yet in the 1590s Guilpin can only fantasize about such future insulation. For it *is* a fantasy, as the imagery of imprisonment and entombment indicate, as the suggestive adjective 'lushious' hints, and as the playful pun on being 'grave' with Plato betrays. However much the persona protests (and it comes to feel like *too* much), they know sequestration from the city is deathly. This ambivalence is caught in the succeeding question: 'What more variety of pleasures can / An idle Citty-walke affoord a man?' This question is at once affirmative about the existence of pleasure in the city, yet uncertain that such pleasures can be realized; it is both ironic and inviting. Ironic because, as the poem goes on, we learn the city is 'troublesome and tedious', its 'peopled streets' are overcrowded, and these crowds make the city seem chaotic. In this poem, London assaults the senses: being in the city makes it hard to work out what is what, and to describe what's going on. There are 'rude sounds' and 'many noyses', and all make up a 'hotch-potch' that challenges language and the satirist's ability to explicate, as they resort to para-words. Moreover, there is something strange and estranging about these noises. The city barely registers the sound of civilized English language. Instead, the mixed-up hybrid ruckus of London resounds like 'harsh *Barbary*', a description evoking barbarism, savagery and also unsettlingly outlandish places. This echoes an alienating city full of alien tongues.

Yet the suggestiveness of the question remains. And despite all these negative associations and effects, for the persona, there's something fascinating and stimulating about the experience of being in the city. We start to see this in the next few lines, describing the noise, but also people's reactions to it: 'Compos'd of seuerall mouthes, and seuerall cries, / Which to mens eares turne both their tongs & eies'. There may be alienating 'Chaos' but people are captivated by it. They tune their senses to it, and in turn add their own 'tongs' to the mix. The persona might be disgusted by the city's forms, beginning by being physically and intellectually removed from it, but they too are captivated, and they too contribute to its stimulating noise:

> There squeaks a cart-wheele, here a tumbrel rumbles;
> Heere scolds an old Bawd, there a Porter grumbles.
> Heere two tough Car-men combat for the way …

With their repetitive use of 'Heere', itself punningly echoing 'hear' to connect sites and sounds, these lines ingeniously record the city: the 'squeaks' are onomatopoeic, and the alliteration rumbles like cartwheels, or clashing car-men. Simply put, the persona vividly reproduces the sounds of the city even as they attack them. Furthermore, Guilpin amplifies the echoic clamour by reproducing others' attacks on it: these lines imitate complaints about the noise of carts and wagons in Rome that Juvenal made in *Satire* 3.[62]

The negative, satirical description does continue, but more than ever this works to create a sense that the persona is simply resisting being seduced by the city's sounds and sensations. So the persona isn't *just* appalled by the city's chaotic noise. They are also dismayed by its deceits, loose morals and criminality, embodied in 'coney-catching knaues' and sexual deviants like 'Whores' and 'bawds'. But the 'Bedles … and Sergeants' who are supposed to be stopping criminality and deviance are corrupt too: legal and illegal figures all combine in a noisome 'Iigge'. There is no law and order, just a hypocritical – and raucous – show. This nauseates the moral observer and auditor:

> Yet might all this, this too bad be excusd,
> Were not an Ethicke soule much more abusd,
> And her still patience choakt by vanitie,
> With vnsufferable inhumanitie:
> For whose gall is't that would not ouerflow,
> To meete in every streete where he shall goe,
> With folly maskt in diuers semblances?

Yet even here, the persona can't resist perpetuating or giving voice to the very things they claim to despise. For example, they go on to employ the phrase 'nipt a bong' to signify thieves' slang for 'picked a pocket' or 'pinched a purse'. But they use the term unglossed. This betrays their own facility with the language, and their ability to reproduce it. Equally significantly, using this unglossed phrase also suggests readers would have known and understood this language: it does not require translation in a polyglot city.

If, as this poem itself articulates, the city was a chaotic, jumbled, hotch-potch place, then the acoustic, linguistic, spatial and moral boundaries between the illicit, illegal, immoral underworld, and the seemingly licit, legal and moral mainstream were permeable. And so the ambiguities of the poem, and of the city itself, reverberate. We learn that the city is 'the mappe of vanities', full of idiots obsessed with material possessions: 'The marte of fooles'. Guilpin revives his bestializing imagery to remind us that city-dwellers are so stupid and degenerate they are barely human – more like 'apes' of '*Affricke Tabraca*'. So the persona makes foreigners animals, yet also makes city-dwellers strange, and makes their city, the nation's capital, foreign.

But such lines and images also herald the city's vitality: London might be full of savage fools, but it teems with life, ingenuity, enterprise and busy-ness, commercial, linguistic, material and aural. And if London alienates visitors and locals alike, it is also exotic and exciting – a place of wonderful novelty as well as weirdness. This ambiguity registers in the poem's imperative, active verbs. Rather than warning us off, the persona invites their readers to get involved in the city they describe. We are instructed to 'walk in Poules', 'see yonder' and 'obserue'. Clearly visual verbs are potent here: this collection is subtitled, after all, 'A Shadowe of Truth', suggesting how the satire can represent only a dark fragment of reality, because that reality is itself so confusingly dark. Yet auditory imperatives figure too: 'Satire V' generates resonant homophones, we are asked to listen to someone who 'bellows rimes like thunder', and implored to be 'soft' to eavesdrop on vanities. Is this then as much an *echo* of truth as a shadow? Whichever – we are compelled to visit the city, move through its streets, participate in its scenes and revel in the diversity of its 'sundry kindes of shapes', its many people and voices, just as the persona is doing. This, perhaps, is the final irony. The satire may condemn the city, but it couldn't exist without it. London makes this poem necessary and possible, just as the poem re-presents the sounds of the city. And while the city threatens the integrity of personal, satiric, linguistic, social and national identities, as discrimination declines and one wallows in the filthy and fatuous appetites of the many, *not* to engage in the city is fatal.

Conclusion

'Satire V' sustains this ambiguous relation to the city and its noises in its ending:

> Enough of these then, and enough of all,
> I may thanke you for this time spent; but call
> Henceforth I'le keepe my studie …
> Now let vs home, I'me sure tis supper time,
> The horn hath blowne, haue done my merry rime.

The Inns of Court blew their horns to call scholars and lawyers to meals. A reassuringly regular sound therefore relieves the persona's appetites and lets them pull themselves out of the streets (and together). Yet this sound – and the sites that made it – are part of the noise of the city. Earlier satirists knew this. In *Satire* 7, Juvenal describes how lawyers 'talk big' with chests like bellows exhaling lies and spit.[63] Later writers would return to this theme. In Thomas Middleton and Thomas Dekker's *The Roaring Girl* (1611), Holborn is designated a 'wrangling' (noisy and argumentative) street, 'because lawyers' walk there.[64] And in the period's most sound-fixated play, Jonson's *Epicoene*, the trope of noisy lawyers recurs:

DAUPHINE: Ha' you spoke with a lawyer, sir?
MOROSE: Oh no! There is such a noise i' the court, that they have frighted me home with more violence than I went! Such speaking and counter-speaking …[65]

This brings new significance to Guilpin's association with the Inns and the legal profession: he is thus subject to and maker of these kinds of noises, attracted and repulsed by forces he recognizes in the city and himself, and in the sounds of the city his poetry evokes. Yet Guilpin's close connection to a site of sonic and literary production, and his reputation as a maker of satiric noise (even to drown out the sounds of a sinful, idiotic city), would prove problematic. Early in June 1599, John Whitgift, Archbishop of Canterbury and Richard Bancroft, the Bishop of London, issued a ban on certain named authors, texts and genres, resulting in book burnings at the Stationers' Hall, opposite St. Paul's.[66] Guilpin's satires were explicitly included, alongside works by other writers associated with the Inns, such as John Marston and John Davies. This was an outright attempt to regulate what was becoming seen as the city's tumultuous literary culture. The noise generated by satirists like Guilpin had grown too loud.

Nonetheless, in 1601, it would seem Guilpin contrived a characteristically piercing response to the authorities' treatment of his work. There's an unrepentant sound of self-justification in 'The Whipper of the Satyre his pennance in a white Sheet: OR, The Beadles Confutation': satire does more good than harm, and helps those unwilling or unable to 'giue ears' to the 'heauenly voice' and 'dulcet noise' of 'sacred Teachers' to 'learne aright'. Despite the tin-eared assaults of religious and civil authorities, what Guilpin terms 'good *Satyre*' actually 'Sings to the soule a blessed lullabie'. This raises the issue of whether Guilpin intends 'good' to be an ironic adjective here, and whether effective satire actually lulls a soul to sleep. However we resolve such concerns, it is evident Guilpin audibly sustains the value of satire in terms of sonic power:

> No no, since kinde perswasions will not do,
> Sung from the tongue of dulcet pietie,
> Let irefull *Fury* whip and scourge them to,
> Sounding their soules perpetuall miserie.[67]

He was right to do so. Though Guilpin was forced to adopt a defensive position, the power exerted over producers of cultural noise was never total. Ultimately, as ever in this city, satiric sound silenced in one location would echo in another: the theatres.[68]

Notes

1 Everard Guilpin, *Skialetheia, Or A Shadowe of Truth in Certain Epigrams and Satyres* (1598), ed. D. Allen Carroll (Chapel Hill: University of North Carolina Press, 1974), 'Satire V', p. 86.
2 Jacques Attali, *Noise: The Political Economy of Music*, trans. Brian Massumi (Minneapolis and London: University of Minneapolis Press, 2003), p. 3.
3 Matthew Hodgart, *Satire* (Verona: World Publishing, 1969), p. 129; Alvin Kernan *The Cankered Muse: Satire of the English Renaissance* (New Haven: Yale University Press, 1959), pp. 7–8.
4 Bruce R. Smith, *The Acoustic World of Early Modern England: Attending to the O-Factor* (Chicago and London: University of Chicago Press, 1999), p. 49.
5 Smith, *Acoustic World*, p. 52.
6 Ibid., pp. 61–2.
7 Emily Cockayne, *Hubbub: Filth, Noise & Stench in England 1600–1770* (New Haven and London: Yale University Press, 2007), pp. 112–13.
8 Laura Wright, 'Speaking and Listening in Early Modern London', in Alex Cowan and Jill Steward (eds), *The City and the Senses: Urban Culture Since 1500* (Aldershot: Ashgate, 2007), pp. 60–74, at p. 63. For other similarly useful perspectives on the contradictions surrounding and inherent to linguistic nationalism, with a focus on London, see also Dick Leith, 'Tudor London: Sociolinguistic Stratification and Linguistic Change', *Anglo-American Studies*, 4 (1984): 59–72; Julia G. Ebel, 'Translation and Cultural Nationalism in the Reign of Elizabeth', *Journal of the History of Ideas*, 30 (1969): 593–602; Richard Helgerson, 'Language Lessons: Linguistic Colonialism, Linguistic Postcolonialism, and the Early Modern English Nation', *Yale Journal of Criticism*, 11/1 (Spring, 1998): 289–99; and Emma Smith, '"Signes of a Stranger":

The English Language and the English Nation in the Late Sixteenth Century', in Philip Schwyzer and Simon Mealor (eds), *Archipelagic Identities: Literature and Identity in the Atlantic Archipelago 1550–1800* (Hampshire: Ashgate, 2004), pp. 169–79.

9 Wright, 'Speaking and Listening', p. 73.

10 John Eliot, *The Parlement of Pratlers* (1593; London: Franfrolico Press, 1928), p. 27.

11 Smith, *Acoustic World*, p. 54.

12 Laura Gowing '"The Freedom of the Streets": Women and Social Space, 1560–1640', in Paul Griffiths and Mark S.R. Jenner (eds), *Londinopolis: Essays in the Cultural and Social History of Early Modern London* (Manchester and New York: MUP, 2000), pp. 130–51, at pp. 134–6. See Mary Thomas Crane, 'Illicit Privacy and Outdoor Spaces in Early Modern England', *Journal for Early Modern Cultural Studies*, 9/1 (2009): 4–22; Lena Cowen Orlin, 'Boundary Disputes in Early Modern London', in Lena Cowen Orlin (ed.), *Material London, ca. 1600* (Philadelphia: University of Pennsylvania Press, 2000), pp. 344–76; Lena Cowen Orlin, *Locating Privacy in Tudor London* (Oxford: Oxford University Press, 2007); and Lena Cowen Orlin, *Privacy Matters and Public Culture in Post-Reformation England* (Ithaca: Cornell University Press, 1994).

13 David Garrioch, 'Sounds of the City: The Soundscapes of Early Modern Towns', *Urban History*, 30/1 (2003): 5–25, at p. 8.

14 Katherine Eisaman Maus, *Inwardness and Theater in the English Renaissance* (Chicago and London: University of Chicago Press, 1995), pp. 29–31, 75.

15 Richard Leppert, *The Sight of Sound: Music, Representation and the History of the Body* (Berkeley, LA and London: University of California Press, 1993), pp. xix–xx.

16 Jill Steward and Alexander Cowan, 'Introduction', in Jill Steward and Alexander Cowan (eds), *The City and the Senses: Urban Culture Since 1500* (Aldershot: Ashgate, 2007), pp. 1–22, at p. 13.

17 Garrioch, 'Sounds': 5–6.

18 Ibid.: 15.

19 Garrioch, 'Sounds': 18.

20 Quoted in Kate Aughterson (ed.), *The English Renaissance: An Anthology of Sources and Documents* (London: Routledge, 2002), pp. 386–7.

21 Wes Folkerth, *The Sound of Shakespeare* (London: Routledge, 2002), p. 73.

22 Smith, *Acoustic World*, pp. 66–8.

23 Richard Dering, *The Cries of London* (n.p., 1599); ed. Denis Stevens (Pennsylvania: Pennsylvania State University Press, 1964), pp. 7, 11, 30, 35–6.

24 This and all subsequent quotes in the paragraph are from Dering, *Cries*, pp. 5–6, 20, 9–11, 27–8.

25 Dering, *Cries*, pp. 18–19.

26 Ibid., p. 15.

27 Carroll, 'Introduction', p. 7. On the cultural milieu of the Inns of Court, with some references to Guilpin's place in it, see also Philip J. Finkelpearl, *John Marston of the Middle Temple* (Cambridge, MA: Harvard University Press, 1969), pp. 3–33.

28 Lawrence Manley, *Literature and Culture in Early Modern London* (Cambridge: Cambridge University Press, 1995), pp. 372–3.

29 Manley, *Literature and Culture*, p. 379.

30 Ibid., pp. 374–5.

31 Ibid., p. 393.

32 Ibid., p. 375.

33 Ibid., p. 422.

34 Manley, *Literature and Culture*, p. 426.

35 Ibid., p. 427.

36 Guilpin, 'Satire VI', p. 93.

37 Guilpin, 'Epigram 47', p. 52.

38 Guilpin, 'Epigram 7', p. 41; see Juvenal, *Satires* in *Juvenal and Persius*, trans. Susanna Morton Braund (Cambridge, MA and London: Harvard University Press, 2004), pp. 174–5.

39 Guilpin, 'Satire V', p. 87.
40 Juvenal, *Satires*, pp. 260–61, 270–71.
41 Guilpin, 'Epigram 29', p. 47; 'Epigram 46', p. 51; 'Epigram 50', p. 52.
42 Guilpin, 'Epigram 4', p.40.
43 Guilpin, 'Epigram 8', p. 41.
44 Guilpin, 'Satyre Preludium', pp. 59–60.
45 Guilpin, 'Satire VI', p. 89.
46 Guilpin, 'Epigram 68', pp. 56–7.
47 Guilpin, 'Satire VI', p. 90.
48 Guilpin, 'Satire III', p. 73.
49 See Laura Yungblut, *Strangers Settled Here Amongst Us: Policies, Perceptions & the Presence of Aliens in Elizabethan England* (London and New York: Routledge, 1996), pp. 78, 128, 147–8.
50 See Steve Rappaport, *Worlds Within Worlds: Structures of Life in Sixteenth-Century London* (Cambridge: Cambridge University Press, 1989), p. 76, and Ian W. Archer, 'Shakespeare's London', in David Scott Kastan (ed.), *A Companion to Shakespeare* (Oxford: Blackwell, 1999), pp. 43–56, at p. 44.
51 Juvenal, *Satires*, pp. 170–71.
52 Guilpin, 'Epigram 2', p. 39.
53 Guilpin, 'Epigram 70', p. 58.
54 Ibid., p. 58.
55 Smith, *Acoustic World*, p. 44.
56 Guilpin, 'Satire I', p. 63.
57 Ibid., p. 77.
58 Ibid., p. 67.
59 Guilpin, 'Satire V', pp. 82–7. All the subsequent quotes in this section, including the indented extracts, are from Guilpin 'Satire V', pp. 82–87.
60 Cockayne, *Hubbub*, p. 127.
61 James D. Mardock, *Our Scene is London: Ben Jonson's City and the Space of the Author* (New York and Abingdon: Routledge, 2008), p. 73. On *Epicoene*, spatial incontinency and aurality, see Karen Newman, *Cultural Capitals: Early Modern London and Paris* (Princeton: Princeton University Press, 2007), pp. 84–91; Ann C. Christensen, '"The Doors Are Made Against You": Domestic Thresholds in Ben Jonson's Plays', *Journal of the Rocky Mountain Medieval and Renaissance Association*, 18 (1997): 153–78; and Adam Zucker, 'The Social Logic of Ben Jonson's *Epicoene*', *Renaissance Drama*, 33 (2004): 37–62.
62 Juvenal, *Satires*, pp. 187–8.
63 Juvenal, *Satires*, pp. 306–7.
64 Thomas Middleton and Thomas Dekker, *The Roaring Girl* (1611), in Simon Barker and Hilary Hinds (eds), *The Routledge Anthology of Renaissance Drama* (Abingdon: Routledge, 2005), pp. 326–77, at p. 357.
65 Ben Jonson, *Epicoene, Or the Silent Woman* (1609), in Simon Barker and Hilary Hinds (eds), *The Routledge Anthology of Renaissance Drama* (Abingdon: Routledge, 2005), pp. 272–325, at p. 316.
66 For studies of this event, see Richard A. McCabe, 'Elizabethan Satire and the Bishops' Ban of 1599', *Yearbook of English Studies* (1981): 188–93, and Cyndia Susan Clegg, *Press Censorship in Elizabethan England* (Cambridge: Cambridge University Press, 1997).
67 Guilpin, 'The Whipper of the Satyre his pennance in a white Sheet: OR, The Beadles Confutation' (London: Thomas Pauier, 1601), *Literature Online* (http://lionchadwyck.co.uk/). Lines 91–6, 241–4, 199–202 (accessed 17 June 2009).
68 See Lynda E. Boose, 'The 1599 Bishops' Ban, Elizabethan Pornography, and the Sexualization of the Jacobean Stage', in Richard Burt and John Michael Archer (eds), *Enclosure Acts: Sexuality, Property, and Culture in Early Modern England* (Ithaca, London: Cornell University Press, 1994), pp. 185–200, and Tom Rutter, '*Patient Grissil* and Jonsonian Satire', *Studies in English Literature*, 48/2 (Spring, 2008): 283–303.

11 The Soundscape of a City in the Nineteenth Century

Olivier Balaÿ

Introduction

This chapter retraces part of the history of the transformation of the 'soundscape of 19th century towns'.[1] It shows, on the basis of the study of the urban fabric of Lyon, France, that developments made in the middle of the century changed the conditions under which sounds were produced and propagated: urban 'repairs', the widening of thoroughfares and the removal of overhangs, corbelled constructions and awnings, led slowly but surely to changes in what was heard, to average intensities, lower frequencies and continuous noises that gradually put an end to high-intensity noises or silence and high-pitched, intermittent, human sounds.[2]

The Acoustic Metamorphoses of the Street

The urbanscape is not only organized in terms of the visual, but also by sound. Sounds propagate in different ways depending on the width of streets and the nature of the materials and decorative elements used for façades. Developing a space necessarily changes its original acoustic make-up. By modifying the width of streets and the height of buildings, and introducing regulations governing the architectural design of façades, the authorities affected the characteristics of the urban soundscape.

The end of the first half of the nineteenth century was a crucial moment in urban transformation. In Lyon, much as in other large towns, the local and national government authorities, who were responsible for defining urban ills and the appropriate remedial policies, made alterations on an unprecedented scale that changed the urban morphology inherited from previous centuries and paved the way for the shaping of the acoustic environment that we find in the contemporary public space.

The streets of eighteenth-century Lyon, as we know, were narrow and lined with high buildings. The façades were decorated with various cavities and projections. Visiting travellers were struck by the considerable height of the buildings lining the narrow streets of the town centre. In Lyon, as one visitor put it, 'you only see a sample of the sky'.[3]

The historian Maurice Garden refers to an expert appraisal dating from 1723–24 and covering the whole of Rue des Prêtres, one of the busiest thoroughfares close to Saint-Georges' gate and the wooden Archevêché bridge. The street was 9 foot wide (3 metres) and there was not enough room for two carts to pass. Two-thirds of the length of Rue du Boeuf (in the Saint-Jean district) was trapped between three buildings, with the same width of 9 feet. Rue du Plat-d'Argent (on the central peninsula) was no more than 14 foot wide (4.2 metres). Rue Royale, which was built in the eighteenth century in what is now the Croix-Paquet district (at the bottom of the Croix-Rousse slopes) was 24 foot wide (7.20 metres). That is 'much more than the old streets', writes Garden.[4]

At the same time, at least three-quarters of the buildings in the Hôtel-Dieu district had at least five storeys. Around the corn market (Cordeliers district), some 41 per cent of buildings had five storeys or more. Around the town hall (on what is now Place des Terreaux) 28 per cent of the buildings boasted six floors. Taking the city as a whole, without its suburbs, 71 per cent of buildings had more than four storeys.[5]

From 1823, with successive additions to street by-laws, the Lyon authorities banned various devices that absorbed and dispersed sound, and, inadvertently, established street acoustics that were less satisfactory in terms of sound clarity. Several by-laws banned dyers, drapers, cleaners and the like from setting up poles and hanging out fabric. Such drapes were only permitted if they were located at least 6 metres above the ground and did not jut out more than 80 centimetres into the street. Encorbelled additions were banned, and existing elements were gradually demolished as structures were rebuilt or repaired. Under article III of a police by-law issued on 9 December 1823, awnings were only allowed on squares and quays, never in the streets. However, permission would be granted subject to agreement by the occupants of upper floors. There was no question of awnings being supported by wooden columns driven into the paving stones. This was still allowed in 1813 and proved a considerable hindrance to traffic to judge by an advisor to the *Préfet*, who wrote: 'On Rue du Plat-d'Argent a dyer who has hung out fabric is taking up more than half the width of the street'.[6] Under the awnings, fixed displays jutting out by more than 20 centimetres (also known as *montres*) were still authorized in 1813. But by 1825 the rules had been changed. Cafes were only allowed to set up chairs and benches on squares and quays, on condition that they did not encroach on public thoroughfares by more than 1.5 metres. Elsewhere placing chairs outside cafes was forbidden.

The drawings in Figure 11.1 show how, in the same street, sound propagation changes as a consequence of the drapes and the size of outcrops, encorbelled elements and signs located on the façades. For example the fabric of awnings absorbs low and high frequencies. Cornices diffuse vocal frequencies well. Glass canopies, authorized in streets more than 10 metres wide and above shops, protect dwellings on upper floors.

The cross-section views showing street dimensions regroup this data with the rules for the height of various elements found in a by-law dated 15 November 1849. The drawings show how the profile of the street changes the sound level,

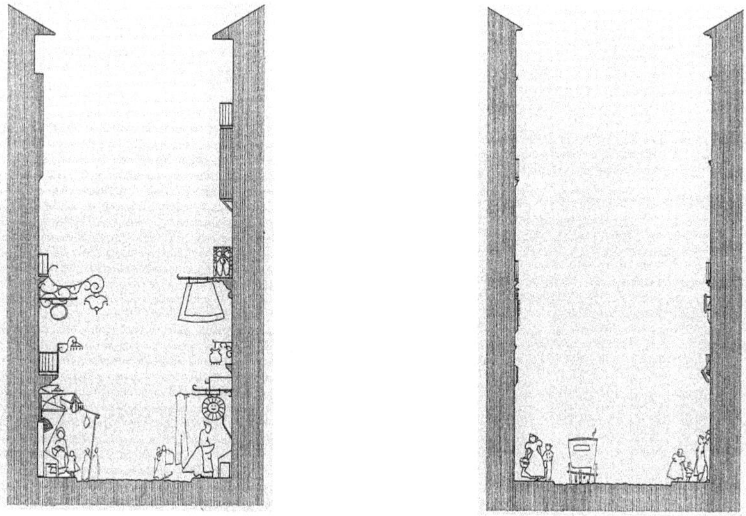

La rue traditionnelle au début du XIXe siècle : étalages et tentes absorbent les sons. Dessin de Bruno Delescure.

35. La rue selon les règlements de voiries du milieu du XIXe siècle : absorption et diffusion des sons disparaissent. Dessin de Bruno Delescure.

Figure 11.1 At the beginning of the nineteenth century, hangings and tents absorbed sound. In the middle of the nineteenth century absorption and diffusion of sounds disappeared.

Source: Drawing by Bruno Delescure.

due to acoustic reflection on the walls, regardless of the nature of the noise and the number of sources (one or more voices, for example). Acoustic analysis, which we will develop in greater detail below, can be used to specify the reverberation parameters for each case. For a given sound level for city activity, the reverberation of a narrow street heightens the perception of human density. Reflected sound is multiplied by the closeness of the walls, creating conditions in which sources are more difficult to locate and isolate for the listener.

In the wider streets less sound is reflected. Sounds do not reverberate for so long because the streets open more onto the sky, into which sound may escape. The opening onto the sky increases from 12 per cent in a narrow street (five metres wide) to 20 per cent in a street twice as wide. The difference in the length of time streets reverberate should not be neglected (about 0.5 to 1 second). Depending on whether the street is narrow or wide, the human voice will linger differently (the sound envelope of the voice is amplified) or there is an echo.

Urban Transformation Changes the Sound Atmosphere

But there is more to the sound history of the town than just the existence of sound sources and auditory presences. Something intermediate intervenes, reshaping the sound source, boosting or weakening its range, filtering its spectrum and shaping its duration. It is, then, the *morphology* and *material* of the place that has the biggest impact on how the different ways of sounding linger on.

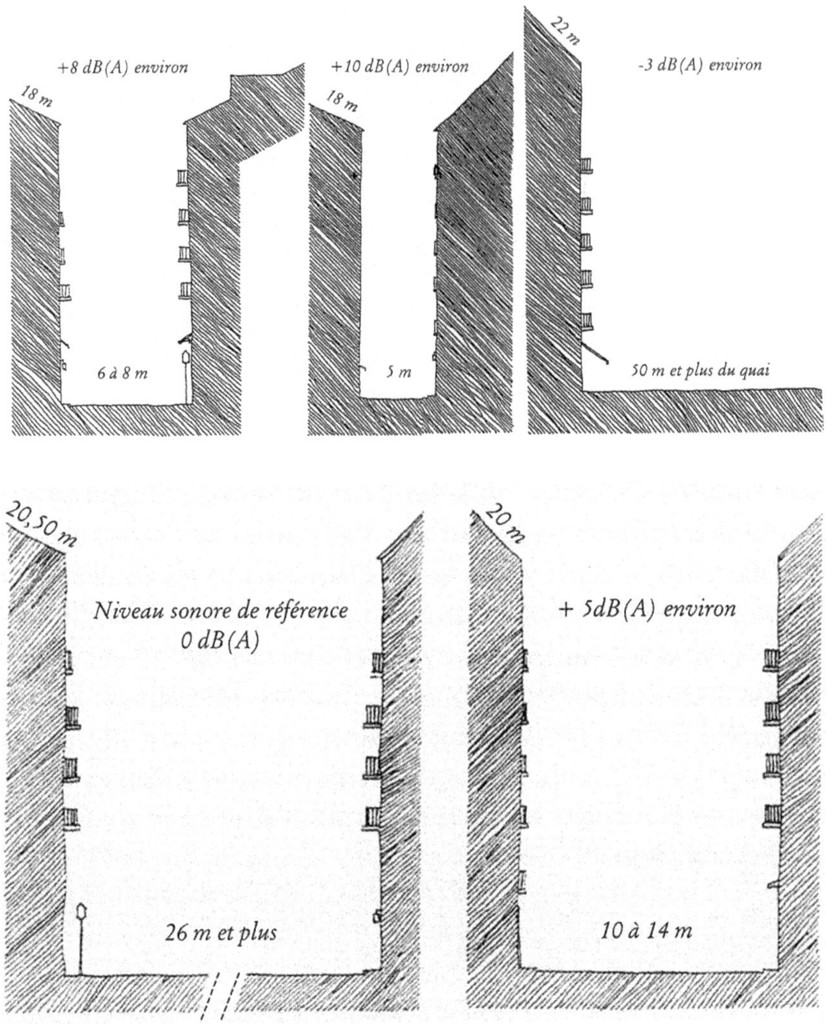

Figure 11.2 From the left to the right, graphics showing how a sound level increases with the narrowness of the streets.

Source: Drawing by Bruno Delescure.

Reorganization of the Central Districts

The map of Lyon's town centre in Figure 11.3 reveals other acoustic changes that we have measured using a sonometer.[7]

Before the major redevelopment work carried out in the mid-nineteenth century the streets did not run in straight lines, as they do today. Curved or angled streets were characteristic of the pre-Haussmann town. Typical examples were Rue Mercière, Rue du Plâtre and, above all, Rue du Bessard (now straightened out and renamed Rue de Constantine). After the redevelopment work, directed

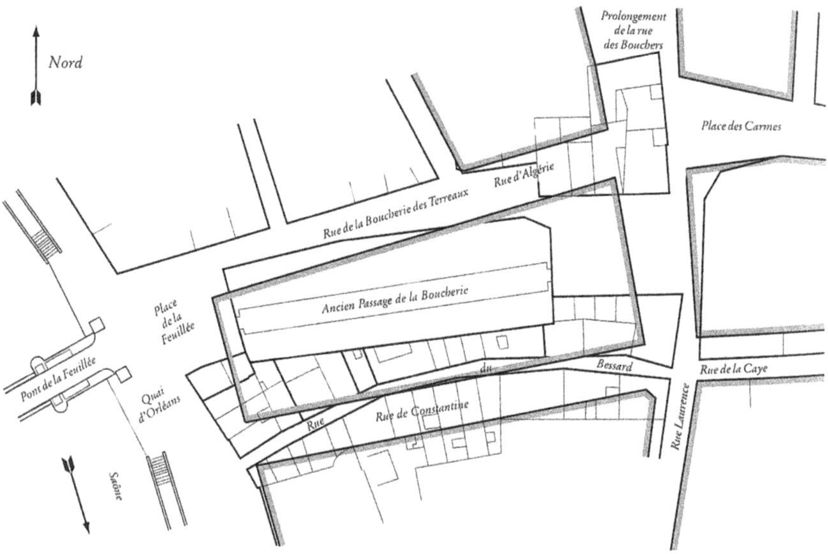

38. Le quartier de la Boucherie des Terreaux.
Plans comparatifs des tracés de rues avant et après le remodelage de 1840-1841.

Figure 11.3 The Terreaux meat-market area, before and after redevelopment.
Source: Drawing by Olivier Balaÿ.

by Prefect Vaïsse, only Rue Port-du-Temple retained a bend along its full length. These alterations affected the Terreaux district, after the Perrache abattoir opened, enabling the old meat-market to be closed and the neighbourhood to be reorganized. The architect René Dardel was in charge of this operation from 1838 to 1846. As may be seen from the map (Figure 11.3), the narrow stretch of Rue de la Boucherie-des-Terreaux and the bend in Rue Bessard were replaced by new streets, Rue d'Algérie and Rue de Constantine, with their neat parallel façades.

Berthelet de Barbot (1773) confirms this layout: 'The waters of two rivers bathe the shores of the peninsula. It is packed with houses separated by streets, three of which run parallel to the rivers, deviating slightly. These streets are crossed by others running obliquely from east to west, but never from the Rhône to the Saône. Most of them are winding, forming more or less open angles with the main streets'.[8] The configuration of these 'angles' was completely different after the work carried out in the mid-nineteenth century, as we shall show.

From 1853 to 1864 Lyon was governed by the Prefect Vaïsse. He resolutely transformed the town, opening up major thoroughfares. Vaïsse wanted to 'conserve life and movement in the town centre'. His objective was to 'prevent the centre from moving'.[9]

Over 11 years a series of streets (designated by their present-day names) were carved out and built: Rue de la République (former Rue Impériale),[10] Rue Édouard-Herriot (former Rue de l'Impératrice), Rue de Brest, Rue Grenette, Rue Childebert and Rue Terme. Even the Rhône and Saône quays were either built or rehabilitated.

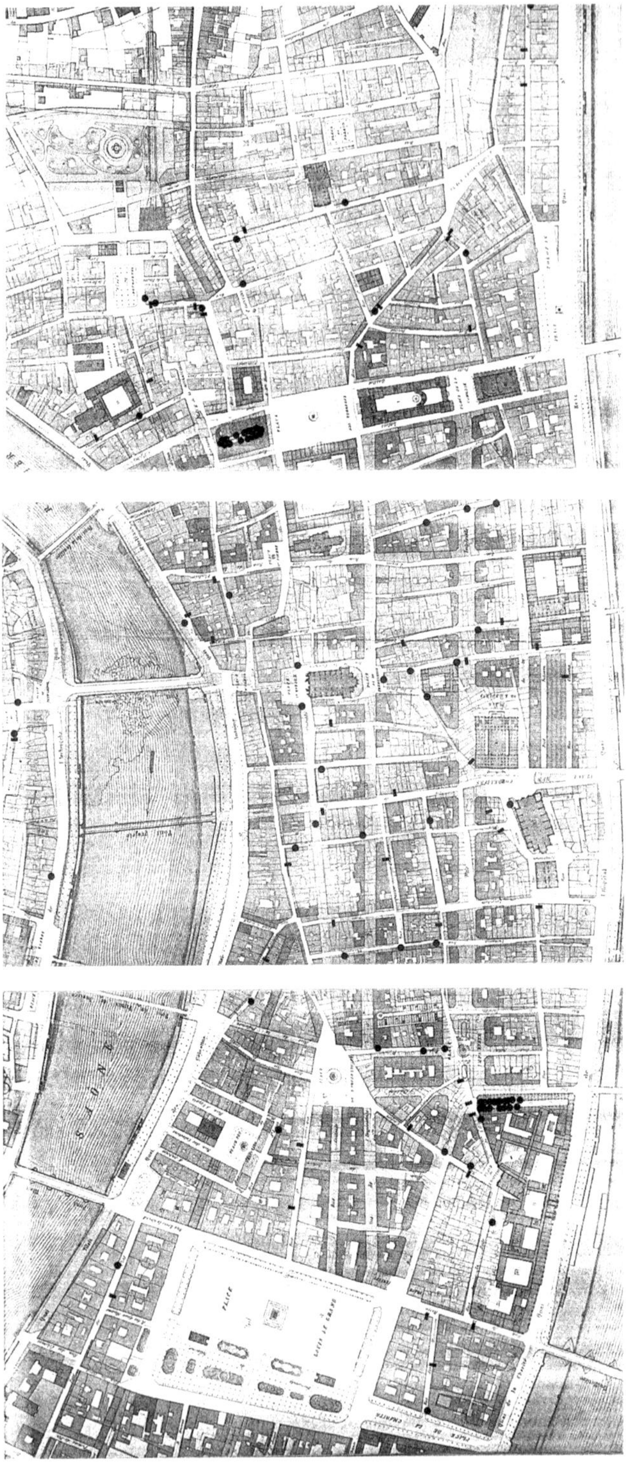

Figure 11.4 Layout of the streets of Lyon's town centre, before the alterations of 1853–64.

Source: Collection Bibliothèque municipale de Lyon, 5143. © BM de Lyon. Annotations by Olivier Balaÿ.

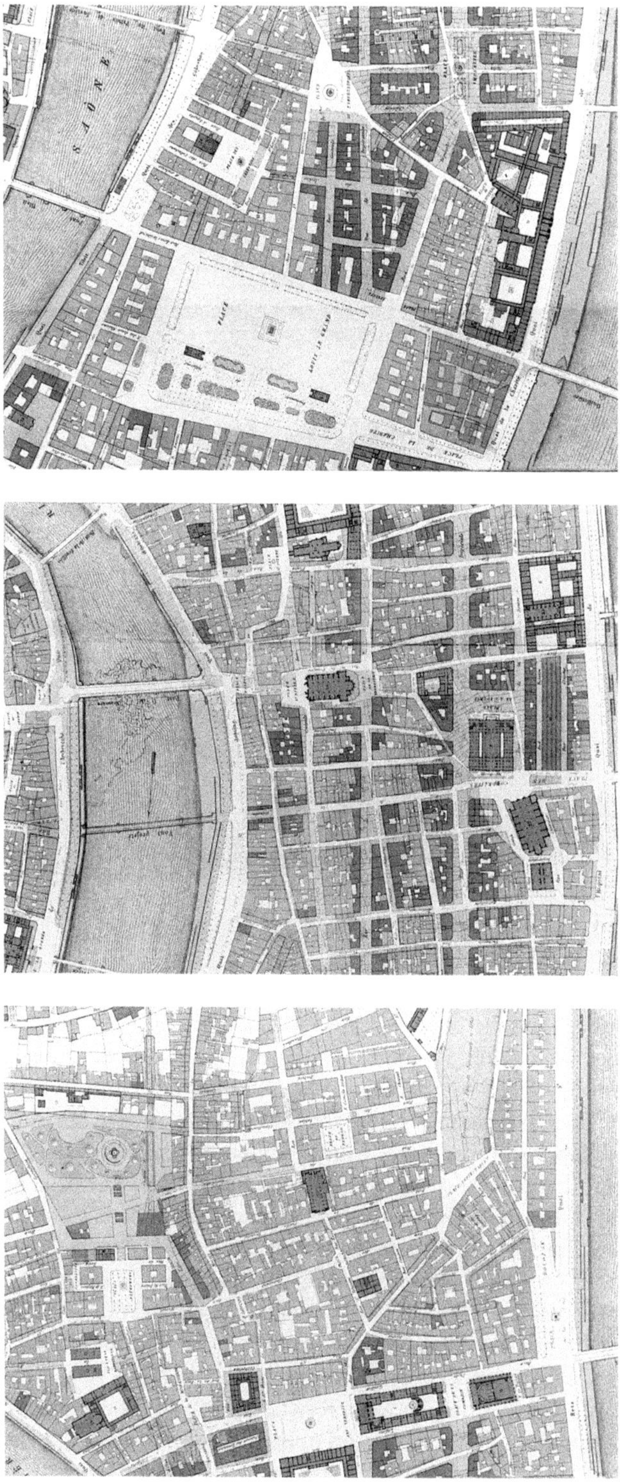

Figure 11.5 Layout of the streets of Lyon's town centre, after the alterations of 1853–64. Gustave Bonnet, Lyon 1863, Don 1937, Inv. 37.456.

Source: Collection Bibliothèque municipale de Lyon, 5143. © BM de Lyon.

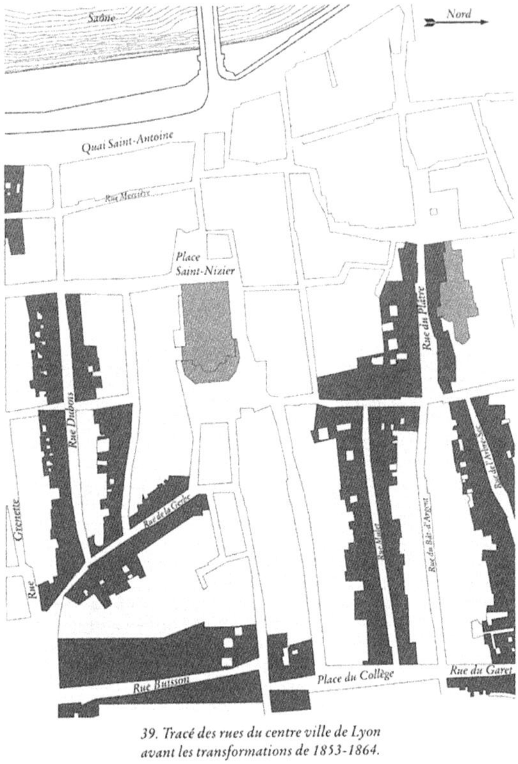

Figure 11.6 Acoustic consequences of the new layout of the streets in the centre of Lyon, before the transformations of 1853–64.

Source: Drawing by Olivier Balaÿ.

To summarize, redevelopment work proceeded as follows.[11] At the northern and southern ends of the peninsula the Saône and Rhône were connected by two large thoroughfares, one running past the town hall (Hôtel de Ville), Place de la Comédie and Place des Terreaux, the other running past Place Bellecour and the Hôtel-Dieu. Three new streets, which should ideally have been parallel, connected the two large thoroughfares: Rue Centrale, built between 1846 and 1848 under the supervision of R. Dardel and B. Poncet; Rue Impériale (now Rue de la République), supervised by the same architects, 22 metres wide; and Rue de l'Impératrice (now Rue Édouard-Herriot) planned by J. Bonnet, with work starting in 1860. Along Rue Impériale, on which work started in 1855, openings into cross streets were funnel-shaped.[12] A great deal of work was done to correct the alignment of intersecting streets. This was particularly noticeable in Rue Thomassin, Rue Grenette, Rue du Bât-d'Argent and Rue de l'Arbre-Sec. Rue du Bât-d'Argent was substantially widened. The junction between Rue Grenette and Rue Basse-Grenette was reduced to no more than the extension of the first of these streets. There are still narrow streets in the old town centre, becoming even narrower at the extremities,

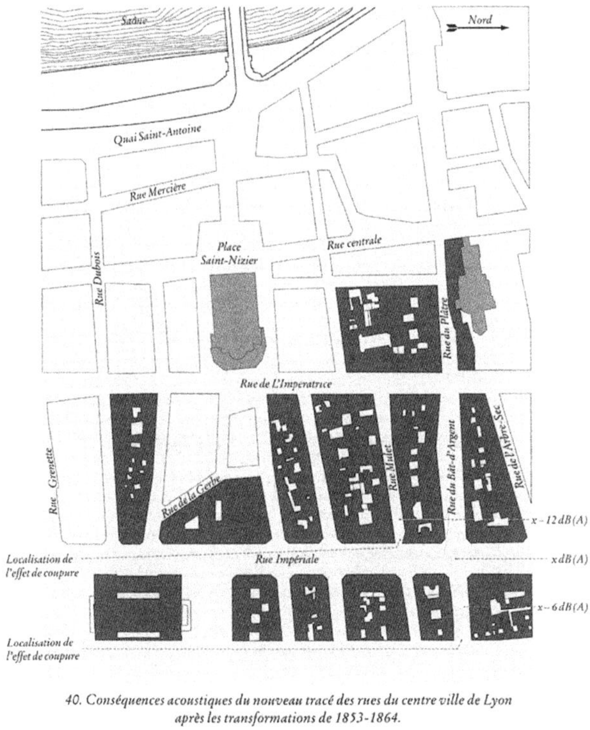

Figure 11.7 Acoustic consequences of the new layout of the streets in the centre of Lyon, after the transformations of 1853–64.

Source: Drawing by Olivier Balaÿ.

for example Rue Mulet at the point where it joins the former Rue de l'Impératrice, Rue du Plâtre at the intersection with Rue Centrale and also Rue du Garet, but they are much shorter, which also has an impact on the way they sound.

Acoustic Consequences of the New Street Layout

The 'strangled' configuration of crossroads, as Maurice Garden puts it, limits the penetration of noise from the crossroads, as shown by the acoustic study carried out in 1992 (-12dB(A)).[13] On the other hand, the acoustic study also highlights the extent to which a closed junction accentuates our perception of a break in the sound as we walk along the street, on leaving or entering the crossroads. When a street connects to the crossroads from the narrower side of the junction, there is a sudden change in the sound environment. Or, in other words, a city-dweller from the first half of the nineteenth century had the distinct impression of a change in sound atmosphere on entering a street in the old centre of Lyon.

Urban background noise penetrates much more (+6dB(A)) into a short, wide street, than into one that is long and narrow. This phenomenon is amplified by a funnel-shaped connecting section opening onto the crossroads. Nowadays, much as for people in the nineteenth century, this means that sounds seem flatter to passers-by in the street: high-intensity sounds and silences have less presence when mixed with sounds from the crossroads. The city-dweller's perception of the sound identity of the street – a perception that is informed by the habits of its occupants – is altered. The new urban layout changed the old street network's tendency to restrict the amount of noise entering a narrow thoroughfare from a crossroads, allowing sounds that make it possible to identify the presence and nature of the street's local forms of social exchange. More detailed acoustic observation has also shown that the new urban fabric, with its much shorter, wider streets makes the micro-social sounds produced in workshops or dwellings opening onto the street almost inaudible.

Sound Effects Left Over from Nineteenth-Century Redevelopment

It is impossible to appreciate the full extent of the sound consequences of the urban and architectural morphology, without some understanding of the intensity of the sounds and noises being reflected. Unfortunately we have no idea of the level of sound in the streets during the period we are studying. Nor do we have any way of comparing the acoustic measurements that we can carry out now with other more remote observations.[14] The oldest study observed average variations ranging from 46.5 to 56.5 decibels in the streets of Paris at night[15] and a value of 60 decibels for daytime noise on Place Saint-Augustin. A comparable study, carried out at the end of the first third of the twentieth century shows that car horns and police whistles reach a level of 84.5 decibels, the noise at certain times and places becoming continuous. This level of sound is commonplace in the centre of Lyon nowadays. But we have no comparable information for nineteenth-century Lyon. All we know is that measurements of the average noise taken on the spot in the summer of 1996 reveal an acoustic level that is practically equivalent to the level in Paris in 1935 (61 to 62dB(A) Leq by day, 55dB(A) Leq by night).

Consequently, to complete this historical study, the sound effect[16] concept was used, as currently employed by Cresson. Some of the variants of such sound effects can be used for a backward-looking analysis of the sound consequences of urban redevelopment in the nineteenth century and for a plausible description of the most striking differences in the quality of sounds between the urban fabric of the *Ancien Régime* and its Haussmannian counterpart. In terms of quality and quantity sound effects enable us to approximate the past properties of a street and gain a better understanding of how city-dwellers of the past heard their neighbourhood.

Cutting Sound Effect

In practical terms the cutting sound effect corresponds to a sudden drop in intensity associated with an abrupt change in the spectral envelope or the reverberation of noise.

The cutting sound effect perceived when moving from a junction into a street, as discussed above, did not disappear altogether in the streets laid out in the nineteenth century, but it was less pronounced than in eighteenth-century streets. Lower frequencies take advantage of the opening out onto the junction to enter the street. The cutting effect is pushed back up the street, becoming less noticeable, above all when combined with the change in the perception of local social life mentioned above. As soon as one entered an eighteenth-century street this effect produced a more favourable sense of the emergent sound of the daily life of the street.

Masking Sound Effect

In towns sound masking takes two main forms. Either a more powerful sound masks another one or a building interposes itself between the subject and the source of the sound.[17] As one enters a street, moving away from a junction, the intensity of the sound of the crowd or traffic from the junction (ambient noise) diminishes in proportion to the width of the street. It also diminishes as a function of the layout of the junction. Depending on whether the junction with the street is closed or open, the variation in the intensity of the masking sound effect, as measured, is about 6dB(A).

The greater the masking effect, the less passers-by can differentiate sounds, localize their source and gauge their distance. When the ambient noise is relatively low the occupant of a dwelling can hear sounds close-to but also remote emergent sounds. If the ambient noise is louder that is no longer possible. This means the outer limits of one's 'home' come closer to the subject when the spatial configuration only allows them to perceive sounds in the immediate proximity. Conversely, they are pushed back, when it is possible to distinguish remote sounds.

Reverberation Sound Effect

In simple terms reverberation corresponds to the addition to direct sound, between the source and the subject, of one or more reflections off surfaces in the surrounding space, resulting in what is commonly referred to as an echo. The longer the reflections retain their energy, the longer the reverberation lasts. Physical evaluation of a reverberation takes into account several criteria, the main ones being the duration of reflections, the colour of the timbre and how clear or intelligible the original sound is.[18]

Reverberation in a narrow street is an essential part of how the neighbourhood is perceived soundwise, and that was just as true in the past. This note, written by Mrs. Thrale about the narrow streets of Paris, is particularly impressive. Mrs. Thrale, the wife of a London brewer and member of parliament, noted in her diary (in 1775) that the streets of Paris (much as the streets of Lyon mentioned by Maurice Garden) 'were so high that they caused an echo and each sound bounced around so much that it became deafening'.[19] From the window of her lodgings in Rue Jacob Mrs. Thrale could hear 'continuous quarrels, carriages overturned and confusion'.

In the broader streets of the second half of the nineteenth century reverberation did not last so long, as we have already pointed out, and human sounds were much less noticeable. This is enough to completely alter one's perception of the environment and neighbourhood. In Lyon, to this day, one may experience this phenomenon.

Mixing Sound Effect

This refers to the joint penetration of simultaneous, neighbouring sources. For the mixing sound effect to occur the various sounds need to be at similar levels of intensity. This effect is primarily found in transition areas likely to receive sound atmospheres from different places. It is of particular relevance to the circular nature of sound between the inside of dwellings and the street. Bearing in mind Maurice Garden's comments on eighteenth-century Lyon streets, it is quite clear that the perception of sound in a narrow street may be at one and the same time stable and dynamic. It is stable because the circularity of sounds between the inside and outside is constant, or because the circularity lasts or repeats itself for long enough to be perceived as such. It is dynamic because sound production is random and it is the movement in sound that the subject retains: 'Song and laughter, shouts and scraps of conversation can be heard at all hours, particularly in summer, when the open windows let the household sounds escape'. Children, apprentices 'can not remain closed all day severely in the workshops. They escape at every opportunity, always ready to run, always willing to take advantage of any distraction from their parents and their employers. ... The slightest noise in the street attracts them: everyone gathers at their window at the slightest excuse, though they say they saw nothing if the incident turns nasty'.[20]

In the contrasting case of a broad street in which the sound atmosphere is continuous a series of sound events completely changes the overall perception of the surroundings. The circularity of sounds between the inside and outside is not stabilized. Or to take another example, large public squares produce a fairly high level of sound stability, but powerful emergent sounds are nevertheless heard and may consequently intrude on the subject's perception.

As we can see, the eighteenth- and nineteenth-century urban models both produce their particular form of remanence. Auditory perception and the various ways of making space sound differ with variations in morphology: the nature of sound exchanges between the street and dwellings and the forms of appropriation are transformed. The shape and dimensions of streets, and the relative 'smoothness' of facades contribute to differentiating the conditions under which ambient sound propagates. In this attempt to reconstitute the way one would have heard sounds in the nineteenth-century town, this approximation of the ways in which individuals, from another age, perceived sounds depending on the space and social presence, in their homes and in the street, is an inaugural step. It shows how the urban sound atmosphere, and consequently the sense of comfort, has evolved. In a more open spatial morphology, human voices are diluted, the presence of local social life is almost inaudible, neighbourhood relations are transformed and the perception of a background noise made up of city activities (to which various means of transport have been added) becomes dominant, increasing the sense of a sound intrusion. Lastly, for city-dwellers walking in the street, there is less contrast in urban acoustics. Even today these throwbacks to the sounds of the nineteenth century are still influencing our behaviour and the way we hear the city.

Notes

1 Olivier Balaÿ, *L'espace sonore de la ville au XIXe siècle* (Bernin (Isère): À la Croisée, Collection Ambiances, Ambiance, 2003).

2 Other authors have already noted this change: see R. Murray Schafer's *Le paysage sonore* (Paris: Jean-Claude Lattès, 1979). Pascal Amphoux asked, in similar terms, how to combat 'the loss of sound differentiation' in contemporary towns; see Pascal Amphoux, *Aux écoutes de la ville* (Lausanne: IREC, Cresson, 1991). Let us hope that this short chapter, which explains in detail the reasons for this phenomenon, also convinces the reader that town planning is a way of combating the loss of sound differentiation in our towns.

3 Maurice Garden, *Lyon et les lyonnais au XVIIIe siècle* (Paris: Les belles lettres, 1970), p. 32. All translations are my own unless otherwise stated.

4 Ibid., pp. 26–7.

5 Garden, *Lyon*, p. 33. This distribution dates from 1791.

6 Ibid., p. 72.

7 Bibliothèque municipale de la Part-Dieu, plan n°2.S.189: surveyor's drawing of the meat market in the Terreaux district, scale 1:130, eighteenth century. Comparative drawings for the rehabilitation of the neighbourhood n°185, 186, 187 (1840–41).

8 Berthelet de Barbot, *Topographie médicale de Lyon et de son territoire* (Lyon, 1773), p. 5.

9 Claude-Marius Vaïsse, *Projet de traité pour l'exécution de divers travaux d'utilité générale* (Lyon, 1855), p. 22.

10 *Bibliothèque municipale de la Part-Dieu*, plan n°2.S.393: project for Rue Impériale and third project for the Terreaux neighbourhood, scale 1:200, 1854. Plans n°2.S.398 and 2.S.399 of Rue Impériale, Lyon, scale 1:250, 1853.

11 See Figure 11.6.

12 Vaïsse, *Projet*, p. 22.

13 See Figure 11.6 for its position.

14 On the origin of the decibel, see the article by Father Pierre Chavasse, 'Les origines techniques et "lyriques" du décibel', *Revue d'Acoustique*, 28 (1974): 53–6. The forerunners of the decibel go back to 1925. It was certified in 1927.

15 Louis Astier and Paul Baron, 'Bruits produits par les distributions urbaines d'énergie électrique et bruits dus à la circulation dans les rues', preface by François Cellerier', *Le génie civil*, 5 (1935): 28–9. In addition the authors point out that the noise from building sites and road mending is so loud that it covers the noise of substations producing electrical energy (the level reached during peak output in winter: about 76dB, and the level reached at night of about 55dB). At this time engineers had already invented silencer systems that could cut the noise by 20 to 25 decibels (p. 23). In a contemporary study (Grégoire Chelkoff et al., *Entendre les espaces publics* (Grenoble: Plan urbain, Cresson, 1988)), the sound level on the open space in front of the Pompidou Centre (Beaubourg) in Paris was estimated as between 58 and 67dB (A) (decibels A), the measurement range being due to the variation in activities (small drama groups, music, etc.) in the background noise. However, these noises cannot really be compared, the unit value not being the same. To the best of our knowledge no acoustician has so far tried to compare them.

16 Jean-François Augoyard and Henry Torgue (with all the team at Cresson) (eds), *Sonic Experience: A Guide to Everyday Sounds* (Montreal: McGill-Queen's University Press, 2006), translation of *À l'écoute de l'environnement, Répertoire des effets sonores* (Marseille: Parenthèses, 1995).

17 Masking sound effect: 'acoustic effect defined as the existence of a noise which on account of its level and frequency completely or partly erases another weaker noise. This effect, which is easily demonstrated using measuring devices, is less frequent from a physiological point of view due to the higher differential sensitivity of the ear'. Augoyard and Torgue, *Sonic Experience*, p. 66.

18 Propagation effect, by which some frequencies resonate for a short time, but long enough for the discrepancy between the direct wave and the reflected wave to be audible. The notion of reverberation is linked to a definition of measurement: the time taken for the sound to decrease by 60 decibels. If the baseline sound enters into a particular acoustic relation with the background noise, it is referred to as a drag effect: the time during which the sound continues after stopping in other frequential regions, before being drowned by the background noise. See Augoyard and Torgue, *Sonic Experience*, p. 111.

19 Cited in Annik Pardailhe-Galabrun, *La naissance de l'intime* (Paris: PUF, 1988), p. 226.

20 Garden, *Lyon*, pp. 405ff.

12 Porosity and Modernity

Lisbon's Auditory Landscape from 1864 to 1908

João Silva

The intense changes that transformed Lisbon's urban fabric between 1864 and 1908 were a key agent in reshaping the political economy of sound in the city.[1] During the second half of the nineteenth century, Lisbon was getting to grips with modernity in a period when not only was the term 'modern' starting to be seen in a positive light, but modernity itself was also being commodified.[2] Furthermore, changes in the entertainment market and the replacement of several elements of urban everyday life that were perceived as representatives of an earlier 'Romantic' Lisbon (such as pleasure gardens and theatres) with spaces that echoed transnational articulations of modernity (such as wide boulevards) were essential to the emerging sociability networks of that time. Moreover, these developments played a key role in the new forms of entertainment available, and the repertoires produced for this market were rapidly integrated in the city's soundscape, from the stage to the streets and to domestic spaces. In this sense, urban change is deeply associated with the new sounds that permeated the capital and with a dynamic relationship between the sonic materials and the spaces they crossed.

Lisbon's Modernity: The Building of a New Capital

I start this chapter by outlining the profound changes that transformed Lisbon's urban fabric during the period and, in particular, relate them to a wide leisure market.[3] This market included a varied set of spaces and goods, including theatres, taverns, *retiros*, *hortas*, cafés, shops and music stores (that sold instruments, sheet music, or phonograms). Moreover, the music that circulated through these settings was transformed, according to the spaces and contexts it traversed, to maximise its reach. For example, opera arias originally sung in the theatre were also published as sheet music for domestic consumption, transcribed for wind band or issued as a piano roll or record. Thus, its ability to be adapted for a wide variety of settings – both public and private – promoted the ubiquitous circulation of music in Lisbon. Indeed, the popular theatre stage provided a focal point in the city's music life, and was of particular significance to the dissemination of musical repertoires throughout Lisbon in a variety of forms and commodities. The entertainment market consisted of a set of venues where people would spend their leisure time, including, in particular, theatres and taverns. This circuit was central to the reshaping of Lisbon's soundscape because the music originally presented in such venues was also arranged and performed in other urban contexts and became a ubiquitous presence in the city's auditory landscape. The activity of non-subsidized theatrical endeavours tended to be ephemeral, a characteristic of Lisbon's market for cultural goods during the period. In this sense, it is possible to trace the transformation of the entertainment circuit by analysing changes in both theatrical repertoire and venues.

One of the most significant reformulations of urban space that was to reshape Lisbon and its entertainment spaces, and one of the most symbolic, was the opening of the Avenida da Liberdade, a wide avenue that linked the city centre to the new urbanized areas further north. This required the destruction of the Passeio Público, a public park built in the eighteenth century.[4] Despite constant work to make the park more attractive, it was not until the second third of the nineteenth century that the Passeio Público finally became a venue for aristocratic and bourgeois sociability. In his study on Romanticism in Portugal, José-Augusto França associates the everyday life of the aristocracy and of the bourgeoisie of Portugal with the idea of 'Passeio' (promenade). França describes 'Passeio' as an activity that is enacted in venues such as theatres, cafés, the circus, the bullring and the parliament and includes all aspects of public entertainment and 'publicness'.[5] In this sense, Passeio consists not only of the physical spaces frequented by the economically privileged strata of Lisbon's society, but also the social spaces where the process of seeing and being seen (key in the establishment and reinforcement of sociability networks) is played out. Thus, read in the context of França's 'Passeio', the Passeio Público stands as a metonymical representation of aristocratic and bourgeois lifestyles in Lisbon during the second third of the nineteenth century, a period when the Portuguese Constitutional Monarchy was stabilized after a long period of intense turmoil that included a civil war between 1828 and 1834.

From the Liberal Revolution of 1820 until 1846, theatrical activity in Lisbon was focused mainly on three theatres built in the eighteenth century: the Real Teatro de S. Carlos (a theatre established in 1793 to perform mostly opera in Italian, run by an impresario, although subsidized by the state);[6] the Teatro da Rua dos Condes (established in 1765 and mostly focused on the presentation of drama and, occasionally, musical events); and the Teatro do Salitre (established in 1783, presenting the same genres as the Teatro da Rua dos Condes). The opening of the Teatro Nacional D. Maria II in 1846 was a key moment in the development of Lisbon's theatrical activity, reshaping Lisbon's cultural market and providing a venue for the presentation of drama.[7] The same year saw the opening of the Teatro do Ginásio, a theatre mostly devoted to drama, but which, as in most of the other theatres, presented several types of entertainment. Alongside drama, for instance, Teatro do Ginásio also staged several comic operas.[8] Another venue for the study of theatrical activity was the short-lived Teatro de D. Fernando, in use between 1849 and 1859. Smaller theatres were also established in Lisbon during the 1850s, such as the Teatro da Floresta Egípcia or the Teatro do Calvário, most of them of ephemeral duration.[9] Relating theatrical life with the established sociability networks of the time, a journalist writing in 1853 noted that Lisbon's city life spanned only from the Teatro do Ginásio (in Trindade) to the Teatro de D. Maria II (in Rossio), a relatively small area within the city.[10]

Associated with the demolition of the Passeio Público was the construction of the Avenida da Liberdade on the same plot, a process that started in 1879, and which enabled the expansion of the urban centre of Lisbon. Just as the Passeio Público metonymically stood for everyday bourgeois life, so the Avenida da Liberdade is emblematic of its time. The construction of the avenue had been an intention of the city council since at least 1859.[11] Nevertheless, it is a work that reveals the emergence of new planning strategies for the city, especially after the appointment of the engineer Frederico Ressano Garcia (1847–1911) as head of the technical department of the city's council in 1874. Ressano Garcia had studied in Lisbon and in Paris and was responsible for planning and constructing the new and expanded areas of Lisbon.[12]

During the period in which Ressano Garcia led the city council's technical department, Lisbon was deeply changed by the reconfiguration of its urban landscape, an important process that was reflected in the auditory landscape of the city. New neighbourhoods were built in the city and were linked to new and broad avenues (such as the Avenida da Liberdade and the Avenida das Picoas). Furthermore, a network of streetcars (horse-drawn and, later, electric) was created to facilitate the circulation of people inside the city.[13] Another aspect of the development of the transport system was the construction of several funiculars and elevators (such as the Elevador de Santa Justa, the Ascensor da Bica, the Elevador do Lavra and the Elevador da Glória – the last two leading to the new Avenida da Liberdade). These new facilities seemed to epitomize the trends of modernity realized in the new planning of Lisbon, especially by the integration of iron as a visible construction material in a modern city (see the Elevador de Santa Justa, for example) and by the definitive altering of Lisbon's auditory landscape.[14]

Another important development in the transport network was the development of the railway system leading to Lisbon, especially the Sud-Express (that linked Paris to Lisbon and started operating in 1877) and the regional railways (such as the West railway and the Sintra railway, that started operating in 1887). As a consequence of the development of rail travel and of its integration in a wider transport planning strategy, a new railway station, the Estação da Avenida (Avenida Station, nowadays called the Rossio Railway Station), was inaugurated in 1890.[15] Situated at the southern end of the Avenida da Liberdade, close to the city centre, it was the terminal for several national and international routes. Moreover, alongside the screeching of the railways and the bells of the trams, the sounds of the automobile became part of Lisbon's auditory landscape in the last decade of the nineteenth century.[16]

While studying in Paris, Garcia had come into contact with new perspectives on urbanism operating in that city, especially its reconfiguration promoted by Georges-Eugène Haussmann (1809–91) between 1853 and 1868. In this sense, the Avenida da Liberdade can be compared to the wide Parisian boulevards built according to Haussmann's principles of urban regeneration.[17] The process of so-called 'Haussmanization' in Paris had a tremendous impact on urban planning across Europe at the time, and some of its aspects were adopted in the reconfiguration of Lisbon.[18]

Another significant process in the development of Lisbon during the period was the several attempts to improve the city's port (which included the shipyards and the customs' facilities) and its surrounding areas. Being a littoral capital with a long shoreline, the need for a rationalization of this space emerged during several periods of the city's history, although with more visibility in the second half of the nineteenth century, when several projects were presented to its council.[19] The project that started its implementation in 1887 contemplated the construction of several specific docks (for commercial and fishing boats), piers, industrial buildings and railways, concentrating on the Western part of Lisbon's shoreline.[20] On the one hand, this rationalization of space was a key aspect for the capital of an empire that depended on its port for an important part of its trading. On the other hand, several authors have associated this modernization with the construction of new buildings that contributed to a physical separation between the city and the river.[21] Urban interventions of the magnitude of the construction of the Avenida da Liberdade and of the improvement of the city's port also contributed to the auditory landscape of Lisbon for several years, embedding the sounds associated with the construction works in everyday life.

The modern cityscape as 'a site of disambiguation and rationalisation of social space through either privatisation and atomisation (in the case of Berlin and Paris) or collectivisation (in the case of Moscow) of built space and hence also atomisation or collectivisation of the subjects who dwelt within these spaces' is a key aspect in the work of Walter Benjamin.[22] Benjamin points out a tendency for a stricter separation between public and private spaces in Paris starting from the reign of Louis-Philippe I (1830–48).[23] Furthermore, this separation displays the rationalization of space as a tendency inherent in the development of modern cities. Subsequently, during the Second Empire, Paris was transformed in order to suit the needs of an imperial capital of an industrial state, much to the concerted efforts of Napoleon III and of Haussmann.[24] During that time:

> Haussmann's urbanistic ideal was one of views in perspective down long street-vistas. It corresponded to the tendency which was noticeable again and again during the 19th century, to ennoble technical exigencies with artistic aims. The institutions of the worldly and spiritual rule of the bourgeoisie, set in the frame of the boulevards, were to find their apotheosis.[25]

Furthermore, Benjamin associated Haussmann with the 'rendition into stone' of what he called the phantasmagoria,[26] an image a 'commodity-producing society … produces of itself … and that it customarily labels as its culture'.[27] When discussing the commodity-form in *Capital*, Marx states that: 'It is nothing but the definite social relation between men themselves which assumes here, for them, the fantastic [the German term Marx uses is *phantasmagorische*] form of a relation between things'.[28] Thus, according to Derrida, Marx analyses 'not only the phantomalization of the commodity-form but the phantomalization of the social bond'.[29]

In *The Arcades Project*, Benjamin expands on Marx's concept of phantasmagoria, specifically those elements of the theory that relate to the commodity-in-the-market. Benjamin shifts his emphasis to a concept of phantasmagoria associated with the commodity-*in-display*, thus stressing its representational value.[30] Shifting the focus from the spectacle of commodities to the process that relates capital to labour, David Harvey has argued that Haussmann's projects were, indeed, a contribution to solving the issue of the concentration of unemployed surpluses of both capital and labour power.[31] Therefore, at that time, urbanization in Paris can be seen as a way of absorbing those surpluses, acting as a vehicle for social stabilization.[32]

During the period on which this chapter focuses, gas was still the main fuel for public lighting in the city, but the late nineteenth and the early twentieth centuries saw electric-powered public illumination reaching a significant part of Lisbon's centre.[33] The shift of public lighting in Lisbon from gas to electricity was uneven throughout the city, first benefiting both the new areas and the traditional commercial and entertainment districts. Consequently, the first spaces to have a definitive installation of electrical lighting (from 1889) were: Chiado, Rua do Ouro, Praça D. Pedro IV (Rossio), Praça do Município, Praça dos Restauradores and Avenida da Liberdade.[34] This planned use of lighting created a kind of delimited corridor for a specific kind of publicness in Lisbon, putting sites such as shops, theatres, public buildings and cafés on display (an aspect I discussed earlier in this chapter).[35] Furthermore, several commercial enterprises helped the reshaping of the Chiado from the last decade of the nineteenth century onwards, especially the first multi-storey department stores in Lisbon (analogous to the French *magasins* of that time): the Grandes Armazéns do Chiado (established in 1894) and the Armazéns Grandella (established in 1907).[36]

The establishment of several stores focused on the commercialization of talking machines and recorded sound expanded the market for musical commodities in Lisbon. Starting from the middle of the nineteenth century, several commercial spaces dedicated to selling instruments and sheet music were established in Lisbon, mainly through the action of European immigrants, such as the German Eduardo Neuparth (1784–1871, owner of the Armazém de Música e Instrumentos de Eduardo Neuparth) or the Italian João Baptista Sassetti (1813–89, founder of the da Sassetti e Comp.ª).[37] At the beginning of the twentieth century, the establishment of the stores of the Companhia Franceza do Gramophone (a company affiliated with the The Gramophone Company, Ltd) and of the Simplex company can be seen not only as an expansion of the market for musical products, but also as a symptom of the incorporation of modern technologies/commodities of sound reproduction into everyday life.

The Theatres of Modern Lisbon

The entertainment business was not immune to the changes operating in the urban landscape during the period 1864–1908. With entertainment being a volatile activity, several theatres opened and closed in various areas of Lisbon, such as the

Teatro do Salitre, which changed its name to Teatro das Variedades in 1858, and was demolished in 1879 during the construction of the Avenida da Liberdade.[38] Another space demolished for the building of the new avenue was the Circo Price, which presented mainly circus, although there are references to operettas and *zarzuelas* also being staged there.[39] The Teatro da Rua dos Condes closed down in 1882 and a provisional theatre was built on that same plot, the Teatro Chalet da Rua dos Condes, which lasted only three seasons.[40] During the last years of the Teatro da Rua dos Condes, the *revistas* (revue theatre) were the most important source of income for the theatre and that same repertoire was also hegemonic in its successor, the short-lived Teatro Chalet.[41] In 1888, a new theatre, the Teatro Condes, was opened in the same space as the older theatres and also concentrated on presenting operettas and *revistas*. It can be argued that this process of the sequential opening of three theatres on the same ground is related to the operative dynamics of a spatial *habitus*.[42] Of the theatres working in the first 40 years of the nineteenth century, only the Real Teatro de S. Carlos, the Teatro de D. Maria II and the Teatro do Ginásio (where the first *revistas* were presented in the 1850s) survived and maintained regular activity throughout the nineteenth century.

In the last third of the nineteenth century, a significant number of theatres were built in Lisbon and in other towns and villages, which perhaps indicates an expansion of theatrical activity in Portugal during this period. Despite the ephemerality of the theatre business, a few of these new venues were key for the expansion of entertainment in Lisbon, especially by becoming prominent venues for the presentation of new genres of musical theatre, which were then becoming hegemonic: the operetta and the *revista* (revue theatre). Lisbon's theatrical scene during this period relied, to a great extent, on translations. In this sense, the presentation of operettas (mostly associated with the French tradition) and *zarzuelas* in Portuguese came to be central to the entertainment market in Lisbon. The performance of these new theatrical genres contributed to the reshaping of Lisbon's auditory landscape because the songs were rapidly adapted and performed in the city's streets and parlours.

Some of the theatres that opened in Lisbon during this period were: the Teatro do Príncipe Real; the Teatro da Trindade; the Teatro Taborda; the Real Coliseu de Lisboa; the Teatro D. Amélia; the Teatro da Avenida; the Teatro da Alegria; and the Coliseu dos Recreios. The Teatro do Príncipe Real opened in Mouraria in 1865 and focused on melodrama and, occasionally, operetta.[43] The Teatro da Trindade opened in 1867 as a venue that could accommodate both drama, comedy and comic operas in the main hall (mainly presenting operetta and *revista*) and concerts and balls in another hall (the Salão).[44] The Teatro Taborda, situated in the Costa do Castelo, and focusing especially on drama, was inaugurated in 1870.[45] The Real Coliseu de Lisboa (mostly dedicated to circus, but also presenting opera, comic opera, operetta and *zarzuela*) opened in 1887. Furthermore, the first projections of film in Lisbon took place in this theatre (from 1896 onwards) and this new type of entertainment was soon shown in other theatres, such as the Teatro D. Amélia, and in venues built mainly for its presentation, such as the Animatógrafo do Rossio (established in 1907).[46] In 1888, the Teatro da Avenida opened, one of the first buildings on the Avenida da Liberdade focused especially on performing comedies and *revistas*.[47] The short-lived Teatro da Alegria (near the Avenida da Liberdade and dedicated to spoken theatre) was built on the same grounds on the Rua Nova da Alegria where a provisional theatre had once stood, and opened in 1890.[48] Later that year, the biggest venue in Lisbon, the Coliseu dos Recreios (an octagonal hall with a glass and iron dome), was opened. The Coliseu premiered with a comic opera by an Italian company, although it mainly specialized in the production of circus shows.[49] Nevertheless, it accommodated several theatrical genres (such as operetta, *zarzuela* and Italian opera), public concerts, film (with Edison's projectoscope) and wrestling.[50] In 1894, an operetta was the inaugural offering at the Teatro D. Amélia, located in Chiado, but the theatre would mainly present drama, comedy and cinema.[51] By interlocking theatres and spaces of circulation and sociability, it is possible to grasp that most of the recently built theatres were still in the areas traditionally associated with leisure (such as Chiado or Trindade), but the expansion of the city with the Avenida da Liberdade also enlarged the theatrical circuit (especially with the Teatro Condes, the Teatro da Alegria, the Teatro da Avenida and the Coliseu dos Recreios).

Furthermore, theatrical repertoire was also changing at this time, especially due to the presentation of naturalist and symbolist plays in several of Lisbon's theatres, performed both by Portuguese and foreign companies.[52] From the 1860s onwards, the Real Teatro de S. Carlos started to regularly present works by Gounod, Halévy, Auber, Thomas, Massenet, Délibes and Bizet. During the 1890s, performances of operas associated with Italian *verismo* and the incorporation of works by German composers (especially Wagner's musical dramas; *Lohengrin* premièred in that theatre in 1883) were key events in Lisbon's operatic seasons.[53] Nevertheless, and considering that the paradigm was changing at the beginning of the twentieth century, the vast majority of these operas were presented in Italian, which reflects the Italian scope of the S. Carlos.

Porosity in Lisbon: Sound and Everyday Life

This section will examine several aspects of Lisbon's urban fabric that contributed to the structuring of everyday life in the city. Focusing on the concept of porosity, I will examine practices such as fado and its sites of performance, the cries of street vendors and pedlars, the sound of church bells and the use of music in the public spaces of the city, with a particular emphasis on the ensemble *banda*.

Extending the analogy of the then recently built areas of Lisbon with Benjamin's Paris, I would also like to propose a parallel between the old neighbourhoods of Lisbon and Benjamin's and Lacis's reading of Naples.[54] In their short essay, 'Naples', Benjamin and Lacis stress the porosity of the architecture of that Southern European city, in which 'building and action interpenetrate in the courtyards, arcades, and stairways'.[55] In this sense, the forced socialization imposed by poverty and the 'passion for improvisation' are linked to 'the scope to become a theater of new, unforeseen constellations'.[56] For the authors, another important aspect of Naples was the interpenetration of public and private spheres, a distinction they stress when comparing Naples with its Northern European counterparts. So, in Naples 'each private attitude or act is permeated by streams of communal life'.[57] Furthermore, 'buildings are used as a popular stage. They are all divided into innumerable, simultaneously animated theaters. Balcony, courtyard, window, gateway, staircase, roof are at the same time stage and boxes'.[58] Hence, to a rationalized concept of urban space where the public and the private is separated, characteristic of modern industrial cities, the writers see Naples as a chaotic and almost archaic city, not yet subject to the rationalization embedded in the capitalist development.[59] On that specific point, 'the structuring boundaries of modern capitalism – between public and private, labour and leisure, personal and communal – have not yet been established'.[60] However, the presence of multi-layered and shifting concepts of publicness and domesticity cannot be consistently and directly translated into a polarity between public and private realms. Boym argues that porosity is a quality present in all cities, consisting in a spatial metaphor for a 'variety of temporal dimensions embedded in physical space' that reflects multiple 'layers of time and history, social problems, as well as ingenious techniques of urban survival' and creates 'a sense of urban theatricality and intimacy'.[61]

Lisbon's older districts, such as Alfama, Bairro Alto, Madragoa and Mouraria stand out as examples of Benjamin's and Lacis's notion of porosity. In these neighbourhoods, the concentration of dwellings and the cohabitation of a significant number of people in small areas were indicative of that forced socialization. Taverns played a key role in the sociability of such districts, marking a space for the performance of *fado* (especially in what might be called its vernacular form). A socially diversified spectrum of frequenters gathered in the taverns and alcohol was constantly mixed with violence.[62] Furthermore, the sounds of the taverns bled into the streets, a fact that displays the porosity inherent to those spaces of the city.[63]

Apart from taverns and brothels, the *hortas* and the *retiros* situated on the rural outskirts of the city also played a relevant role in the sociability of the bohemian segments of Lisbon's population. According to José Machado Pais, these spaces of sociability mixed aristocrats, coach drivers, prostitutes, *fadistas*, journalists and bullfighters, who mingled spontaneously in a non-stratified way.[64] In his text on Marseilles, Walter Benjamin describes the outskirts as 'the state of emergency of a city, the terrain on which incessantly rages the great decisive battle between town and country'.[65] Furthermore, these interstitial spaces between the country and the city and, within the city itself, were privileged places for the sociability of the unstructured *communitas* of Lisbon's bohemians.[66] Other events in which that *communitas* emerged were related to bulls, an aspect associated with the performance of a specific type of masculinity.[67] Both bullfighting and the *espera de touros* (an event where men would taunt a bull that was released in a delimited space) became also places for the sociability of several segments of the population during the period, especially those integrated in the bohemian circles previously mentioned.[68]

As a consequence of the industrialization of Lisbon and its outskirts, the creation of workers' accommodation from the 1870s onwards contributed to the reshaping of the auditory landscape of the city at that time. The reconfiguration of the urban landscape in neighbourhoods such as Alcântara, Graça, Xabregas ou Poço do Bispo, which were situated in the vicinities of industrial activity through the construction of *pátios* and *vilas*, is a significant element in this process.[69] Initially, workers were accommodated in previously existing spaces, such as converted palaces or convents, where the individual rooms were rented and the environment was obviously unwholesome due to the inadequacy of the facilities.[70] Later, several private agents (entrepreneurs, industrialists or co-operatives) started to purchase lands specifically for the building of workers' accommodation.[71] On this aspect, physical and cultural space merged in a narrative of class, sociability and *habitus*.[72] Due to migration, these neighbourhoods became privileged sites for the contact between Portuguese regional cultures and for the emergence of specific traits of a vernacular culture, frequently aestheticized in periodicals and literature.[73]

Another important concern regarding urban planning in Lisbon at this time was public health.[74] The improvement of sanitation through the renewal of the sewer network and the establishment of several indoor markets, such as the Mercado da Ribeira Nova (inaugurated in 1882) and the Mercado de Alcântara (inaugurated in 1905), for example, can be related to public health policy. Nevertheless, this attempt to centralize the commerce of fresh products collided with what was, at that time, the crystallized *habitus* of trading them in the street. Fresh goods were mainly concentrated in outdoor markets (such as the Praça da Figueira), fairs or in the carts or baskets of the pedlars (especially inbound from the rural outskirts of Lisbon known as the *região saloia*), who contributed with their cries to the auditory landscape of the city.

This issue of the perambulating pedlars is deeply associated not only with the city's auditory landscape, but also with the structuring of everyday life through sound.[75] Due to its specific qualities, sound played a key role in announcing the presence of the pedlars. Furthermore, each product had its own specific cry, the *pregão* (in some instances added by a horn or a flute), that informed the customers of who was coming to sell their goods. This practice is recorded since at least the sixteenth century and, in very few cases (such as lottery sellers), remains to the present. Furthermore, Júlio de Castilho states that the *pregões* (cries) were seasonal and varied according to the neighbourhood, due to the fixed routes of the sellers.[76] In that sense, the seasonality embedded in the commerce of fresh products and its advertisement throughout the streets of Lisbon contributed also to the structuring of everyday life on a yearly basis. During the spring, pedlars advertised strawberries and tender vegetables, whereas during the autumn they would sell boiled chestnuts, for example.[77] Conversely, the produce advertised also varied according to the time of the day; Castilho presents milk as being a morning good and associates products such as roasted broad beans or lupini beans with the leisure Sunday afternoons of the working classes.[78]

Apart from the vegetables and dairy products previously mentioned, street pedlars (in which bakers and butchers were included) sold a wide variety of goods. For example, characters such as the *ardina* (paperboy), the coal seller, the *petrolino* (who sold a variety of household goods, such as paraffin or olive oil), the knife-sharpener and the lottery-seller showed their daily presence in the city of Lisbon, filling its auditory landscape with sound.[79]

Migration also played an important role throughout the history of Lisbon. Although I have already mentioned the role of Central European immigration in the establishment of several music shops in Lisbon, I will now stress several specificities regarding certain migrant communities.[80] As the capital of an empire that included colonies in Asia, Africa and South America, Lisbon was home to several communities associated with this colonial process (especially populations of Brazilian and African descent). These communities contributed to Lisbon's multicultural character, and their presence was reflected in its auditory landscape. Furthermore, Graça Índias Cordeiro relates the inclusion of migrants in the work marketplace with their integration into city life, an aspect I will now develop.[81]

An interesting case of immigration is the *galego* (Galician), who exerted a strong presence in Lisbon especially from the nineteenth century onwards. In Lisbon, male *galegos* did all sorts of work that involved carrying weights (such as carrying the injured from the brawls in which the *fadistas* were involved), but they became particularly associated with water-selling in such a way that *galego* became almost a synonym for that activity.[82] In that activity, their role both as a supplier of household water and as key agents in fire extinction was extremely important for the everyday city life.

Another interesting case of migration was the *varina*, mostly a female migrant, originally from the littoral region of Ovar, who sold fresh fish in the streets of Lisbon using specific *pregões*. Cordeiro states that some of the professions stated earlier became associated with the so-called 'popular types' of the city of Lisbon, and gives the *varina* as an example.[83] According to the same author, their ubiquitous visibility, their specific sound markers, their role in supplying the city with products from the river and the sea (displaying a proximity between the seller and that specific urban environment), their conspicuous and exuberant group sociability and their working in the public space contributed to that association.[84]

A ubiquitous presence in the auditory landscape of Lisbon was the sound of church bells. In his study regarding church bells in the nineteenth-century French countryside, Alain Corbin points to several ways in which bells were used to impart 'a rhythm to the ordinary functions of the community', acting as an 'auditory synchronizer' of everyday life.[85] In that sense, church bells were used to summon the community for religious services, to punctuate religious and secular festivities, to mark the passage of chronological time and to ring the alarm, for example. Thus, 'a subtle auditory rhetoric was developed' and integrated in the everyday life of Lisbon.[86] As an example of the usage of bells in Lisbon, I will use some references from Maria Rattazzi's book *Portugal de relance* ('Portugal at a glance'), a set of satirical letters written by Rattazzi about her travels in Portugal. Originally written in French (bearing the title *Le Portugal à vol d'oiseau*), the book was translated into Portuguese and published in Lisbon in 1881.[87] Despite its satirical vein (in which a deeply ethnocentric perspective surfaces) and the poor documentation of the author, some of the first-hand aspects of this account may be useful when dealing with the tolls and peals of the bells in Lisbon.

In her account, Rattazzi comments upon the ubiquity of churches in Lisbon and, consequently, on the sounds of their bells. Moreover, she associates bell ringing with several religious solemnities (such as christenings, funerals and other celebrations) and with the sounding of fire alarms.[88] Apart from the standard tolls and peals, Rattazi emphasizes the usage of melodies such as the national anthem, operetta arias and urban songs in Lisbon's bell ringing.[89] Bearing in mind that a significant part of the sheet music market of the time was based on theatrical music, this points to a porosity between stage, street and domestic space in Lisbon. In this sense, a common repertoire unifies several of the spaces in the city, which displays an auditory continuity amongst them.

Music in the open air was also an important facet of Lisbon's everyday life. Buskers, beggars and street musicians were a frequent presence in the city's streets. A ubiquitous presence in Lisbon's auditory landscape was the mechanical organ (also called barrel organ). For example, in 1890 the music periodical *Amphion* published an article that depicted the sound of the barrel organ as a scourge that afflicted the city of Lisbon and several novels include scenes in which the organ grinder would perform well-known operatic extracts, illustrating both the porosity between the stage and the street and the prominence of theatrical music in other performance contexts.[90]

The magazine *O Occidente* published a small article on itinerant musicians, observing that groups of wandering musicians were frequently seen in Portuguese towns and villages.[91] According to the article, these groups of buskers consisted mainly of a singer and a guitarist or a fiddler and predominantly performed songs drawn from the successful comic operas of the time and their double entendre *coplas*. This view is reinforced by Alberto Pimentel in *Vida de Lisboa*, in which he writes that the most successful theatrical songs were adopted by pedlars and children and immediately reproduced in the city's streets.[92] The involuntary exposure of the inhabitants of Lisbon to the omnipresent sounds of church bells and of the barrel organ, as well as the cries of the pedlars, helps to situate the resulting sonic realm in what Anahid Kassabian designates as 'ubiquitous musics', 'the musics that are always there, beyond our control, slipping under our thresholds of consciousness', a sort of soundtrack of everyday life.[93]

A more organized grouping started to perform in Lisbon's public spaces: the *banda* (an ensemble of wind and percussion instruments). The constitution of these ensembles varied, but they generally included flutes, reed instruments, brass instruments and percussion. Furthermore, some of them were affiliated with military and police institutions, while others were associated with voluntary societies (some of them established by the agents connected to the workers movement). The *bandas* usually performed on Sundays and bank holidays in theatre halls, balls, gatherings promoted by several societies and public gardens (such as the Jardim da Estrela, the Praça D. Fernando and the Avenida da Liberdade).[94] These ensembles performed a heterogeneous repertoire that extended from marches, waltzes, polkas and anthems to transcriptions of classical pieces, such as opera overtures.

The impact of this movement in the city's musical life cannot be overstated, due to their frequent presence in public spaces and gatherings. Furthermore, several military and police bands were among the first Portuguese ensembles to be recorded by international companies such as The Gramophone and Typewriter Ltd and Pathé.

The rise of new forms of political demonstrations in public spaces towards the last third of the nineteenth century also contributed to the reshaping of Lisbon's auditory landscape. Especially associated with forms of republican and socialist propaganda that emerged in the period, the rally became an important event in the urban space. This form of demonstration enabled the congregation of a significant number of people and was mainly held in public spaces, such as Lisbon's avenues. The bustling of the crowd, the speeches and, sometimes, the performance of a *banda* helped to delimit and demarcate a specific place and time in the city.

Between 1864 and 1908 Lisbon was undergoing a significant period of modernization and several major developments that helped to reshape the auditory landscape were introduced. The weaving of sounds produced by tramways and automobiles in the city's sonic fabric added several layers to the constant bustle of the capital. Despite the trend for the rationalization of space and time associated with modern urban planning, Lisbon (and other cities) retained a significant presence of elements of the past. Modern amenities such as markets and boulevards coexisted with the labyrinth of streets of the older districts and the cries of the pedlars were intersected by the sounds of the phonograph and the automobile. This porosity between different spaces and times is central to understanding Lisbon as a complex and changing sonic site where the shifting relations between music, sound and noise helped to structure everyday life until well into the twentieth century.

Conclusion

In this chapter, I have tried to show, in a way similar to that used by James Clifford in his approach to Michel de Certeau's theory of everyday life, how: '"space" is never ontologically given. It is discursively mapped and corporeally practiced'.[95] This aspect of space brings to mind Henri Lefebvre who concentrated 'not on space as an a priori or ontological entity, but on the processes and strategies of producing space, which are by definition historical'.[96] About his groundbreaking book *The Production of Space*, Deborah Pellow observes that Lefebvre 'emphasizes the symbolic meaning and significance of particular spaces, in effect how spaces are culturalized, but also how culture is spatialized, how practices are lived in space'.[97] In this sense, spatial and cultural practices are not only inseparable, but are inherently intertwined with each other, a process this study of Lisbon's urban reconfiguration in the period 1864–1908 evidences.

In his work *The Urban Revolution*, Lefebvre constructs a synchronic analysis of space in contemporary societies divided into three levels: the global, the mixed and the level of habiting.[98] Although the author was theorizing about contemporary society, due to the necessary historicity implicit in his thought, these analytical levels can be applied to other historical moments, especially if one pays close attention to their historical specificities. In Lefebvre's theory, the global level is presented by its association with the exercise of power (by politicians, for instance) and it relates to what he terms institutional space.[99] By assuming that the global level 'projects itself into part of the built domain: buildings, monuments, large-scaled urban projects', it is possible to relate the planning and construction of the Avenida da Liberdade, for example, with this level.[100]

The mixed level is, for Lefebvre, the specifically urban, an intermediary level between the global and the private, where elements related to 'institutions and higher-level entities' are removed from the cityscape.[101] On this specific point, Lefebvre's examples relate, in a particular way, to contemporary society, because he includes, for example, avenues and a few institutional buildings such as city councils and parish churches on that level. Regarding a city with the dimension of Lisbon and the centralization of local and national institutions in the period 1864–1908, there is a complex permeability between the global and the mixed levels, a kind of porosity between them.

Finally, the level of habiting is associated by Lefebvre with the private realm, mainly housing.[102] Again, the blurring of the threshold between public and private spaces presented in Lisbon points to a bleeding of the mixed level into the level of habiting and vice-versa. In the case of Lisbon during the second half of the nineteenth century, it is possible to observe a porosity not only of spaces, but also of levels. Therefore, a hypothetical reconstruction of the produced land/soundscapes in the several levels of the city of Lisbon can only be multi-layered and complex.

Notes

1 Excerpted by permission from the publishers from João Silva, *Entertaining Lisbon: Music Theater, and Modern Life in the Late 19th Century* (New York: Oxford University Press, 2016), 49–74.
2 Raymond Williams, *Keywords* (London: Fontana Press, 1988), pp. 208–9 and Walter Benjamin, *The Arcades Project* (Cambridge, MA: Harvard University Press, 1999).
3 For an analysis of Lisbon's soundscape in the year of 1890 see Rui Leitão, 'A ambiência musical e sonora da cidade de Lisboa no ano de 1890', Master's thesis (Universidade Nova de Lisboa, 2006).
4 About the constitution of the Passeio Publico as a Romantic garden see Maria Alexandra Salgado Quintas, 'Do Passeio Público à Pena: um percurso do jardim romântico', Master's thesis (Universidade Técnica de Lisboa, 2001).
5 José-Augusto França, *O Romantismo em Portugal: Estudo de factos socioculturais* (Lisbon: Livros Horizonte, 1974), p. 364.
6 About the production system of that theatre from 1834 to 1854 see Luísa Cymbron, 'A ópera em Portugal 1834–1854: o sistema produtivo e o repertório nos teatros de S. Carlos e de S. João', PhD thesis (Universidade Nova de Lisboa, 1998).
7 Luiz Francisco Rebello, *O teatro romântico (1838–1869)* (Lisbon: Instituto de Cultura e Língua Portuguesa, 1980), pp. 31–48 and Maria de Lourdes Lima dos Santos, 'Para a análise das ideologias da burguesia', *Análise Social*, 14/53 (1978): 39–80.
8 António de Sousa Bastos, *Diccionario do theatro portuguez* (Lisbon: Imp. Libânio da Silva, 1908), p. 343.

9 This survey of regular theatrical activity in Lisbon relied heavily on advertisements published in the press. This method is, of course, limited because it excludes spaces that did not advertise and those associated with amateur theatrical activity.

10 *A Revolução de Setembro*, 24 January 1853: 1.

11 Isabel Maria Rodrigues, 'As avenidas de Ressano Garcia', *Boletim Lisboa Urbanismo* 13 (Sept./Oct. 2000): 20–23, and 14 (Nov./Dec. 2000): 30–34.

12 Raquel Henriques da Silva (ed.), *Lisboa de Frederico Ressano Garcia 1874–1909* (Lisbon: Câmara Municipal de Lisboa/Fundação Calouste Gulbenkian, 1989).

13 Rodrigues, 'As avenidas de Ressano Garcia'.

14 For a study of the acoustic changes introduced by modernity in terms of architecture see Emily Thompson, *The Soundscape of Modernity: Architectural Acoustics and the Culture of Listening in America 1900–1933* (Cambridge, MA: MIT Press, 2004).

15 Maria Helena Lisboa, *Os engenheiros em Lisboa, urbanismo e arquitectura (1850–1930)* (Lisbon: Livros Horizonte, 2002), pp. 101–40.

16 José Carlos Barros Rodrigues, *O automóvel em Portugal: 100 anos de história* (Lisbon: CTT, 1995). For a newspaper narrative on urban transport in Lisbon during this period, see Alfredo Mesquita, 'Do omnibus ao automóvel', *Illustração Portugueza*, 24 (September 1908): 233–8.

17 The idea of Paris as the epicentre of European cultural life and a model to emulate is very much present in Portugal during this period. See, for example, Álvaro Manuel Machado, *O 'francesismo' na literatura portuguesa* (Lisbon: Instituto de Cultura e Língua Portuguesa, 1984). On the association of the Parisian boulevard with the rise of a mass culture in Paris in which entertainment and the spectacle are embedded see Vanessa R. Schwartz, *Spectacular Realities: Early Mass Culture in fin-de-siècle Paris* (Berkeley: University of California Press, 1998).

18 About the processes of Haussmannization see, for example, Schwartz, *Spectacular Realities*, pp. 16–26. Aspects of that paradigm of urbanization were also incorporated in the reconfiguration of Rio de Janeiro in the first decade of the twentieth century. See Jaime Benchimol, *Pereira passos: um Haussmann Tropical* (Rio de Janeiro: Secretaria Municipal de Cultura, Turismo e Esportes, 1992) and Glória Kok, *Rio de Janeiro no tempo da Av. Central* (São Paulo: Bei Comunicação, 2005).

19 For a detailed discussion of these projects see Júlio de Castilho, *A Ribeira de Lisboa: descripção histórica da margem do Tejo desde a Madre de Deus até Santos-o-Velho* (Lisbon: Imprensa Nacional, 1893), pp. 639–701.

20 This is as according to the blueprint of 1887, made by João Veríssimo Mendes Guerreiro, and extant in the National Library of Portugal (shelfmark C.C. 316 A.).

21 Ana Martins Barata, 'A ordenação do espaço litoral de Lisboa, 1860–1940', *Scripta Nova*, 13/296.4 (2009): http://www.ub.es/geocrit/sn/sn-296/sn-296-4.htm (accessed 12 March 2010).

22 Peter Schmiedgen, 'Interiority, Exteriority and Spatial Politics in Benjamin's Cityscapes', in Andrew Benjamin and Charles Rice (eds), *Walter Benjamin and the Architecture of Modernity* (Prahran: re.press, 2009), pp. 147–58, at p. 147.

23 Walter Benjamin, 'Paris – Capital of the Nineteenth Century', *New Left Review*, 1/48 (1968): 77–88, at p. 83.

24 Bill Risebero, *Modern Architecture and Design: An Alternative History* (Cambridge, MA: MIT Press, 1982), pp. 79–84. Nevertheless, the strict identification of Haussmann with modernity is problematic. Despite his modernization of the surface of the city, Matthew Gandy argues that the underground infrastructures, such as the sewage system, stood as a problematic issue during Haussmann's urban reform. See Matthew Gandy, 'The Paris Sewers and the Rationalization of Urban Space', *Transactions of the Institute of British Geographers*, 24/1 (1999): 22–44.

25 Benjamin, 'Paris – Capital of the Nineteenth Century': 86.

26 Walter Benjamin, 'Exposé de 1939', in Benjamin, *The Arcades Project*, p. 24.

27 Benjamin, *The Arcades Project*, p. 669. For a more detailed discussion on Benjamin's concept of phantasmagoria see Margaret Cohen, 'Le Diable à Paris: Benjamin's

Phantasmagoria', in Margaret Cohen, *Profane Illumination: Walter Benjamin and the Paris of Surrealist Revolution* (Berkeley: University of California Press, 1995), pp. 217–69.

28 Karl Marx, *Capital* (London: Penguin Books Ltd, 1992), p. 165.

29 Jacques Derrida, *Specters of Marx* (London/NY: Routledge, 1994), p. 159.

30 Susan Buck-Morss, *The Dialectics of Seeing: Walter Benjamin and the Arcades Project* (Cambridge, MA: MIT Press, 1991), pp. 81–2.

31 David Harvey, *Paris, Capital of Modernity* (London/NY: Routledge 2003), p. 118.

32 David Harvey, 'The Right to the City', *New Left Review*, 53/September-October (2008): 23–40. For an interesting discussion about surplus from Spinoza to Lacan see Kiarina Kordela, *$urplus (Spinoza, Lacan)* (New York: SUNY Press, 2007).

33 Abílio Fernandes, *Lisboa e a electricidade* (Lisboa: EDP, 1992), pp. 15–63.

34 Fernandes, *Lisboa e a electricidade*.

35 In 1907, there were 75 cafés and 51 restaurants in Lisbon. See Manuela Rêgo (ed), *1907: no advento da República*. (Lisbon: Biblioteca Nacional, 2007), p. 11.

36 Joaquim Palminha Silva, 'Armazéns Grandella: como nasceram e o que foram', *História*, 112 (1988): 4–27 and João Mário Mascarenhas (ed.), *Grandella, o grande homem* (Lisbon: Câmara Municipal de Lisboa, 1994).

37 Leonor Losa, João Silva and Gonçalo Antunes de Oliveira, 'A edição de música impressa enquanto agente de mediatização do Fado: o caso do Fado do 31', *Etno-Folk: Revista Galega de Etnomusicoloxía*, 12 (2008): 55–68, at p. 56.

38 *O Occidente*, 15 September 1879: 138–40, Bastos, *Diccionario*, pp. 362–3 and Francisco da Fonseca Benevides, *O Real Theatro de S. Carlos de Lisboa desde a sua fundação em 1793 até à actualidade: Estudo histórico* (Lisbon: Typ. Castro Irmão, 1883), pp. 316–17.

39 *O Occidente*, 11 April 1883: 83 and 85 and Bastos, *Diccionario*, p. 194.

40 Bastos, *Diccionario*, pp. 325, 359–61.

41 Ibid., and Luiz Francisco Rebello, *História do teatro de revista em Portugal*, vol. 1 (Lisbon: Publicações D. Quixote, 1984), p. 84.

42 Pierre Bourdieu, *Outline of a Theory of Practice* (Cambridge, MA: Cambridge University Press, 1977).

43 Bastos, *Diccionario*, p. 355.

44 Tomaz Ribas, *O teatro da Trindade: 125 anos de vida* (Oporto: Lello & Irmão, 1993), p. 13 and Luiz Francisco Rebello, *O teatro naturalista e neo-romântico (1870–1910)* (Lisbon: Instituto de Cultura Portuguesa, 1978), pp. 53–4.

45 Bastos, *Diccionario*, p. 371.

46 Marina Tavares Dias, *Lisboa desaparecida* (Lisbon: Quimera Editores, 1987), pp. 143–6.

47 Bastos, *Diccionario*, pp. 312, 318.

48 Ibid., p. 316 and *O Occidente*, 1 June 1890: 123.

49 Bastos, *Diccionario*, pp. 309–10. Another possible entertainment for the population of Lisbon was the bullfight. These have been frequently presented in Lisbon since, at least, the last years of the eighteenth century (first in Salitre, then in the Campo de Santana). In 1892, a modern and specially designated space for its practice was opened, the Praça de Touros do Campo Pequeno.

50 Mário Moreau, *Coliseu dos Recreios: um século de história* (Lisbon: Quetzal Editores, 1994), pp. 17–20, 25, 38, 205–6.

51 Bastos, *Diccionario*, pp. 332–3. For several reviews of presentations in the Teatro de D. Amélia (among other chronicles) see Fialho d'Almeida, *Actores e autores* (Lisbon: Livraria Clássica Editora, 1925). This spatial distribution of theatrical genres was presented both in Bastos, *Diccionario*, p. 35 and in Rebello, *O teatro naturalista*, pp. 53–4. About the presentation of film in the Teatro de D. Amélia see Dias, *Lisboa desaparecida*, p. 145.

52 Rebello, *O teatro naturalista*.

53 Bastos, *Diccionario*, pp. 261–2 and Mário Moreau, *O teatro de S. Carlos: dois séculos de história*, vol. 1 (Lisbon: Hugin Editores, 1999), pp. 88–157.

54 Walter Benjamin and Asja Lacis, 'Naples', in Walter Benjamin, *Reflections: Essays, Aphorisms, Autobiographical Writings* (New York and London: Schoken Books, 2007), pp. 163–73.
55 Benjamin and Lacis, 'Naples', pp. 165–6.
56 Simón Parker, *Urban Theory and the Urban Experience: Encountering the City* (London and New York: Routledge, 2004), p. 16 and Benjamin and Lacis, 'Naples', p. 166.
57 Benjamin and Lacis, 'Naples', p. 171.
58 Ibid., p. 167.
59 Jennifer Robinson, *Ordinary Cities: Between Modernity and Development* (London and New York: Routledge, 2006), p. 33.
60 Buck-Morss, *The Dialectics of Seeing*, p. 26.
61 Svetlana Boym, *The Future of Nostalgia* (New York: Basic Books, 2001), p. 77.
62 José Machado Pais, 'A prostituição na Lisboa boémia nos inícios do século XX', *Análise Social*, 19/77–79 (1983): 939–60, at pp. 943–4.
63 Cesário Verde, *O livro de Cesário Verde* (Lisboa: Typographia Elzeveriana, 1887), p. 67.
64 Pais, 'A prostituição': 942–3.
65 Walter Benjamin, 'Marseilles', in Benjamin, *Reflections*, pp. 135–6.
66 Victor Turner, *The Ritual Process: Structure and Anti-Structure* (Hawthorne, NY: Aldine de Gruyter, 1997), pp. 132–40. For a discussion of the concept of 'community' see Roberto Esposito, *Communitas: The Origin and Destiny of Community* (Palo Alto, CA: Stanford University Press, 2009).
67 Miguel Vale de Almeida, 'Marialvismo: Fado, Touros e Saudade como Discursos da Masculinidade, da Hierarquia Social e da Identidade Nacional', *Trabalhos de Antropologia e Etnologia*, 37/1–2 (1997): 41–66.
68 Pinto de Carvalho, *História do fado* (Lisbon: Empreza da História de Portugal, 1903), pp. 28–9 and Pais, 'A prostituição': 944.
69 On these types of housing and urbanism in Lisbon see Nuno Teotónio Pereira, 'Pátios e vilas de Lisboa, 1870–1930: a promoção privada do alojamento operário', *Análise Social*, 29/127 (1994): 509–24.
70 Ibid.: 511.
71 Pereira, 'Pátios e vilas de Lisboa': 511.
72 Henri Lefebvre, *The Production of Space* (Oxford: Blackwell Publishers, 1991).
73 See, for instance, Fialho D'Almeida, *Contos* (Porto and Braga: Livraria Chardron, 1881) or Abel Botelho, *Amanhã!* (Porto: Livraria Chardron, 1901).
74 Lisboa, *Os engenheiros*, pp. 151–77.
75 Graça Índias Cordeiro, 'Trabalho e Profissões no imaginário de uma cidade: sobre os tipos populares de Lisboa', *Etnográfica*, 5/1 (2001): 7–24, at p. 16.
76 Júlio de Castilho, *Lisboa antiga* (Lisbon: Antiga casa Bertrand, José Bastos, 1904), p. 216.
77 Ibid., pp. 217–18.
78 Ibid., pp. 218–19 and pp. 229–33.
79 One of the foremost guitarists at that time was Luís Carlos da Silva (1859–1934), nicknamed *Petrolino* possibly due to his profession. See Alberto Pimentel, *A triste canção do Sul: Subsídios para a História do Fado* (Lisbon: Livraria Central, 1904), p. 62.
80 Rui Cidra, 'Música e Migração', in Salwa Castelo-Branco (ed.), *Enciclopédia da Música em Portugal no Século XX*, vol. 3 (Lisbon: Círculo de Leitores, 2010), pp. 774–5.
81 Cordeiro, 'Trabalho e Profissões': 11.
82 Pais, 'A prostituição': 941, and Cordeiro, 'Trabalho e Profissões': 11.
83 Cordeiro, 'Trabalho e Profissões': 8.
84 Ibid.: 16–17.

85 Alain Corbin, *Village Bells: Sound and Meaning in the Nineteenth-century French Countryside* (London: Papermac, 1999), p. xi.
86 Ibid.
87 Maria Rattazzi, *Portugal de relance* (Lisbon: Livraria Zeferino, 1881).
88 Rattazzi, *Portugal de relance*, pp. 28–9.
89 Ibid.
90 Leitão, 'A ambiência musical e sonora', p. 178 and Eça de Queirós, *O primo Bazilio, episódio domestico* (Porto/Braga: Livraria Chardron, 1878).
91 *O Occidente*, 25 November 1896: 258–9.
92 Alberto Pimentel, *Vida de Lisboa* (Lisbon: Parceria António Maria Pereira, 1900), p. 52.
93 Anahid Kassabian, 'Popular', in Bruce Horner and Thomas Swiss (eds), *Key Terms in Popular Music and Culture* (Malden, MA: Blackwell Publishing, 1999), pp. 117–23, at p. 117.
94 Leitão, 'A ambiência musical e sonora', pp. 130–35.
95 James Clifford, *Routes: Travel and Translation in the Late Twentieth Century* (Cambridge, MA: Harvard University Press, 1997), p. 54.
96 Stefan Kipfer, Kanishka Goonewardena, Christian Schmid and Richard Milgrom, 'On the Production of Henri Lefebvre', in Kanishka Goonewardena, Stefan Kipfer, Richard Milgrom and Christian Schmid (eds), *Space, Difference, Everyday Life: Reading Henri Lefebvre* (London and New York: Routledge, 2008), pp. 1–24, at p. 9.
97 Deborah Pellow, 'The Architecture of Female Seclusion in West Africa', in Denise Lawrence-Zuniga and Setha M. Low (eds), *The Anthropology of Space and Place: Locating Culture* (Cambridge: Blackwell Publishers, 2003), pp 160–83, at p. 161 and Lefebvre, *The Production of Space*.
98 Henri Lefebvre, *The Urban Revolution* (Minneapolis, MN: University of Minnesota Press, 2003), p. 78.
99 Lefebvre, *The Urban Revolution*, p. 79.
100 Ibid.
101 Ibid., p. 80.
102 Lefebvre, *The Urban Revolution*, pp. 80–81.

Bibliography

Abbate, Carolyn, *In Search of Opera* (Princeton: Princeton University Press, 2001).

Agnew, Vanessa, *Enlightenment Orpheus: The Power of Music in Other Worlds* (New York: Oxford University Press, 2008).

Alpers, Paul, '"The Philoctetes Problem" and the Poetics of Pastoral', *Representations*, 86, Special Issue: 'Music, Rhythm, Language' (Spring, 2004): 4–19.

Altman, Rick (ed.), *Sound Theory/Sound Practice* (New York: Routledge, 1992).

Amerus, *Practica artis musice*, ed. C. Ruini ([Rome]: American Institute of Musicology, 1977).

Amphoux, Pascal, *Aux écoutes de la ville* (Lausanne: IREC, Cresson, 1991).

Ancina, G., *Tempio armonico della Beatissima Vergine* (Rome: N. Mutii, 1599).

Anon, 'Dalcroze Eurhythmics Have Come to New York', *New York Times*, 25 January 1914: SM5.

Antegnati, C., *Salmi a otto voci* (Venice, 1592).

Anthony, James R., *French Baroque Music from Beaujoyeulx to Rameau*, rev. ed. (Portland, OR: Amadeus, 2003).

Appadurai, Arjun, *Modernity at Large: Cultural Dimensions of Globalization* (Minneapolis: University of Minnesota Press, 1996).

Archer, Ian W., 'Shakespeare's London', in David Scott Kastan (ed.), *A Companion to Shakespeare* (Oxford: Blackwell, 1999), pp. 43–56.

Astier, Louis and Paul Baron, 'Bruits produits par les distributions urbaines d'énergie électrique et bruits dus à la circulation dans les rues', preface by François Cellerier, *Le génie civil*, vol. 5 (1935): 28–9.

Attali, Jacques, *Noise: The Political Economy of Music*, trans. Brian Massumi (Minneapolis and London: University of Minnesota Press, 1985; reprinted 1999).

Aughterson, Kate (ed.), *The English Renaissance: An Anthology of Sources and Documents* (London: Routledge, 2002).

Augoyard, Jean-François and Henry Torgue (eds), *Sonic Experience: A Guide to Everyday Sounds* (Montreal: McGill-Queen's University Press, 2006); translation of *À l'écoute de l'environnement, Répertoire des effets sonores* (Marseille: Parenthèses, 1995).

Aurelian of Réôme, *Musica disciplina*, ed. L. Gushee ([Rome]: American Institute of Musicology, 1975).

Bagnall-Yardley, Ann, *Performing Piety: Musical Culture in Medieval English Nunneries* (New York: Palgrave Macmillan, 2006).

Bal, Mieke, *Quoting Caravaggio: Contemporary Art, Preposterous History* (Chicago: University of Chicago Press, 1999).

Balaÿ, Olivier, *L'espace sonore de la ville au XIXe siècle* (Bernin (Isère): À la Croisée, Collection Ambiances, Ambiance, 2003).

Barata, Ana Martins, 'A ordenação do espaço litoral de Lisboa, 1860–1940', *Scripta Nova*, 13/296.4 (2009): <http://www.ub.es/geocrit/sn/sn-296/sn-296–4.htm> (accessed 12 March 2010).

Bartky, Ian, *Selling the True Time, Nineteenth Century Timekeeping in America* (Stanford: Stanford University Press, 2000).

Bartlet, M. Elizabeth C., 'Beaumarchais and Voltaire's "Samson"', *Studies in Eighteenth-Century Culture*, 11 (1982): 33–49.

Bastos, António de Sousa, *Diccionario do theatro portuguez* (Lisbon: Imp. Libânio da Silva, 1908).

Beethoven, Ludwig van, *Letters*, vol. III, collected, trans. and ed. Emily Anderson (London: Macmillan; New York: St. Martin's Press, 1961).

Bellarmino, Roberto, *De controuersiis christianae fidei, aduersus huius temporis haereticos: tribus tomis comprehensae* (Ingolstadt, 1581).

———, *De aeterna felicitate sanctorum* (Antwerp: Ex officina plantiniana, 1617).

Benchimol, Jaime, *Pereira passos: um Haussmann Tropical* (Rio de Janeiro: Secretaria Municipal de Cultura, Turismo e Esportes, 1992).

Benevides, Francisco da Fonseca, *O Real Theatro de S. Carlos de Lisboa desde a sua fundação em 1793 até à actualidade: Estudo histórico* (Lisbon: Typ. Castro Irmão, 1883).

Benjamin, Walter, 'Paris – Capital of the Nineteenth Century', *New Left Review*, 1/48 (1968): 77–88.

———, *The Arcades Project* (Cambridge, MA: Harvard University Press, 1999).

———, *Reflections: Essays, Aphorisms, Autobiographical Writings* (New York and London: Schoken Books, 2007).

Benjamin, Walter, and Asja Lacis, 'Naples', in Walter Benjamin, *Reflections: Essays, Aphorisms, Autobiographical Writings* (New York and London: Schoken Books, 2007), pp. 163–73.

Berlioz, Hector, *A Treatise on Modern Instrumentation and Orchestration*, trans. Mary Cowden Clarke (London: Novello, Ewer and Co., 1856).

Berry, Mary, and Franklyn Gellnick, 'Cistercian Monks', *Grove Music Online*, <http://www.grove.com> (accessed 9 April 2010).

Bhabha, Homi, *The Location of Culture* (New York and London: Routledge, 1994).

Bianconi, Lorenzo, *Il Seicento* (Turin: EDT, 1991).

Biddle, Ian, 'Visitors or the Political Ontology of Noise', *Radical Musicology*, Volume 4 (2009), <http://www.radical-musicology.org.uk/> (accessed May 12 2014).

Birla, Ritu, *Stages of Capital: Law, Culture, and Market Governance in Late Colonial India* (Durham: Duke University Press, 2009).

Blackton, Charles, 'The Action Phase of the 1915 Riots', *Journal of Asian Studies*, 29/2 (February 1970): 235–54.

Blondel, François, *Cours d'architecture* (Paris: Lambert Roulland, 1675).

Bolton, Thaddeus L., 'Rhythm', *The American Journal of Psychology*, 6 (January 1894): 145–238.

Bond, George, *The Buddhist Revival in Sri Lanka* (Columbia: University of South Carolina Press, 1988).

Bonus, Alexander Evan, 'The Metronomic Performance Practice: A History of Rhythm, Metronomes, and the Mechanization of Musicality', PhD Dissertation (Case Western Reserve University, 2010).

Boose, Lynda E., 'The 1599 Bishops' Ban, Elizabethan Pornography, and the Sexualization of the Jacobean Stage', in Richard Burt and John Michael Archer (eds), *Enclosure Acts: Sexuality, Property, and Culture in Early Modern England* (Ithaca and London: Cornell University Press, 1994), pp. 185–200.

Boring, Edwin G., *A History of Experimental Psychology* (New York: Appleton-Century Company, 1929).

Born, Georgina, and David Hesmondhalgh, 'Introduction', in Georgina Born and David Hesmondhalgh (eds), *Western Music and its Others: Difference, Representation, and Appropriation in Music* (Berkeley: University of California Press, 2000), pp. 1–58.

Botelho, Abel, *Amanhã!* (Porto: Livraria Chardron, 1901).

Botstein, Leon, 'Toward a History of Listening', *The Musical Quarterly*, 82/3–4, Special Issue: 'Music as Heard' (Autumn–Winter, 1998): 427–31.

Bourdieu, Pierre, *Outline of a Theory of Practice* (Cambridge: Cambridge University Press, 1977).

Bowles, Edmund A., 'Tower Musicians in the Middle Ages', *Brass Quarterly*, 5 (1961–62): 91–103.

Boym, Svetlana, *The Future of Nostalgia* (New York: Basic Books, 2001).

Brenner, Clarence D., *The Théâtre Italien: Its Repertory, 1716–1793* (Berkeley: University of California Press, 1961).

Brooks, Jeanice, *Courtly Song in Late Sixteenth-Century France* (Chicago and London: University of Chicago Press, 2000).

Brooks, William, Buford Norman and Jeanne Morgan Zarucchi (eds), *Alceste, suivi de La Querelle d'Alceste: Anciens et Modernes avant 1680* (Paris: Droz, 1994).

Brower, Harriette Moore, *Self-Help in Piano Study* (New York: Frederick A. Stokes Co., 1920).

Brown, Bruce Alan, *Gluck and the French Theatre in Vienna* (Oxford: Clarendon Press, 1991).

Bruggisser-Lanker, Therese, 'Die Stadtpfeifer von Nürnberg im 16. Jahrhundert. Ikonographische und quellenkritische Hinweise zur Aufführungspraxis', *Schweizer Jahrbuch für Musikwissenschaft*, n.s.10 (1990): 43–72.

Brunner, Wolfgang, 'Städtisches Tanzen und das Tanzhaus im 16. Jahrhundert', in Alfred Kohler and Heinrich Lutz (eds), *Alltag im 16. Jahrhundert: Studien zu Lebensformen in mitteleuropäischen Städten* (Wiener Geschichte der Neuzeit. Bd.14; Wien: Verlag für Geschichte und Politik, 1987), pp. 45–64.

Buci-Glucksmann, Christine, *La folie du voir: Une esthétique du virtuel* (Paris: Galilée, 2002).

Buck-Morss, Susan, *The Dialectics of Seeing: Walter Benjamin and the Arcades Project* (Cambridge, MA: MIT Press, 1991).

Bulst, Neithard, 'Feste und Feiern unter Auflagen. Mittelalterliche Tauf-, Hochzeits-, und Begräbnisordnungen in Deutschland und Frankreich', in Detlef Altenburg, Jörg Janut and Hans-Hugo Steinhoff (eds), *Feste und Feiern im Mittelalter. Paderborner Symposium Mediävistenverbandes* (Sigmaringen: Thorbecke, 1991), pp. 39–51.

Buratti, Adele, 'L'azione pastorale dei Borromeo a Milano e la nuova sistemazione urbanistica della città', in Adele Buratti (ed.), *La città rituale. La città e lo Stato di Milano nell'età dei Borromeo* (Milan: Franco Angeli, 1982), pp. 9–56.

C.A.B., 'George Grove's Analyses of Beethoven', *The Musical World*, 3 November 1888.

Caldwell, John, 'Relations Between Liturgical and Vernacular Music in Medieval England', in Susan Rankin and David Hiley (eds), *Music in the Medieval English Liturgy* (Oxford: Clarendon Press, 1993), pp. 285–93.

Carpenter, William, *Principles of Mental Physiology*, Fourth edition (New York: D. Appleton and Company, 1883).

Casali, Lodovico, *Generale invito alle grandezze, e maraviglie della musica* (Modena, 1629).

Castelvecchi, Stefano, 'From *Nina* to *Nina*: Psychodrama, Absorption and Sentiment in the 1780s', *Cambridge Opera Journal*, 8/2 (1996): 91–112.

Castilho, Júlio de, *A Ribeira de Lisboa: descripção histórica da margem do Tejo desde a Madre de Deus até Santos-o-Velho* (Lisbon: Imprensa Nacional, 1893).

———, *Lisboa antiga* (Lisbon: Antiga casa Bertrand, José Bastos, 1904).

Cattell, James McKeen, 'The Psychological Laboratory at Leipsic', *Mind*, 13 January 1888.

Chakrabarty, Dipesh, *Provincializing Europe: Postcolonial Thought and Historical Difference* (Princeton: Princeton University Press, 2000).

Chandler, James, 'The Languages of Sentiment', *Textual Practice*, 22/1 (2008): 21–39.

Charlton, David, *Grétry and the Growth of Opéra-comique* (Cambridge: Cambridge University Press, 1986).

Charpentier, Gustave, *Louise*, libretto. English translation by Henry Grafton-Chapman. Liner Notes, Sony Classical 46429.

Charrak, André, *Raison et perception: Fonder l'harmonie au XVIIIe siècle* (Paris: Librairie philosophique J. Vrin, 2001).

Chavasse, Father Pierre, 'Les origines techniques et "lyriques" du décibel', *Revue d'Acoustique*, 28 (1974): 53–6.

Chelkoff, Grégoire et al., *Entendre les espaces publics* (Grenoble: Plan urbain, Cresson, 1988).

Chion, Michel, *The Voice in Cinema*, trans. and ed. Claudia Gorbman (New York: Columbia University Press, 1999).

Christensen, Ann C., '"The Doors Are Made Against You": Domestic Thresholds in Ben Jonson's Plays', *Journal of the Rocky Mountain Medieval and Renaissance Association*, 18 (1997): 153–78.

Christensen, Thomas, *Rameau and Musical Thought in the Musical Enlightenment* (Cambridge: Cambridge University Press, 1993).

Christiani, Adolph Friedrich, *The Principles of Expression in Pianoforte Playing* (New York: Harper & Brothers, 1885).

Chua, Daniel, 'Vincenzo Galilei, Modernity and the Division of Nature', in Suzannah Clark and Alexander Rehding (eds), *Music Theory and Natural Order from the Renaissance to the Early Twentieth Century* (Cambridge: Cambridge University Press, 2001), pp. 17–29.

———, 'Untimely Reflections on Operatic Echoes: How Sound Travels in Monteverdi's *L'Orfeo* and Beethoven's *Fidelio* with a Short Instrumental Interlude', *The Opera Quarterly*, 21/4 (2006): 573–96.

Chudinova, Irina A., 'Song, Bells, Ritual: The Topography of St. Petersburg's Ecclesiastical Musical Culture', *Russian Studies in History*, 45/3 (Winter, 2006–7): 31–59.

Cidra, Rui, 'Música e Migração', in Salwa Castelo-Branco (ed.), *Enciclopédia da Música em Portugal no Século XX* (Lisbon: Círculo de Leitores, forthcoming).

Clegg, Cyndia Susan, *Press Censorship in Elizabethan England* (Cambridge: Cambridge University Press, 1997).

Clifford, James, *Routes: Travel and Translation in the Late Twentieth Century* (Cambridge: Harvard University Press, 1997).

Clouzot, Martine, *Images de musiciens (1350–1500). Typologie, figurations et pratiques sociales* (Turnhout and Tours: Brepols-CESR, 2007).

Cockayne, Emily, *Hubbub: Filth, Noise & Stench in England 1600–1770* (New Haven and London: Yale University Press, 2007).
Cohen, Margaret, 'Le Diable à Paris: Benjamin's Phantasmagoria', in Margaret Cohen, *Profane Illumination: Walter Benjamin and the Paris of Surrealist Revolution* (Berkeley: University of California Press, 1995), pp. 217–69.
Colledge, Edmund (ed.), *The Latin Poems of Richard Ledrede, C.F.M., Bishop of Ossory, 1317–1360* (Toronto: Pontif. Inst. of Mediaeval Studies, 1974).
Colton, Lisa, 'The Articulation of Virginity in the Medieval *Chanson de nonne*', *Journal of the Royal Musical Association*, 133/2 (November, 2008): 159–88.
Cooke, James Francis, *Great Pianists on Piano Playing: Study Talks with Foremost Virtuosos* (Reprint, Philadelphia: Theodore Presser Co., 1917).
Cooley, Timothy J., 'Casting Shadows in the Field: An Introduction', in Gregory Barz and Timothy J. Cooley (eds), *Shadows in the Field: New Perspectives for Fieldwork in Ethnomusicology* (New York: Oxford University Press, 1997), pp. 3–19.
Corbin, Alain, *Village Bells: Sound and Meaning in the Nineteenth-Century French Countryside* (New York: Columbia University Press, 1998).
Cordeiro, Graça Índias, 'Trabalho e Profissões no imaginário de uma cidade: sobre os tipos populares de Lisboa', *Etnográfica*, 5/1 (2001): 7–24.
Cowart, Georgia J., *The Triumph of Pleasure: Louis XIV and the Politics of Spectacle* (Chicago: University of Chicago Press, 2008).
Crampton, Georgia Ronan (ed.), *The Shewings of Julian of Norwich* (Kalamazoo: Medieval Institute Publications, 1994).
Crane, Julia E., *Music Teachers Manual*, Seventh Edition (Potsdam: Elliot Fay & Sons, 1915).
Crane, Mary Thomas, 'Illicit Privacy and Outdoor Spaces in Early Modern England', *Journal for Early Modern Cultural Studies*, 9/1 (2009): 4–22.
Crombie, Alistair C., *Styles of Scientific Thinking in the European Tradition: The History of Argument and Explanation Especially in the Mathematical and Biomedical Sciences and Arts*, 3 vols (London: Gerald Duckworth & Company, 1995).
Crossley, Ceri, *French Historians and Romanticism: Thierry, Guizot, the Saint-Simonians, Quinet, Michelet* (London: Routledge, 1993).
Cymbron, Luísa, 'A ópera em Portugal 1834–1854: o sistema produtivo e o repertório nos teatros de S. Carlos e de S. João', PhD Dissertation (Universidade Nova de Lisboa, 1998).
Cyr, Mary, 'Towards a New Chronology of Rameau's Cantatas', *The Musical Times*, 124 (1983): 539–41.
D'Accone, Frank A., *The Civic Muse: Music and Musicians in Siena During the Middle Ages and the Renaissance* (Chicago and London: The University of Chicago Press, 1997).
d'Almeida, Fialho, *Contos* (Porto and Braga: Livraria Chardron, 1881).
———, *Actores e autores* (Lisbon: Livraria Clássica Editora, 1925).
Dahlhaus, Karl, *Foundations of Music History* (Cambridge: Cambridge University Press, 1983).
Dainotto, Roberto M., *Europe (In Theory)* (Durham: Duke University Press, 2007).
Darnton, Robert, *The Great Cat Massacre and Other Episodes in French Cultural History* (New York: Basic Books, 1984).
———, 'Mademoiselle Bonafon and the Private Life of Louis XV: Communication Circuits in Eighteenth-Century France', *Representations*, 87/2 (2004): 102–24.
Daston, Lorraine and Peter Galison, *Objectivity* (New York: Zone Books, 2007).

de Almeida, Miguel Vale, 'Marialvismo: Fado, Touros e Saudade como Discursos da Masculinidade, da Hierarquia Social e da Identidade Nacional', *Trabalhos de Antropologia e Etnologia*, 37/1–2 (1997): 41–66.
de Barbot, Berthelet, *Topographie médicale de Lyon et de son territoire* (Lyon, 1773).
de Carvalho, Pinto, *História do fado* (Lisbon: Empreza da História de Portugal, 1903).
de Granada, Louis, *Of Prayer and Meditation*, trans. Richard Hopkins (Paris, 1582).
de Morand, Pierre, *Théâtres et oeuvres diverses de M. de Morand*, 3 vols (Paris: Sébastien Jorry, 1751).
de Souza, Armand, *One Hundred Days in Ceylon: Under Martial Law 1915* (Colombo: Young Socialist Publications, 1916, reprinted 2006).
Deansley, Margaret (ed.), *The Incendium Amoris of Richard Rolle of Hampole* (Manchester: Manchester University Press, 1915).
Deleuze, Gilles, *The Fold: Leibniz and the Baroque* (Minneapolis: University of Minnesota Press, 1993).
Delumeau, Jean, *Une histoire du paradis – 3. Que reste-t-il du paradis?* (Paris: Fayard, 2000).
Dering, Richard, *The Cries of London* (n.p., 1599); ed. Denis Stevens (Philadelphia: Pennsylvania State University Press, 1964).
Derrida, Jacques, *Specters of Marx* (London/NY: Routledge, 1994).
Di Lorenzo, Andrea, 'Gaudenzio Ferrari e la cupola di Saronno', in Luigi Lazzaroni and Alfredo Fusetti (eds), *Il concerto degli angeli* (Milan: Silvana, 1994), pp. 24–33.
Dias, Marina Tavares, *Lisboa desaparecida* (Lisbon: Quimera Editores, 1987).
Dierig, Sven, 'Engines for Experiment: Laboratory Revolution and Industrial Labor in the Nineteenth-Century City', *Osiris*, 18 (2003): 116–34.
Donati, Ignazio, *Sacri Concentus* (Venice: G. Vincenzi, 1612).
Donia, Robert J., *Islam under the Double Eagle: The Muslims of Bosnia and Hercegovina, 1878–1914* (Boulder: East European Monographs, 1981).
———, *Sarajevo: A Biography* (London: Hurst & Company, 2006).
Downey, Peter, 'The Renaissance Slide Trumpet: Fact or Fiction?', *Early Music*, 12 (1984): 26–33.
Drexel, Jeremias, *Infernus damnatorum carcer* (Munich: C. Leysser, 1631).
———, *Caelum beatorum civitas* (Antwerp: J. Cnobbaert, 1635).
———, *A Pleasant and Profitable Treatise of Hell*, trans. not given (London, 1668).
Duffin, Ross W., 'The "*trompette des menestrels*" in the 15th-century "*alta capella*"', *Early Music*, 17 (1989): 397–402.
Duncan, Bruce, 'Remembering Schiller: The Centenary of 1859', *Seminar: A Journal of Germanic Studies*, 35/1 (1999): 1–22.
Duverney, Joseph-Guichard, *Traité de l'organe de l'ouïe. Nouvelle édition* (Leiden: Joh. Arn. Langerak, 1731).
———. *A Treatise of the Organ of Hearing* (1737; New York: AMS Press, 1973).
Ebel, Julia G., 'Translation and Cultural Nationalism in the Reign of Elizabeth', *Journal of the History of Ideas*, 30 (1969): 593–602.
Elders, Willem, *Symbolic Scores* (Leiden, New York, Köln: Brill, 1994).
Eliot, John, *The Parlement of Pratlers* (1593; Franfrolico Press: London, 1928).
Elliott, Richard, 'Popular Music and/as Event: Subjectivity, Love and Fidelity in the Aftermath of Rock 'n' Roll', *Radical Musicology*, 3 (2008), <http://www.radical-musicology.org.uk/> (accessed January 12 2015).
Erlmann, Veit (ed.), *Hearing Cultures: Essays on Sound, Listening and Modernity* (Oxford and New York: Berg, 2004).

———, *Reason and Resonance: A History of Modern Aurality* (New York: Zone Books, 2014).

Ernst, Fritz, 'Die Spielleute im Dienste der Stadt Basel im ausgehenden Mittelalter (bis 1550)', *Basler Zeitschrift für Geschichte und Altertumskunde*, 44 (Basel, 1945): 79–236.

Esposito, Roberto, *Communitas: The Origin and Destiny of Community* (Palo Alto: Stanford University Press, 2009).

Fader, Don, '"The Cabale du Dauphin", Campra, and Italian Comedy: The Courtly Politics of French Musical Patronage Around 1700', *Music & Letters*, 86/3 (2005): 380–413.

Fanon, Frantz, *Les Damnés de la Terre* (Paris: Éditions Maspero, 1961), trans. Constance Farrington as *The Wretched of the Earth* (New York: Grove Press, 1963).

Fein, Susanna Greer, '*Haue Mercy of Me* (Psalm 51): An Unedited Alliterative Poem from the London Thornton Manuscript', *Modern Philology*, 86/3 (1989): 223–41.

Ferino-Pagden, Sylvia (ed.), *Arcimboldo: artista milanese tra Leonardo e Caravaggio* (Milan: Skira, 2011).

Fernandes, Abílio, *Lisboa e a electricidade* (Lisboa: EDP, 1992).

Fernando, P.T.M., 'The British Raj and the 1915 Communal Riots in Ceylon', *Journal of Asian Studies*, 29/2 (February 1970): 245–55.

Filippi, Daniele V., *Selva armonica. La musica spirituale a Roma fra Cinque e Seicento* (Turnhout: Brepols, 2008).

Finck, Henry T., *Success in Music and How it is Won* (New York: Charles Scribner's Sons, 1909).

Fink, Robert, '"Rigoroso (♪ = 126)": "The Rite of Spring" and the Forging of a Modernist Performing Style', *Journal of the American Musicological Society*, 52 (1999): 299–362.

Finkelpearl, Philip J., *John Marston of the Middle Temple* (Cambridge: Harvard University Press, 1969).

Folkerth, Wes, *The Sound of Shakespeare* (London: Routledge, 2002).

Fournel, Victor, *Les Cris de Paris: Types et physionomies d'autrefois* (Paris: Librairie de Firmin-Didot, 1887).

França, José-Augusto, *O Romantismo em Portugal: Estudo de factos socioculturais* (Lisbon: Livros Horizonte, 1974).

Gandy, Matthew, 'The Paris Sewers and the Rationalization of Urban Space', *Transactions of the Institute of British Geographers*, 24/1 (1999): 22–44.

Garden, Maurice, *Lyon et les lyonnais au XVIIIe siècle* (Paris: Les belles lettres, 1970).

Gardey, Loius, *Voyage du sultan Abd-ul-Aziz de Stamboul au Caire* (Paris: E. Dentu, 1865).

Garrioch, David, 'Sounds of the City: The Soundscape of Early Modern European Towns', *Urban History*, 30/1 (2003): 5–25.

Gemeiner, Carl Theodor, *Regensburgische Chronik*, ed. Heinz Angermeier (Munich: C.H. Beck, repr. 1987).

Gerbino, Giuseppe, *Music and the Myth of Arcadia in Renaissance Italy* (Cambridge: Cambridge University Press, 2014).

Gethin, Rupert, Review of *Vain Debates*, in *Bulletin of the School of Oriental and African Studies, University of London*, 62/3 (1999): 574–5.

Gibson, Kirsten, '"So to the Wood went I": Politicizing the Greenwood in Two Songs by John Dowland', *Journal of the Royal Musical Association*, 132 (2007): 221–51.

Giussano, G.P., *Vita di S. Carlo Borromeo* (Rome, 1610).

Gluck, Christoph Willibald, *Orfeo ed Euridice*, Sämtliche Werke, eds Anna Amalie Abert and Ludwig Finscher, series I, vol. 1 (Kassel: Bärenreiter, 1963).

———, *Orphée et Euridice = Orpheus und Eurydike: (Pariser Fassung von 1774): tragédie opéra (drame héroïque) in drei Akten*, in Ludwig Finscher (ed.), *Christoph*

Willibald Gluck: Sämtliche Werke, ser. 1: Musikdramen, 11 vols (Kassel, New York: Bärenreiter, 1967).

Gombrich, Richard, 'The Consecration of a Buddhist Image', *The Journal of Asian Studies*, 26/1 (November 1966): 23–36.

Goodman, Anthony, *Margery Kempe and Her World* (London: Longman, 2002).

Gottsched, Johann Christoph, *Versuch einer kritischen Dichtkunst* in *Ausgewählte Werke*, vol. 2 (Berlin and New York: Walter de Gruyter, [1737] 1973).

Gouk, Penelope, *Music, Science, and Natural Magic in Seventeenth-century England* (New Haven and London: Yale University Press, 1999).

Gould, Stephen Jay, *The Hedgehog, the Fox, and the Magister's Pox: Mending the Gap between Science and the Humanities* (New York: Knopf, 2003).

Gowing, Laura, '"The Freedom of the Streets": Women and Social Space, 1560–1640', in Paul Griffiths and Mark S.R. Jenner (eds), *Londinopolis: Essays in the Cultural and Social History of Early Modern London* (Manchester and New York: MUP, 2000), pp. 130–51.

Green (now Coffey), Helen, 'Stadtpfeifer and Varende Lewte: Secular Musical Patronage in the Imperial Cities of Germany during the Reign of Maximilian I (1486–1519)', DPhil Dissertation (University of Oxford, 2006).

———, 'Defining the City "Trumpeter": German Civic Identity and the Employment of Brass Instrumentalists *c.*1500', *Journal of the Royal Musical Association*, 136 (2011): 1–31.

Green, Marcus, 'Gramsci Cannot Speak: Presentations and Interpretations of Gramsci's Concept of the Subaltern', *Rethinking Marxism*, 14/3 (Fall 2002): 1–23.

Grew, Sydney, *The Art of the Player-Piano* (London: Kegan Paul, Trench, Trubner & Co., 1922).

Gstrein, Rainer, 'Tanzmusik-Ensembles zur Zeit und am Hofe Kaiser Maximilians I'., *Jaarboek van het Vlaamse Centrum voor Oude Muziek*, 3 (1987): 79–96.

Guerrini, Anita, 'The "Virtual Menagerie": The *Histoire des animaux* Project', *Configurations*, 14/1–2 (2006): 29–42.

Guilpin, Everard, *Skialetheia, Or A Shadowe of Truth in Certain Epigrams and Satyres* (1598), ed. D. Allen Carroll (Chapel Hill: University of North Carolina Press, 1974).

———, 'The Whipper of the Satyre his pennance in a white Sheet: OR, The Beadles Confutation' (London: Thomas Pauier, 1601), *Literature Online* <http://lionchadwyck.co.uk/> (accessed 17 June 2009).

Hallays, André, *Les Perrault* (Paris: Librairie Académie Perrin, 1926).

Hammerstein, Reinhold, *Die Musik der Engel. Untersuchungen zur Musikanschauung des Mittelalters* (Bern: Francke, 1990).

Hangi, Antun, *Život i običaji Muslimana.*, 2nd enlarged edition (Sarajevo: Daniel A. Kajon, 1906).

Hanioğlu, M. Şükrü, *A Brief History of the Late Ottoman Empire* (Princeton: Princeton University Press, 2008).

Harris, Ellen T., *Handel as Orpheus: Voice and Desire in the Chamber Cantatas* (Cambridge: Harvard University Press, 2001).

Harrison, Frank Ll. (ed.), *Motets of English Provenance, Polyphonic Music of the Fourteenth Century*, vol. 15 (Monaco: L'Oiseau-Lyre, 1980), pp. 197–8.

Harris-Warrick, Rebecca, and Carol G. Marsh, 'The French Connection', in Rebecca Harris-Warrick and Bruce Alan Brown (eds), *The Grotesque Dancer on the Eighteenth-Century Stage: Gennaro Magri and His World* (Madison: University of Wisconsin Press, 2005), pp. 173–98.

Harvey, David, *Paris, Capital of Modernity* (London and New York: Routledge 2003).
———, 'The Right to the City', *New Left Review*, 53/September-October (2008): 23–40.
Heartz, Daniel, *From Garrick to Gluck: Essays on Opera in the Age of Enlightenment*, John A. Rice (ed.) (New York: Pendragon, 2004).
Hegel, Carl (ed.), 'Etliche Geschichten 1488–1491', in Historische Commission bei der Königlichen Akademie der Wissenschaften, *Die Chroniken der fränkischen Städte. Nürnberg*, vol. 5, Die Chroniken der deutschen Städte, vol. 11 (Leipzig: S. Hirzel, 1874).
Helgerson, Richard, 'Language Lessons: Linguistic Colonialism, Linguistic Postcolonialism, and the Early Modern English Nation', *Yale Journal of Criticism*, 11/1 (Spring, 1998): 289–99.
Herbert, Trevor, *The Trombone* (New Haven and London: Yale University Press, 2006).
Herrmann, Wolfgang, *The Theory of Claude Perrault* (London: A. Zwemmer, 1973).
Herzfeld, Michael, *The Poetics of Manhood: Contest and Identity in a Cretan Mountain Village* (Princeton: Princeton University Press, 1985).
Hilton, Walter, *Scale of Perfection, Book 1*, ed. Thomas H. Bestul (Kalamazoo: Medieval Institute Publications, 2000).
Hobsbawm, Eric, 'Introduction: Inventing Tradition', in Eric Hobsbawm and Terence Ranger (eds), *The Invention of Tradition* (Cambridge: Cambridge University Press, 1983), pp. 1–14.
Hodgart, Matthew, *Satire* (Verona: World Publishing, 1969).
Hofmann, Joseph, 'Playing the Piano Successfully', *The Ladies' Home Journal* (March 1905): 2–14.
Holbach, Maude M., *Bosnia and Herzegovina: Some Wayside Wanderings* (London and New York: J. Lane, 1910).
Holsinger, Bruce W., 'The Vision of Music in a Lollard Florilegium: *Cantus* in the *Rosarium Theologie* (Cambridge, Gonville and Caius College MS 354/581)', *Plainsong and Medieval Music*, 8/2 (1999): 95–106.
———, *Music, Body, and Desire in Medieval Culture: Hildegard of Bingen to Chaucer* (Stanford, CA: Stanford University Press, 2001).
Howard, Patricia (ed.), *C.W. von Gluck: Orfeo* (Cambridge: Cambridge University Press, 1981).
———, *Gluck: An Eighteenth-Century Portrait in Letters and Documents* (Oxford: Clarendon Press, 1995).
Hughes, Andrew, and Margaret Bent (eds), *The Old Hall Manuscript*, 3 vols in 4, *Corpus Mensurabilis Musicae* 46 ([Rome]: American Institute of Musicology, Hänssler-Verlag, 1969–73).
Hunter, Mary, *The Culture of Opera Buffa in Mozart's Vienna* (Princeton: Princeton University Press, 1999).
Huot, Sylvia, *Allegorical Play in the Old French Motet* (Stanford: Stanford University Press, 1997).
Ingham, Ethel, 'Lessons at Hellerau', in Maynard Small (ed.), *The Eurhythmics of Jaques-Dalcroze* (Boston: Small, Maynard, and Company, 1913).
Ingham, Percy B., 'The Jaques-Dalcroze Method', in Maynard Small (ed.), *The Eurhythmics of Jaques-Dalcroze* (Boston: Small Maynard and Company, 1913), pp. 31–45.
Isherwood, Robert S., *Farce and Fantasy: Popular Entertainment in Eighteenth-Century Paris* (New York: Oxford University Press, 1986).
Jahn, Fritz, 'Die Nürnberger Trompeten- und Posaunenmacher im 16. Jahrhundert', *Archiv für Musikwissenschaft*, 7 (1925): 23–52.
James, Henry, *Roderick Hudson* (Boston and New York: Houghton Mifflin Co., 1917).

James, William, 'The Perception of Time', in Charles M. Sherover (ed.), *The Principles of Psychology* (New York: Henry Holt and Co., 1890) vol. 1, Chap XV; reprinted in, *The Human Experience of Time, the Development of its Philosophic Meaning* (Reprint, Evanston: Northwestern University Press, 2001).
Jaques-Dalcroze, Emile, 'The Initiation into Rhythm', in *Rhythm, Music and Education*, trans. Harold F. Rubinstein (New York: G. P. Putnam's Sons, 1921), pp. 79–92.
Jayawardana, Kumari, 'Economic and Political Factors in the 1915 Riots', *Journal of Asian Studies*, 29/2 (February 1970): 223–33.
Jezernik, Božidar, *Wild Europe: The Balkans in the Gaze of Western Travellers* (London: Saqi Books, 2004).
Johnson, James H., *Listening in Paris: A Cultural History* (Berkeley: University of California Press, 1995).
———, 'Listening and Silence in Eighteenth- and Nineteenth-Century France', in Mark M. Smith (ed.), *Hearing History: A Reader* (Athens: University of Georgia Press, 2004), pp. 169–83.
Jonson, Ben, *Epicoene, Or the Silent Woman* (1609), in Simon Barker and Hilary Hinds (eds), *The Routledge Anthology of Renaissance Drama* (Abingdon: Routledge, 2005), pp. 272–325.
Juvenal, *Satires*, in *Juvenal and Persius*, trans. Susanna Morton Braund (Cambridge and London: Harvard University Press, 2004).
Kane, Brian, 'L'Objet Sonore Maintenant: Pierre Schaeffer, Sound Objects and the Phenomenological Reduction', *Organised Sound*, 12/1 (April 2007): 15–24.
Kapferer, Bruce, *A Celebration of Demons: Exorcism and the Aesthetics of Healing in Sri Lanka* (Washington: Smithsonian Books, 1983).
Kassabian, Anahid, 'Popular', in Bruce Horner and Thomas Swiss (eds), *Key Terms in Popular Music and Culture* (Malden, MA: Blackwell Publishing, 1999), pp. 117–23.
———, *Ubiquitous Listening: Affect, Attention, and Distributed Subjectivity* (Berkeley and Los Angeles: California University Press, 2013).
Kastner, Georges, *Les Voix de Paris* (Paris: G. Brandus, Dufour, 1857).
Kearney, Robert, 'The 1915 Riots in Ceylon: A Symposium', *Journal of Asian Studies*, 29/2 (February 1970): 219–22.
Kelkel, Manfred, *Naturalisme, vérisme et réalisme dans l'opéra: de 1890 à 1930* (Paris: Libr. philosophique J. Vrin, 1984).
Kendrick, Robert L., *Celestial Sirens: Nuns and their Music in Early Modern Milan* (Oxford and New York: Oxford University Press, 1996).
———, *The Sounds of Milan, 1585–1650* (New York: Oxford University Press, 2002).
Kernan, Alvin, *The Cankered Muse: Satire of the English Renaissance* (New Haven: Yale University Press, 1959).
Kinser, Samuel, 'Presentation and Representation: Carnival at Nuremberg, 1450–1550', *Representations*, 13 (Winter 1986): 1–41.
———, 'Why Is Carnival So Wild?', in Konrad Eisenbichler and Wim Husken (eds), *Carnival and the Carnivalesque: The Fool, the Reformer, the Wildman, and Others in Early Modern Theatre* (Amsterdam: Rodopi, 1999), pp. 43–87.
Kintzler, Catherine, *Théâtre et opéra à l'âge classique: Une familière étrangeté* (Paris: Fayard, 2004).
———, *Poétique de l'opéra français de Corneille à Rousseau* (1991; Paris: Minerve, 2006).
Kipfer, Stefan, Kanishka Goonewardena, Christian Schmid and Richard Milgrom, 'On the Production of Henri Lefebvre', in Kanishka Goonewardena, Stefan Kipfer, Richard

Milgrom and Christian Schmid (eds), *Space, Difference, Everyday Life: Reading Henri Lefebvre* (London and New York: Routledge, 2008), pp. 1–24.

Kippur, Stephen A., *Jules Michelet: A Study of Mind and Sensibility* (Albany: State University of New York Press, 1981).

Kirnbauer, Martin, 'Die Nürnberger Trompeten- und Posaunenmacher vor 1500 im Spiegel Nürnberger Quellen', in Walter Salmen (ed.), *Musik und Tanz zur Zeit Kaiser Maximilian I: Bericht über die am 21. und 22. Oktober 1989 in Innsbruck abgehaltene Fachtagung* (Innsbruck: Helbling, 1992), pp. 131–42.

Koetschet, Josef, *Aus Bosniens letzter Türkenzeit* (Vienna: A. Hartleben, 1905).

Kok, Glória, *Rio de Janeiro no tempo da Av. Central* (São Paulo: Bei Comunicação, 2005).

Kordela, Kiarina, *$urplus (Spinoza, Lacan)* (New York: SUNY Press, 2007).

Kraljačić, Tomislav, *Kalajev režim u Bosni i Hercegovini (1882–1903)* (Sarajevo: Veselin Masleša, 1987).

Kreševljaković, Hamdija, *Sarajevo za vrijeme austro-ugarske uprave (1878–1918)* (Sarajevo: Arhiv grada Sarajeva, 1969).

Krug, Rebecca, 'Margery Kempe', in Larry Scanlon (ed.), *The Cambridge Companion to Medieval English Literature* (Cambridge: Cambridge University Press, 2009), pp. 217–28.

Kullak, Franz, *Beethoven's Piano-Playing*, trans. Theodore Baker (New York: G. Schirmer, 1901).

La May, Thomasin and Stewart A. Carter, 'Jean-Georges Kastner', *Grove Music Online*, <http://www.oxfordmusiconline.com> (accessed 13 May 2010).

Lefebvre, Henri, *The Production of Space* (Oxford: Blackwell Publishers, 1991).

———, *The Urban Revolution* (Minneapolis: University of Minnesota Press, 2003).

Legrand, Raphaëlle and Nicole Wild, *Regards sur l'opéra-comique: trois siècles de vie théâtrale* (Paris: CNRS, 2002).

Leith, Dick, 'Tudor London: Sociolinguistic Stratification and Linguistic Change', *Anglo-American Studies*, 4 (1984): 59–72.

Leppert, Richard, *The Sight of Sound: Music, Representation and the History of the Body* (Berkeley, LA and London: University of California Press, 1993).

Lesage [Le Sage], Alain René, *Le théâtre de la foire*, 10 vols (Paris: Étienne Ganeau, 1721).

———, *Oeuvres choisies de le Sage*, vols 13–16 (Paris: Leblanc, 1810).

Lessius, L., *De summo bono* (Antwerp: Ex officina plantiniana, 1616).

Lesure, François (ed.), *Textes sur Lully et l'opéra français* (Geneva: Editions Minkoff, 1987).

Levi, Giovanni, 'On Microhistory', in Peter Burke (ed.), *New Perspectives on Historical Writing*, Second Edition (Cambridge: Polity Press, 2001), pp. 97–119.

Lewis, Katherine, 'History, Historiography and Re-writing the Past', in Sarah Salih (ed.), *A Companion to Middle English Hagiography* (Woodbridge: D.S. Brewer, 2006), pp. 122–40.

Lisboa, Maria Helena, *Os engenheiros em Lisboa, urbanismo e arquitectura (1850–1930)* (Lisbon: Livros Horizonte, 2002).

Liszt, Franz, *Letters of Franz Liszt*, vol. II, ed. La Mara, trans. Constance Bache (New York: Charles Scribner's Sons, 1894).

Lochrie, Karma, *Margery Kempe and Translations of the Flesh* (Philadelphia: University of Pennsylvania Press, 1991).

Loose, Wilhelm (ed.), *Anton Tuchers Haushaltbuch (1507 bis 1517)* (Tübingen: H. Laupp, 1877).

Losa, Leonor, João Silva and Gonçalo Antunes de Oliveira, 'A edição de música impressa enquanto agente de mediatização do Fado: o caso do Fado do 31', *Etno-Folk: Revista Galega de Etnomusicoloxía*, 12 (2008): 55–68.

Loud, John F., 'Andrić on Bosnia: The 1924 Dissertation', in Wayne S. Vucinich (ed.), *Ivo Andrić Revisited: The Bridge Still Stands* (Berkeley: University of California, 1995), pp. 187–99.

Machado, Álvaro Manuel, *O 'francesismo' na literatura portuguesa* (Lisbon: Instituto de Cultura e Língua Portuguesa, 1984).

Mairan, Jean-Jacques Dortous de, 'Sur la propagation du son dans les différens tons qui le modifient', in *Histoire de l'Académie royale des sciences avec les mémoires de mathématique et de physique pour la même année tirés des registres de cette académie – Année 1737* (Paris: L'imprimerie Royale, 1737).

Malalgoda, Kitsiri, *Buddhism in Sinhalese Society, 1750–1900* (Berkeley: University of California Press, 1976).

Malcolm, Noel, *Bosnia: A Short History* (London: Macmillan, 1994).

Manley, Lawrence, *Literature and Culture in Early Modern London* (Cambridge: Cambridge University Press, 1995).

Mardock, James D., *Our Scene is London: Ben Jonson's City and the Space of the Author* (New York and Abingdon: Routledge, 2008).

Marmontel, Jean-François, 'Opéra', *Œuvres complètes*, 19 vols (Paris: Verdière, 1818–20).

Martin, Gregory, *Roma sancta, 1581*, ed. by George Bruner Parks (Rome: Edizioni di Storia e letteratura, 1969).

Martin, Isabelle, *Le théâtre de la foire: des tréteaux aux boulevards* (Oxford: Voltaire Foundation, 2002).

Martin, John, 'The Dance: Listening In', *New York Times*, 31 March 1929.

Marx, Adolf Bernhard, *Introduction to the Interpretation of Beethoven Piano Works*, trans. Fannie Louise Gwinner (Chicago: Clayton F. Summy Co., 1895).

Marx, Karl, *Capital* (London: Penguin Books Ltd, 1992).

Mascarenhas, João Mário (ed.), *Grandella, o grande homem* (Lisbon: Câmara Municipal de Lisboa, 1994).

Maus, Katherine Eisaman, *Inwardness and Theater in the English Renaissance* (Chicago and London: University of Chicago Press, 1995).

May, Thomas Robert, *The John Adams Reader: Essential Writings on an American Composer* (Pompton Plains, NJ: Amadeus Press, 2006).

McCabe, Richard A., 'Elizabethan Satire and the Bishops' Ban of 1599', *Yearbook of English Studies* (1981): 188–93.

McDannel, Colleen and Bernard Lang, *Heaven: A History* (New Haven: Yale University Press, 1988).

McGee, Timothy J., *The Sound of Medieval Song: Ornamentation and Vocal Style According to the Treatises* (Oxford: Clarendon Press, 1998).

Mesquita, Alfredo, 'Do omnibus ao automóvel', *Illustração Portugueza*, 24 September 1908: 233–8.

Meyers großes Konversations-Lexikon: Ein Nachschlagewerk des allgemeinen Wissens, 6th edition (Leipzig and Vienna: Verlag des Bibliographischen Instituts, 1907).

Middleton, Thomas, and Thomas Dekker, *The Roaring Girl* (1611), in Simon Barker and Hilary Hinds (eds), *The Routledge Anthology of Renaissance Drama* (Abingdon: Routledge, 2005), pp. 326–77.

Millett, Bella and Jocelyn Wogan-Brown, *Medieval English Prose for Women* (Oxford: Clarendon Press, 1990).

Monnet, Jean, *Supplément au Roman comique, ou Mémoires pour servir à la vie de Jean Monnet*, 2 vols (Paris, 1772).

Montagnier, Jean-Paul, 'Charles-Hubert Gervais's "Psiché burlesque" and the Birth of the Comic "Cantate française"', *Journal of Musicology*, 17/4 (1999): 520–45.

Moreau, Mário, *Coliseu dos Recreios: um século de história* (Lisbon: Quetzal Editores, 1994).

———, *O teatro de S. Carlos: dois séculos de história*, vol. 1 (Lisbon: Hugin Editores, 1999).

Moretti, Franco, 'Style, Inc.: Reflections on Seven Thousand Titles (British Novels, 1740–1850)', *Critical Inquiry*, 36/3 (2009): 134–58.

Mulahalilović, Enver, *Vjerski običaji Muslimana u Bosni i Hercegovini*. 2nd enlarged edition (Sarajevo: El-Kalem, 1989).

Myers, Charles Samuel, *A Text-book of Experimental Psychology* (New York: Longmans, Green & Co., 1909).

Myers, Herbert W., 'Slide Trumpet Madness: Fact or Fiction?', *Early Music*, 17 (1989): 383–9.

Naumburg, Margaret, 'The Dalcroze Idea, What Eurhythmics is and What it Means', *Outlook*, 17 January 1914: 130.

Ndalianis, Angela, *Neo-Baroque Aesthetics and Contemporary Entertainment* (Cambridge, MA: MIT Press, 2004).

Nevola, Fabrizio, *Siena: Constructing the Renaissance City* (New Haven: Yale University Press, 2007).

Newman, Karen, *Cultural Capitals: Early Modern London and Paris* (Princeton: Princeton University Press, 2007).

Nickel, Ekkehart, *Der Holzblasinstrumentenbau in der freien Reichsstadt Nürnberg* (Schriften zur Musik 8; Munich: Emil Katzbichler, 1971).

Obeyesekere, Gananath, 'Religious Symbolism and Political Change in Ceylon', *Modern Ceylon Studies*, 1/1 (1970): 43–63.

———, *The Cult of the Goddess Pattini* (Chicago: University of Chicago Press, 1984).

Oddie, Richard, 'Other Voices: Acoustic Ecology and Urban Soundscapes', in Ingrid Leman-Stefanovic and Stephen B. Scharper (eds), *The Natural City: Re-envisioning the Built Environment* (Toronto: University of Toronto Press, 2012), pp. 161–73.

Okey, Robin, *Taming Balkan Nationalism: The Habsburg Civilizing Mission in Bosnia, 1878–1914* (Oxford: Oxford University Press, 2007).

Ondaatje, Michael, *Running in the Family* (New York: Random House, 1996).

Opie, Iona and Peter (eds), *The Oxford Dictionary of Nursery Rhymes*, 2nd edition (New York: Oxford University Press, 1997).

Orlin, Lena Cowen, *Privacy Matters and Public Culture in Post-Reformation England* (Ithaca: Cornell University Press, 1994).

———, 'Boundary Disputes in Early Modern London', in Lena Cowen Orlin (ed.), *Material London, ca. 1600* (Philadelphia: University of Pennsylvania Press, 2000), pp. 344–76.

———, *Locating Privacy in Tudor London* (Oxford: Oxford University Press, 2007).

Ornithoparchus, Andreas, *Music active micrologus* (Leipzig: Valentin Schumann, 1517; facsimile, New York: Dover, 1973).

Osborne, Richard (ed.), *Conversations with Karajan* (Oxford: Oxford University Press, 1998).

Ouvrard, René, *Architecture harmonique, ou Application de la doctrine des proportions de la musique à l'architecture* (Paris, 1677).

Ovid, *Metamorphoses*, trans. Charles Martin (New York: Norton, 2004).

Pais, José Machado, 'A prostituição na Lisboa boémia nos inícios do século XX', *Análise Social*, 19/77–79 (1983): 939–60.

Pardailhe-Galabrun, Annik, *La naissance de l'intime* (Paris: PUF, 1988).

Parfaict, Claude, *Dictionnaire de théâtres de Paris*, 7 vols (Paris: Rozet, 1767).

Parfaict, Claude, and François Parfaict, *Mémoires pour servir à l'histoire des spectacles de la foire*, 2 vols (Paris: Briasson, 1743).

Parker, Simón, *Urban Theory and the Urban Experience: Encountering the City* (London and New York: Routledge, 2004).

Parsons, Albert Ross, *The Science of Pianoforte Practice* (New York: G. Schirmer, 1886).

Pellegrini, Carlo, *Museum Historico-legale bipartitum* (Rome, 1665).

Pellow, Deborah, 'The Architecture of Female Seclusion in West Africa', in Denise Lawrence-Zuniga and Setha M. Low (eds), *The Anthropology of Space and Place: Locating Culture* (Cambridge: Blackwell Publishers, 2003), pp 160–83.

Pennanen, Risto Pekka, 'Controlling Sound and Music: Aspects of Musical Life in Sarajevo under Austro-Hungarian Rule (1878–1918)', in Tamara Karača and Senad Kazić (eds), *Collection of Papers: 4th International Symposium 'Music in Society', Sarajevo, October 28–30, 2004* (Sarajevo: Musicological Society of the FBiH, 2005), pp. 114–25.

Pereira, Nuno Teotónio, 'Pátios e vilas de Lisboa, 1870–1930: a promoção privada do alojamento operário', *Análise Social*, 29/127 (1994): 509–24.

Perrault, Claude, *Essais de physique; ou, Receuil de plusieurs traitez touchant les choses naturelles*, 4 vols (Paris: Jean Baptiste Coignard, 1680).

———, *Ordonnance for the Five Kinds of Columns after the Methods of the Ancients*, trans. Indra Kagis McEwen (Santa Monica, CA: Getty Center for the History of Art and the Humanities, 1993).

———, *Les dix livres d'architecture de Vitruve* (1673; Paris: Bibliothèque de l'Image, 1995).

———, *Du bruit et de la musique des anciens: Préface de François Lesure* (Geneva: Éditions Minkoff, 2003).

Petzet, Michael, *Claude Perrault und die Architektur des Sonnenkönigs* (Munich: Deutscher Kunstverlag München Berlin, 2000).

Piccolomini, Enea Silvio, *I commentarii*, ed. L. Totaro (Milan: Adelphi, 1984).

Picker, John M., *Victorian Soundscapes* (Oxford: Oxford University Press, 2003).

Picon, Antoine, *Claude Perrault, 1613–1688, ou la curiosité d'un classique* (Paris: Picard Editeur, 1988).

———, 'The Freestanding Column in Eighteenth-Century Religious Architecture', in Lorraine Daston (ed.), *Things That Talk: Object Lessons from Art and Science* (New York: Zone Books, 2004), pp. 67–100.

Pietzsch, Gerhard, 'Die Beschreibungen deutscher Fürstenhochzeiten von der Mitte des 15. bis zum Beginn des 17. Jahrhunderts als musikgeschichtliche Quellen', *Anuario Musicale*, 15 (1960): 21–62.

Pimentel, Alberto, *Vida de Lisboa* (Lisbon: Parceria António Maria Pereira, 1900).

———, *A triste canção do Sul: Subsídios para a História do Fado* (Lisbon: Livraria Central, 1904).

Piron, Alexis, *Oeuvres complettes d'Alexis Piron*, Jean Antoine Rigoley de Juvigny (ed.), 7 vols (Liège: C. Plomteux, 1776).

Plato, *The Symposium*, trans. Christopher Gill (New York: Penguin, 1999).

Polk, Keith, 'The Schubingers of Augsburg: Innovation in Renaissance Instrumental Music', in Friedhelm Brusniak and Horst Leuchtmann (eds), *Quaestiones in musica. Festschrift für Franz Krautwurst* (Tutzing: Schneider, 1989), pp. 495–503.

———, 'The Trombone, the Slide Trumpet and the Ensemble Tradition of the Early Renaissance', *Early Music*, 17 (1989): 389–97.

——, 'Voices and Instruments: Soloists and Ensembles in the 15th Century', *Early Music*, 18 (1990): 179–98.

——, *German Instrumental Music of the Late Middle Ages: Players, Patrons and Performance Practice* (Cambridge: Cambridge University Press, 1992).

Porter, Theodore, *Trust in Numbers: The Pursuit of Objectivity in Science and Public Life* (Princeton: Princeton University Press, 1995).

Pound, Ezra, *Antheil and the Treatise on Harmony* (Reprint, New York: Da Capo Press, 1968).

Prendergast, Christopher, *Paris and the Nineteenth Century* (Oxford, UK, and Cambridge, MA: Blackwell, 1992).

Prothero, Stephen, *The White Buddhist: The Asian Odyssey of Henry Steel Olcott* (Bloomington: Indiana University Press, 1996).

Queirós, Eça de, *O primo Bazilio, episódio domestico* (Porto/Braga: Livraria Chardron, 1878).

Quintas, Maria Alexandra Salgado, 'Do Passeio Público à Pena: um percurso do jardim romântico', Master's Dissertation (Universidade Técnica de Lisboa, 2001).

Ramanathan, Ponnambalam, *Riots and Martial Law in Ceylon, 1915* (London: St. Martin's Press, 1916).

Rameau, Jean-Philippe, *Complete Theoretical Writings*, ed. Erwin R. Jacobi (n.p.: American Institute of Musicology, 1968).

Rappaport, Steve, *Worlds Within Worlds: Structures of Life in Sixteenth-Century London* (Cambridge: Cambridge University Press, 1989).

Rastall, Richard, *Music in Early English Religious Drama. 1. The Heaven Singing* (Cambridge: Brewer, 1996).

Rattazzi, Maria, *Portugal de relance* (Lisbon: Livraria Zeferino, 1881).

Rearick, Charles, *Beyond the Enlightenment: Historians and Folklore in Nineteenth Century France* (Bloomington: Indiana University Press, 1974).

Rebello, Luiz Francisco, *O teatro naturalista e neo-romântico (1870–1910)* (Lisbon: Instituto de Cultura Portuguesa, 1978).

——, *O teatro romântico (1838–1869)* (Lisbon: Instituto de Cultura e Língua Portuguesa, 1980).

——, *História do teatro de revista em Portugal*, vol. 1 (Lisbon: Publicações D. Quixote, 1984).

Reed, Susan, *Dance and the Nation: Performance, Ritual, and Politics in Sri Lanka* (Madison: University of Wisconsin Press, 2010).

Rêgo, Manuela (ed), *1907: no advento da República*. (Lisbon: Biblioteca Nacional, 2007).

Ribas, Tomaz, *O teatro da Trindade: 125 anos de vida* (Oporto: Lello & Irmão, 1993).

Ribot, Théodule, *German Psychology of To-day: The Empirical School*, trans. James Mark Baldwin (New York: Charles Scribner's and Sons, 1886).

Rieber, Robert W. and David K. Robinson (eds), *Wilhelm Wundt in History: The Making of a Scientific Psychology* (New York: Plenum, 2001).

Risebero, Bill, *Modern Architecture and Design: An Alternative History* (Cambridge: MIT Press, 1982).

Roberts, Michael, 'Noise as Cultural Struggle: Tom-Tom Beating, the British and Communal Disturbances in Sri Lanka, 1880s–1930s', in Veena Das (ed.), *Mirrors of Violence: Communities, Riots and Survivors in South Asia* (Delhi: Oxford University Press, 1990), pp. 240–285.

——, *Exploring Confrontation: Sri Lanka: Politics, Culture and History* (Chur, Switzerland: Harwood Academic Publishers, 1995).

Robinson, Jennifer, *Ordinary Cities: Between Modernity and Development* (London and New York: Routledge, 2006).
Rodger, N.A.M., *The Safeguard of the Sea: A Naval History of Britain Vol. 1, 660–1649* (London: HarperCollins, 1997).
Rodrigues, Isabel Maria, 'As avenidas de Ressano Garcia', *Boletim Lisboa Urbanismo* 13 (September/October 2000): 20–23, and 14 (November/December 2000): 30–34.
Rodrigues, José Carlos Barros, *O automóvel em Portugal: 100 anos de história* (Lisbon: CTT, 1995).
Roger, Jacques, *The Life Sciences in Eighteenth-Century French Thought* (Stanford: Stanford University Press, 1997).
Rogers, John D., *Crime, Justice and Society in Colonial Sri Lanka* (London: Curzon Press, 1987).
Roller, Hans-Ulrich, *Der Nürnberger Schembartlauf. Studien zum Fest- und Maskenwesen des späten Mittelalters* (Magstadt: Horst Bissinger, 1965).
Rosow, Lois, 'Power and Display: Music in Court Theatre', in Tim Carter and John Butt (eds), *The Cambridge History of Seventeenth-Century Music* (Cambridge, New York: Cambridge University Press, 2005), pp. 197–239.
Rousseau, Jean Jacques, *Oeuvres complètes*, vol. 3 (Paris: Gallimard, 1964).
Russell, Jeffrey Burton, *A History of Heaven: The Singing Silence* (Princeton: Princeton University Press, 1997).
Rutter, Tom, '*Patient Grissil* and Jonsonian Satire', *Studies in English Literature*, 48/2 (Spring, 2008): 283–303.
Sachs, Curt, *Rhythm and Tempo* (New York: W.W. Norton & Co., 1953).
Said, Edward, *Orientalism* (New York and London: Routledge and Kegan Paul, 1978).
———, *Humanism and Democratic Criticism* (New York: Columbia University Press, 2004).
Saint-Saëns, Camille, trans. Fred Rothwell, 'The Ideas of M. Vincent d'Indy', *The Musical Times*, 61 (1 April 1920): 244.
Salmen, Walter, *Der fahrende Musiker im europäischen Mittelalter* (Kassel: Johann Philipp Hinnenthal Verlag, 1960).
———, 'Dances and Dance Music *c*.1300–1530', in Reinhard Strohm and Bonnie J. Blackburn (eds), *The New Oxford History of Music* iii, Part 1: *Music as Concept and Practice in the Late Middle Ages* (2nd edn., Oxford: Oxford University Press, 2001), pp. 162–90.
Sanford, E.C., 'Personal Equation', *The American Journal of Psychology*, 2 November 1889.
Santos, Maria de Lourdes Lima dos, 'Para a análise das ideologias da burguesia', *Análise Social*, 14/53 (1978): 39–80.
Schaeffer, P. *Traité des objets musicaux* (Paris : Editions du Seuuil, 1966).
Schafer, R. Murray, *The Soundscape: Our Sonic Environment and the Tuning of the World* (Rochester, VE: Destiny Books, 1994 [1977]).
———, *Le paysage sonore* (Paris: Jean-Claude Lattès, 1979).
———, 'Acoustic space', in David Seamon and Robert Mugerauer (eds), *Dwelling, Place and Environment: Towards a Phenomenology of Person and World* (Dordrecht, Netherlands: Martinus Nijhoff, 1985), pp. 88–98.
Scheler, Max, *Wesen und Formen der Sympathie* (Bonn: F. Cohen, 1923).
Schindler, Anton, *The Life of Beethoven: Including His Correspondence with His Friends, Numerous Characteristic Traits, and Remarks on His Musical Works*, vol. II, ed. Ignaz Moscheles (London: Henry Colburn, 1841).

Schmiedgen, Peter, 'Interiority, Exteriority and Spatial Politics in Benjamin's Cityscapes', in Andrew Benjamin and Charles Rice (eds), *Walter Benjamin and the Architecture of Modernity* (Prahran: re.press, 2009), pp. 147–58.

Schwartz, Vanessa R., *Spectacular Realities: Early Mass Culture in fin-de-siècle Paris* (Berkeley: University of California Press, 1998).

Schwarz, David, *Listening Subjects: Music, Psychoanalysis, Culture* (Duke University Press, 1997).

Scott, David, *Conscripts of Modernity: The Tragedy of Colonial Enlightenment* (Durham: Duke University Press, 2004).

Scripture, E.W., 'Some Principles of Mental Education', *The School Review*, 3 November 1895.

———, *Thinking, Feeling, Doing* (Meadville: The Chautauqua-Century Press, 1895).

Sears, Charles H., 'A Contribution to the Psychology of Rhythm', *The American Journal of Psychology*, 13 (January 1902): 28–61.

Seashore, Carl Emil, 'The Measurement of Musical Talent', *The Musical Quarterly*, 1 January 1915.

———, 'The Sense of Rhythm as a Musical Talent', *The Musical Quarterly*, 4 October 1918.

———, *The Psychology of Musical Talent* (Boston: Silver, Burdett and Company, 1919).

———, 'Artistic Deviation as an Esthetic Principle in Music', *The Scientific Monthly*, 54 (February 1942): 99–109.

Seashore, Carl Emil and Milton Metfessel, 'Deviation from the Regular as an Art Principle', *Proceedings of the National Academy of Sciences of the United States of America*, 11 (15 September, 1925): 538–42.

Shaviro, Steven, 'Attali's Noise', *Pinocchio Theory*, (April 15, 2005) <http://www.shaviro.com> (accessed 30th July 2014).

———, *Post Cinematic Affect* (Ropley, Hants: Zero Books, 2010).

Silva, Joaquim Palminha, 'Armazéns Grandella: como nasceram e o que foram', *História*, 112 (1988): 4–27.

Silva, Kingsley De, *A History of Sri Lanka* (Berkeley: University of California Press, 1981).

Silva, Raquel Henriques da (ed.), *Lisboa de Frederico Ressano Garcia 1874–1909* (Lisbon: Câmara Municipal de Lisboa/Fundação Calouste Gulbenkian, 1989).

Silverman, Kaja, *Male Subjectivity at the Margins* (New York: Routledge, 1992).

Smith, Bruce R., *The Acoustic World of Early Modern England: Attending to the O-Factor* (Chicago and London: University of Chicago Press, 1999).

Smith, Emma, '"Signes of a Stranger": The English Language and the English Nation in the Late Sixteenth Century', in Philip Schwyzer and Simon Mealor (eds), *Archipelagic Identities: Literature and Identity in the Atlantic Archipelago 1550–1800* (Hampshire: Ashgate, 2004), pp. 169–79.

Smith, Grace, 'The Eurhythmics of Jaques-Dalcroze', *Journal of Proceedings and Addresses, National Education Association*, 16–27 August, 1915: 871.

Smith, Mark M., *Listening to Nineteenth-century America* (Chapel Hill, NC: University of North Caroline Press, 2001).

———, (ed.), *Hearing History: A Reader* (Athens: University of Georgia Press, 2004).

———, *Sensory History* (New York: Berg, 2007).

Somaratna, G.P.V., *Kotahena Riot 1883: Religious Riot in Sri Lanka* (Nugegoda: Deepanee Publications, 1991).

Stefani, Gino, *Musica barocca 2. Angeli e Sirene* (Milan: Bompiani, 1988).

Stephanov, Darin, 'Cannon Salvos for the Monarch: Notes on the Ceremonial Usage of Artillery at Nineteenth-Century Ottoman and Russian Accession, Coronation and Other

Public Dynastic Festivities', unpublished conference paper presented at the Bosphorus University's Ottoman Material Culture Workshop (Istanbul, Turkey, September 2006).

Sternberg, Constantin von, *Ethics and Esthetics of Piano-Playing* (New York: G. Schirmer, 1917).

Sternfeld, F.W., *The Birth of Opera* (Oxford: Clarendon Press, 1993).

Stevens, John, and Dennis Libby, 'Carol', *Grove Music Online*, in *Oxford Music Online*, <http://www.grove.com> (accessed March 14 2014).

Steward, Jill and Alexander Cowan, 'Introduction', in Jill Steward and Alexander Cowan (eds), *The City and the Senses: Urban Culture Since 1500* (Aldershot: Ashgate, 2007), pp. 1–22.

Stewart, Susan A., *Poetry and the Fate of the Senses* (Chicago: Chicago University Press, 2002).

Stiefel, Eberhard, 'Musikgeschichte der ehemaligen Reichsstadt Schwäbisch Gmünd', PhD Dissertation (University of Tübingen, 1949).

Stirrat, R.L., 'Constructing Identities in Nineteenth-Century Colombo', in John R. Campbell and Alan Rew (eds), *Identity and Affect: Experiences of Identity in a Globalising World* (London: Pluto Press, 1999), pp. 39–63.

Stoddart, J.M., *Stoddart's Encyclopedia Americana*, vol. II, (New York: J.M Stoddart, 1884).

Stravinsky, Igor, *Poetics of Music in the Form of Six Lessons*, trans. Arthur Knodel and Ingolf Dahl (New York: Vintage Books, 1956).

Sumberg, Samuel L., *The Nuremberg Schembart Carnival* (New York: Columbia University Press, 1941).

Sykes, Jim, 'The Musical Gift: Sound, Sovereignty and Multicultural History in Sri Lanka', PhD Dissertation (University of Chicago, 2011).

Tambiah, Stanley J., *Leveling Crowds: Ethnonationalist Conflicts and Collective Violence in South Asia* (Berkeley: University of California Press, 1996).

Taruskin, Richard, *The Oxford History of Western Music: The Earliest Notations to the Sixteenth Century*, vol. 1 (Oxford: Oxford University Press, 2005).

Thomas, Downing A., *Aesthetics of Opera in the Ancien Régime, 1647–1785* (Cambridge: Cambridge University Press, 2002).

Thompson, Emily, *The Soundscape of Modernity: Architectural Acoustics and the Culture of Listening in America 1900–1933* (Cambridge, MA: MIT Press, 2004).

Thompson, Paul, *The Voice of the Past* (Oxford: Oxford University Press, 1978).

Titchener, Edward Bradford, *Experimental Psychology: A Manual of Laboratory Practice*, vol. I (New York: The Macmillan Company, 1901).

———, *An Outline of Psychology* (New York: The Macmillan Company, 1902).

———, *A Text-book of Psychology* (New York: The Macmillan Company, 1910).

———, *A Beginner's Psychology* (New York: The Macmillan Company, 1916).

Tomlinson, Gary, *Metaphysical Song: An Essay on Opera* (Princeton, NJ: Princeton University Press, 1999).

Trevor, Roy, *My Balkan Tour: An Account of Some Journeyings and Adventures in the Near East, Together with a Descriptive and Historical Account of Bosnia & Herzegovina, Dalmatia, Croatia & the Kingdom of Montenegro* (London: John Lane, 1911).

Tunley, David, *The Eighteenth-Century French Cantata* (Oxford: Clarendon Press, 1997).

Turner, Victor, *The Ritual Process: Structure and Anti-Structure* (Hawthorne, NY: Aldine de Gruyter, 1997).

Uberti, Grazioso, *Contrasto musico: opera dilettevole* (Rome: Grignani, 1630; facs. ed., Lucca: LIM, 1991).

Vaïsse, Claude-Marius, *Projet de traité pour l'exécution de divers travaux d'utilité générale* (Lyon, 1855).

Valentine, Daniel E., *Fluid Signs: Being a Person the Tamil Way* (Berkeley: University of California Press, 1984).

———, *Charred Lullabies: Chapters in an Anthropography of Violence* (Princeton: Princeton University Press, 1996).

Valier, A., *Vita Caroli Borromei* (Verona, 1586).

van Eck, Caroline, 'Giannozzo Manetti on Architecture: the *Oratio de secularibus et pontificalibus pompis in consecratione basilicae Florentinae* of 1436', *Renaissance Studies*, 12/4 (December, 1998): 449–75.

Vaughan, Robert (ed.), *Tracts and Treatises of John de Wycliffe, D.D. with Selections and Translations from his Manuscripts and Latin Works* (London: Blackburn and Pardon, 1845).

Vendrix, Phillipe, *L'opéra-comique en France au XVIIIe siècle* (Liège: Margada, 1992).

Venturoli, Alberto, *Visita alle Sette Chiese* (Rome: Città Nuova, 2006).

Verde, Cesário, *O livro de Cesário Verde* (Lisboa: Typographia Elzeveriana, 1887).

Vierordt, Karl, *Der Zeitsinn nach Versuchen* (Tübingen: H. Laupp'schen Buchhandlung, 1868).

Virgil, *Virgil's Georgics*, trans. Janet Lembke (New Haven: Yale University Press, 2005).

Vogeleis, Martin, *Quellen und Bausteine zu einer Geschichte der Musik und des Theaters im Elsass, 500–1800* (Geneva: Minkoff, 1979).

Vorgrimler, Herbert, *Geschichte der Hölle* (Munich: Fink, 1993).

Waddell, Chrysogonus, 'The Origin and Early Evolution of the Cistercian Antiphonary: Reflections on Two Cistercian Chant Reforms', in M. Basil Pennington (ed.), *The Cistercian Spirit: A Symposium in Memory of Thomas Merton* (Spencer: Cistercian Publications, 1970), pp. 190–223.

Waeber, Jacqueline, *En musique dans le texte: Le mélodrame de Rousseau à Schoenberg* (Paris: Van Dieren, 2006).

Walny, Adolf (ed.), *Bosnischer Bote. Universal-Hand- und Adreßbuch für Bosnien und die Hercegovina* (Sarajevo: Self-published, 1901).

Wathey, Andrew, 'The Peace of 1360–1369 and Anglo-French Musical Relations', *Early Music History*, 9 (1990): 129–74.

Watson, Nicholas, *Richard Rolle and the Invention of Authority* (Cambridge: Cambridge University Press, 1991).

Webster's Revised Unabridged Dictionary (1913) <http://machaut.uchicago.edu/websters> and Wordnik <www.wordnik.com> (accessed January 4 2015).

Wedeen, Lisa, *Peripheral Visions: Publics, Power, and Performance in Yemen* (Chicago: University of Chicago Press, 2008).

Wegs, J. Robert, 'The Marshaling of Copper: An Index of Austro-Hungarian Economic Mobilization during World War I', *Austrian History Yearbook*, 12–13/1 (1976): 189–202.

Weidman, Amanda, *Singing the Classical, Voicing the Modern: The Postcolonial Politics of Music in South India* (Durham: Duke University Press, 2006).

Welker, Lorenz, '"Alta Capella": Zur Ensemblepraxis der Blasinstrumente im 15. Jahrhundert', *Basler Jahrbuch für historische Musikpraxis*, vii (1983): 119–65.

Wickramasinghe, Nira, *Sri Lanka in the Modern Age: A History of Contested Identities* (Honolulu: University of Hawaii Press, 2006).

Wild, Nicole, and David Charlton, *Théâtre de l'opéra-comique Paris: Répertoire 1762–1927* (Sprimont: Margada, 2005).

Wilkins, Nigel, *La musique du diable* (Sprimont: Mardaga, 1999).

Williams, David H., *The Cistercians in the Early Middle Ages* (Trowbridge: Cromwell Press, 1998).

Williams, Raymond, *Keywords* (London: Fontana Press, 1988).

Windeatt, Barry (ed.), *The Book of Margery Kempe* (London: Longman: 2000).

Wise, M. Norton (ed.), *The Values of Precision* (Princeton: Princeton University Press, 1995).

Wolters, Clifton (ed.), *Rolle: The Fire of Love* (Harmondsworth: Penguin, 1972).

Wright, Laura, 'Speaking and Listening in Early Modern London', in Alex Cowan and Jill Steward (eds), *The City and the Senses: Urban Culture Since 1500* (Aldershot: Ashgate, 2007), pp. 60–74.

Wundt, Wilhelm Max, *Lectures on Human and Animal Psychology*, Second German Edition (1892), trans. J.E. Creighton and E.B. Titchener (London: Swan Sonnenschein & Co., 1894).

———, *Ethics: An Investigation of the Facts and Laws of the Moral Life*, 2nd German Edition (1892), trans. Edward Bradford Titchener, Gulliver and Margret Floy Washburn, vol. 1 (London: Swan Sonnenschein & Co., Lim., 1902).

———, *Outlines of Psychology*, Fourth Edition, trans. Charles Hubbard Judd (Leipzig: Wilhelm Engelmann, 1902).

Wyatt, J., *The Repertory of Arts, Manufacturers, and Agriculture*, XXVIII, Second Series (London: J. Wyatt, 1816).

Young, R.F. and G.P.V. Somaratne, *Vain Debates: The Buddhist-Christian Controversies of Nineteenth-Century Ceylon* (Vienna: Publications of the De Nobli Research Library, 1996).

Yungblut, Laura, *Strangers Settled Here Amongst Us: Policies, Perceptions & the Presence of Aliens in Elizabethan England* (London and New York: Routledge, 1996).

Zangwill, O.L., 'Wundt', *The Oxford Companion to the Mind Online*, <http://www.oxfordreference.com> (accessed 22 September 2007).

Zucker, Adam, 'The Social Logic of Ben Jonson's *Epicoene*', *Renaissance Drama*, 33 (2004): 37–62.

Index

For Product Safety Concerns and Information please contact our EU representative GPSR@taylorandfrancis.com
Taylor & Francis Verlag GmbH, Kaufingerstraße 24, 80331 München, Germany

www.ingramcontent.com/pod-product-compliance
Ingram Content Group UK Ltd.
Pitfield, Milton Keynes, MK11 3LW, UK
UKHW020015200926
475702UK00014B/1893
9780367230517